S0-ASH-040

$4595G

Parag Sapre

Parag

INTRODUCTION TO THE THEORY AND DESIGN OF ACTIVE FILTERS

McGraw-Hill Series in Electrical Engineering

Consulting Editor
Stephen W. Director, Carnegie—Mellon University

Networks and Systems
Communications and Information Theory
Control Theory
Electronics and Electronic Circuits
Power and Energy
Electromagnetics
Computer Engineering and Switching Theory
Introductory and Survey
Radio, Television, Radar, and Antennas

Previous Consulting Editors

Ronald M. Bracewell, Colin Cherry, James F. Gibbons, Willis W. Harman, Hubert Heffner, Edward W. Herold, John G. Linvill, Simon Ramo, Ronald A. Rohrer, Anthony E. Siegman, Charles Susskind, Frederick E. Terman, John G. Truxal, Ernst Weber, and John R. Whinnery

Networks and Systems

Consulting Editor
Stephen W. Director, Carnegie—Mellon University

Antoniou: *Digital Filters: Analysis and Design*
Athans, Dertouzos, Spann, and Mason: *Systems, Networks, and Computation: Multivariable Methods*
Balabanian and LePage: *Electrical Science, Book I*
Balabanian and LePage: *Electrical Science, Book II: Dynamic Networks*
Becker and Jensen: *Design of Systems and Circuits for Maximum Reliability or Maximum Production Yield*
Belove and Drossman: *Systems and Circuits for Electrical Engineering Technology*
Belove, Schachter, and Schilling: *Digital and Analog Systems, Circuits and Devices*
Bracewell: *The Fourier Transform and Its Applications*
Calahan: *Computer Aided Network Design*
Cannon: *Dynamics of Physical Systems*
Chirlian: *Basic Network Theory*
Cornetet and Battocletti: *Electronic Circuits by System and Computer Analysis*
Desoer and Kuh: *Basic Circuit Theory*
Fitzgerald, Higginbothan, and Grabel: *Basic Electrical Engineering*
Ghausi: *Principles and Design of Linear Active Circuits*
Hammond and Gehmlich: *Electrical Engineering*
Hayt and Hughes: *Introduction to Electrical Engineering*
Hayt and Kemmerly: *Engineering Circuit Analysis*
Hilburn and Johnson: *Manual of Active Filter Design*
Huelsman: *Active Filters: Lumped, Distributed, Integrated, Digital, and Parametric*
Huelsman: *Digital Computations in Basic Circuit Theory*
Huelsman: *Theory and Design of Active RC Circuits*
Huelsman and Allen: *Introduction to the Theory and Design of Active Filters*
Kalman, Falb, and Arbib: *Topics in Mathematical System Theory*
Lewis, Reynolds, Bergseth, and Alexandro: *Linear Systems Analysis*
Liu and Liu: *Linear Systems Analysis*
Murdoch: *Network Theory*
Papoulis: *The Fourier Integral and Its Application*
Peatman: *Design of Digital Systems*
Ramey and White: *Matrices and Computers in Electronic Circuit Analysis*
Ruston and Bordogna: *Electric Networks: Functions, Filters, Analysis*
Sage: *Methodology for Large Scale Systems*
Schwartz and Friedland: *Linear Systems*
Temes and LaPatra: *Introduction to Circuit Synthesis*
Timothy and Bona: *State Space Analysis*
Truxal: *Introductory System Engineering*
Tuttle: *Circuits*

INTRODUCTION TO THE THEORY AND DESIGN OF ACTIVE FILTERS

L. P. Huelsman

Professor of Electrical Engineering
University of Arizona

P. E. Allen

Associate Professor of Electrical Engineering
Texas A & M University

McGraw-Hill Book Company

New York   St. Louis   San Francisco   Auckland   Bogotá   Hamburg
Johannesburg   London   Madrid   Mexico   Montreal   New Delhi
Panama   Paris   São Paulo   Singapore   Sydney   Tokyo   Toronto

This book was set in Times Roman. The editors were Frank J. Cerra and Madelaine Eichberg; the cover was designed by Scott Chelius; the production supervisor was Leroy A. Young. The drawings were done by J & R Services, Inc.

INTRODUCTION TO THE THEORY AND DESIGN OF ACTIVE FILTERS

Copyright © 1980 by McGraw-Hill, Inc. All rights reserved. Printed in the United States of America. No part of this publication may be reproduced, stored in a retrieval system, or transmitted, in any form or by any means, electronic, mechanical, photocopying, recording, or otherwise, without the prior written permission of the publisher.

4567890 HDHD 898765432

Library of Congress Cataloging in Publication Data

Huelsman, Lawrence P
    Introduction to the theory and design of active
filters.

    (McGraw-Hill series in electrical engineering)
    Bibliography: p.
    Includes index.
    1. Electric filters, Active. I. Allen, Phillip E.,
joint author. II. Title. III. Series.
TK7872.F5H82   621.3815'32   79-20998
ISBN 0-07-030854-3

To our wives:
Jo
and
Margaret;

AND OUR CHILDREN:
David
and
Kurt, Cheryl, and Paul

CONTENTS

Bibliography 393

Appendixes

Index 417

PREFACE

This book is designed to provide the basic material for an introductory senior or first-year graduate course in the theory and design of active filters. Synthesis methods for active filters exhibit many differences from the ones for passive filters. One very important difference is that the former tend to recognize the tremendous advances made in integrated-circuit processing techniques. As a result of these advances, we now find that active elements are frequently cheaper to produce than passive ones. As a consequence, modern synthesis methods lean heavily toward the use of active elements, which, together with resistors and capacitors, virtually eliminate the need to use inductors in many frequency ranges. Such active *RC* circuits present significant advantages over their passive counterparts, with respect to cost, reliability, weight, and the availability of gain; thus they are taking over an ever-increasing share of the total of filter production and application.

The first chapter of the book provides introductory material outlining the scope of the filtering problem, reviewing some of the basic concepts of the network function approach to circuit theory, and defining the procedures of impedance and frequency normalization. In Chapter 2 the subject of approximation is introduced. The treatment includes not only the usual Butterworth, Chebyshev, and Thomson characteristics but also the important elliptic functions. Frequency transformations are covered in detail, and charts are used for ready reference to the results. The charts not only present the traditional pole-zero and polynomial coefficient information but also give tabulations of quadratic factors. These latter are especially useful in many active filter synthesis methods. The subject of sensitivity is introduced in Chapter 3. In addition to treating the various types of sensitivity and their interrelations, discussions of adjoint-network sensitivity determinations and statistical sensitivity measures are given. Chapter 4 begins the discussion of active filters and treats *RC*-amplifier configurations. The development given for some of the more important classes of these configurations is also used to provide an understanding of many fundamental concepts basic to this class of active filters. In Chapter 5 the discussion of *RC*-amplifier filters is extended to include treatments of the state-variable filter, the resonator, and the universal active filter. Design relations are given for each of the various types of networks considered in these two chapters. The sensitivities of the different circuits are discussed. In Chapter 6 the use of passive network simulation methods of active *RC* filter realizations is presented. Such approaches, involving the use of synthetic inductors, generalized impedance converters, and frequency-dependent negative resistors, and utilizing configurations and element values from passive filter realizations, provide extremely low sensitivity filter structures. Design tables for the passive prototypes from which these filters are derived are given in Appendix A. The uses of other direct realization methods such as the leapfrog, the PRB, and the parallel-cascade are also included. In almost all practical active filter techniques, the basic active device which is used is the operational amplifier. A review of the basic properties of operational amplifiers is given in Appendix B. In all the chapters (4, 5, and 6) in which operational amplifiers are applied, the student is shown in detail how deviations from the ideal operational amplifier behavior, such as that produced by finite gain bandwidth, place restrictions on the performance obtainable in specific active filter configurations. Finally, in Chapter 7, some specialized modern topics in active filter research are treated, namely, their use in high-frequency (active *R*) filters and in analog sampled-data (switched-capacitor) filters. In addition, examples of the complete design procedure for some actual industrial filter applications are given.

The material presented in this book has been classroom tested at the University of Arizona, the University of California at Santa Barbara, and Texas A & M University for the past five years. It is suitable either for a one-semester course on active filters or, with the addition of some material on passive synthesis techniques (readily available from many sources), for a two-semester course on active and passive filters.

The dominant theme to be found in this book is the presentation of a unified view of modern active filter synthesis techniques. In developing this theme, the authors have relied heavily on their involvement for many years in the filter field and, rather than just presenting a taxonomy of the various published techniques, instead have selected a much shortened collection for presentation, emphasizing the realizations that have proven their "real world" utility and practicality. At the same time, considerable attention has been given to providing basic theoretical background material that will permit the reader to successfully analyze and evaluate future realizations as they are discovered and published in the literature.

The authors would like to express their appreciation to the many persons who assisted and encouraged them in the preparation of this book, especially Dr. R. H. Mattson, Head of the Department of Electrical Engineering at the University of Arizona; Dr. J. G. Skalnik, Dean of the College of Engineering at the University of California at Santa Barbara; and Dr. W. B. Jones, Jr., Head of the Department of Electrical Engineering at Texas A & M University. Thanks are also expressed to the many students whose patience and enthusiasm made the development of this material a pleasure. Finally, our special thanks to our wives and families, for their forbearance of our absence during the many hours that the manuscript was under preparation.

L. P. Huelsman
P. E. Allen

INTRODUCTION TO THE THEORY AND DESIGN OF ACTIVE FILTERS

ONE

INTRODUCTION

The study of almost any engineering subject, be it electrical, mechanical, hydraulic, thermal, etc., can always be divided into two parts, analysis and synthesis. In *analysis* we are concerned with finding the characteristics or properties of some existing system. The process is illustrated by the flow chart shown in Fig. 1.1-1a. This may be read, "Given a system, find its properties." Frequently, of course, the system may "exist" only as a schematic showing the interconnection of idealized elements. In that case, the schematic defines a *model* of the system, and the analysis then gives the properties of the model. If the system (or model) is completely specified, its properties are, of course, unique. Thus, in an analysis problem, there is *only one solution.*

In *synthesis*, on the other hand, the starting point is a desired set of properties, and the goal is to find a system in actual or (more usually) in modeled form which has those properties. The process is illustrated by the flow chart shown in Fig. 1.1-1b. This may be read, "Given a set of properties, find a system possessing them." In general, as indicated in the figure, there is usually more than one such system. Thus, in a synthesis problem, *the solution is rarely unique.* Because of this nonuniqueness a final step in the synthesis process is usually required, namely, the evaluation of several different systems, all of which have the desired original set of properties, to find out which one is best. Before this can be done, we must define what is meant by the word *best.* Another way of looking at this is that an additional property or properties must be added to those originally specified, in order that a unique choice may be made from among the systems which have been found. Obviously, the synthesis process is considerably more complicated (and challenging) than the analysis one!

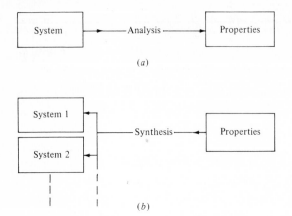

Figure 1.1-1 Flow charts for (a) analysis and (b) synthesis.

In this book our goal is to investigate the synthesis of a specific class of systems, namely, filters for electronic circuits. Such filters occur widely in modern communications and signal-processing fields. Indeed, it is difficult to find even a moderately complex electronic device that does not contain one or more filters. In our study of filters, we will be concerned with finding ways of interconnecting network elements so as to realize circuits which have specific filtering characteristics. To achieve this goal we will need to consider several topics. One of these is the determination of methods for expressing or approximating the properties of filters in such a way as to facilitate their synthesis. This topic is called *approximation*. Another useful topic is the development of methods for expressing the ways in which the properties of a given filter realization change as a result of variation from the design values of its elements. The variation may be caused by aging, temperature changes, component tolerances, etc. This topic is called *sensitivity*. It provides valuable techniques for making a comparative evaluation of filter circuits which, although different in form, have similar properties. Still another topic to be covered here is the study of *active network elements*. Filters designed using such elements often have many advantages over those using only passive ones. In the following chapters, these topics as well as others will be presented in such a way as to give the reader a comprehensive and unified treatment of the modern methods of active filter synthesis.

1.1 NETWORK FUNCTIONS

In this section we briefly review some of the basic concepts of network theory. Additional discussions of these concepts may be found in the texts listed in the Bibliography.

The usual variables that we associate with an electronic circuit are the voltages and currents measured at various points in the circuit. These are "physical" or "real world" variables in the sense that they can be measured with meters, can be displayed on an oscilloscope, and, if they are large enough, can even provide us with a tangible indication of their presence such as a spark or a shock. In most filtering situations, these variables are not constant but have values which change with time. Thus, we write them in the form $v(t)$ and $i(t)$ and refer to them as being in the *time domain*. The time-domain determination of the filtering characteristics for a specific application represents a specification of how some response (or output) variable $r(t)$, which may be a voltage or a current, is produced as the result of some excitation (or input) variable $e(t)$, which may also be either a voltage or a current. Determining the relations between the input and the output variables when they are functions of time, however, is computationally tedious, since it involves the solution of integrodifferential equations and gives little insight into the way in which the various elements of the network affect the relation. Instead, for circuits with linear time-invariant elements, we usually use the Laplace transform to create new (transformed) variables, namely, $R(s)$ for the response variable and $E(s)$ for the excitation variable. These are related to $r(t)$ and $e(t)$ by the Laplace transform

$$R(s) = \mathscr{L}\{r(t)\} = \int_0^\infty r(t)e^{-st}\,dt \qquad E(s) = \mathscr{L}\{e(t)\} = \int_0^\infty e(t)e^{-st}\,dt \qquad (1)$$

where $\mathscr{L}\{\ \}$ may be read " the Laplace transform of." In the above integrands, the exponent in the term e^{-st} must be dimensionless. Therefore, we conclude that the quantity s must have the dimensions of reciprocal time or frequency. Thus the variables $R(s)$ and $E(s)$ are said to be in the *frequency domain*, and s is called the *complex-frequency variable*. Its values are frequently displayed on a two-dimensional *complex-frequency plane*. The relation between $R(s)$ and $E(s)$ is in general given by defining a *network function* $N(s)$ as the ratio of response to excitation, for the case where all network initial conditions are zero. Thus

$$N(s) = \frac{R(s)}{E(s)} \qquad (2)$$

For example, consider the network shown in Fig. 1.1-2. If we define the response variable as $V_2(s)$ and the excitation variable as $V_1(s)$, by routine circuit analysis we find the network function $N(s)$ is

$$N(s) = \frac{V_2(s)}{V_1(s)} = \frac{0.5}{s^3 + 2s^2 + 2s + 1} \qquad (3)$$

Frequency-domain expressions, such as the one given above, are completely general in the sense that they apply to almost any type of time-domain excitation

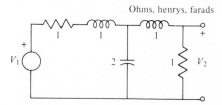

Ohms, henrys, farads

Figure 1.1-2 A circuit which has the network function given in (3).

function. Most filtering requirements, however, are based on the *sinusoidal steady-state* behavior of a network. Thus they assume that the excitation has the form

$$e(t) = \sqrt{2} E_0 \cos(\omega t + \alpha) \tag{4}$$

where E_0 is the root mean square (rms) value of $e(t)$, ω is the frequency in radians per second, and α is the phase in radians. For such an excitation, assuming $N(s)$ is stable,[1] after the transient time-domain components of $r(t)$ have decayed to the point where they are negligible, i.e., after a "steady state" has been reached, $r(t)$ will have the form

$$r(t) = \sqrt{2} R_0 \cos(\omega t + \beta) \tag{5}$$

where R_0 is the rms value of $r(t)$ and β is its phase. The relation between $e(t)$ and $r(t)$ under these conditions is readily found by using complex numbers called *phasors* to represent the sinusoidally varying quantities. For $e(t)$ and $r(t)$ we will use $\mathscr{E}$ and $\mathscr{R}$ to represent the phasors. These are defined as

$$\mathscr{E} = E_0 e^{j\alpha} \qquad \mathscr{R} = R_0 e^{j\beta} \tag{6}$$

The relationship between the phasors $\mathscr{E}$ and $\mathscr{R}$ is directly obtained by replacing the variable s in the network function $N(s)$ in (2) with the variable $j\omega$. Thus we obtain

$$N(j\omega) = \frac{\mathscr{R}}{\mathscr{E}} = \frac{R_0 e^{j\beta}}{E_0 e^{j\alpha}} \tag{7}$$

Frequently we are interested only in the way the rms magnitudes of the excitation and response sinusoids are related. In such a case, taking the magnitudes of both sides of (7), we obtain

$$|N(j\omega)| = \left|\frac{\mathscr{R}}{\mathscr{E}}\right| = \frac{|\mathscr{R}|}{|\mathscr{E}|} = \frac{R_0}{E_0} \tag{8}$$

This magnitude may of course be expressed in logarithmic measure (decibels) by taking $20 \log |N(j\omega)|$. Alternately, we may be interested in the phase difference between the excitation and response sinusoids. In this case, taking the argument of both sides of (7), we may write

$$\arg N(j\omega) = \arg \frac{\mathscr{R}}{\mathscr{E}} = \arg \mathscr{R} - \arg \mathscr{E} = \beta - \alpha \tag{9}$$

As examples of these quantities, for the network of Fig. 1.1-2 and the network function given in (3) we find that

$$|N(j\omega)| = \frac{0.5}{[(1 - 2\omega^2)^2 + (2\omega - \omega^3)^2]^{1/2}} \tag{10}$$

$$\arg N(j\omega) = -\tan^{-1} \frac{2\omega - \omega^3}{1 - 2\omega^2} \tag{11}$$

---

[1] By *stable* we mean that for any bounded excitation $e(t)$, the response $r(t)$ will also be bounded.

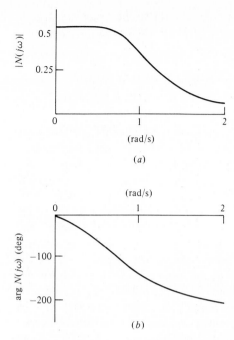

(rad/s)

(*a*)

(rad/s)

(*b*)

Figure 1.1-3 Magnitude and phase plots for the network of Fig. 1.1-2.

Plots of these functions are given in Fig. 1.1-3. Thus we see that the network function $N(s)$ defined in (2) has a very important property, namely, it determines for all frequencies the sinusoidal steady-state response of the network for which it is specified. Almost all filter synthesis methods use the network function as a starting point.

1.2 PROPERTIES OF NETWORK FUNCTIONS

In the preceding section we introduced the concept of a network function. Such functions may be defined in various ways. The simplest situation occurs for the case where the voltage and current variables of a network are only of interest at a single pair of terminals. By convention, the relative reference polarities of the variables are arranged as shown in Fig. 1.2-1, the network is referred to as a *one-port network*, and the network functions are called *driving-point functions*.

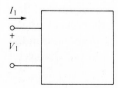

Figure 1.2-1 A one-port network.

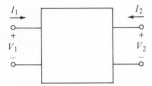

Figure 1.2-2 A two-port network.

Only two types of driving-point network functions are defined for a one-port network. If current is treated as the excitation variable, then the network function is defined as a *driving-point impedance* $Z(s) = V_1(s)/I_1(s)$; while if voltage is considered as the excitation variable, the network function is defined as a *driving-point admittance* $Y(s) = I_1(s)/V_1(s)$. Obviously $Y(s) = 1/Z(s)$. The situation which is of more interest in practical filtering applications is the one where there are four external terminals brought out from the network. These are arranged into pairs called *ports* so as to define four variables, two of voltage and two of current. By convention, the relative reference polarities are arranged as shown in Fig. 1.2-2, and the network is called a *two-port network*. The lower terminals of the two ports are frequently common, in which case the network is sometimes referred to as a *three-terminal network*. In addition to the driving-point functions which may be defined for such a two-port network, any network function which involves variables from both of the two ports is called a *transfer function*. Such functions can be transfer impedances and admittances as well as dimensionless transfer voltage functions such as $V_2(s)/V_1(s)$ or $V_1(s)/V_2(s)$, or dimensionless transfer current functions such as $I_2(s)/I_1(s)$ or $I_1(s)/I_2(s)$. The latter require that a path be available for the flow of the response current. This is usually provided by putting a short circuit or a load across the terminals of the response port.

In general, a network function $N(s)$ has the form of a ratio of a numerator polynomial $A(s)$ to a denominator polynomial $B(s)$. Thus

$$N(s) = \frac{A(s)}{B(s)} \tag{1}$$

The zeros of the numerator polynomial $A(s)$ are referred to as the *zeros of the network function* $N(s)$; that is, they are the values of s, also referred to as the *locations on the complex-frequency plane*, where the magnitude of $N(s)$ is zero. The zeros of the denominator polynomial $B(s)$, on the other hand, are referred to as the *poles of the network function* $N(s)$. They are the values of s, or the locations on the complex-frequency plane, where the magnitude of $N(s)$ is infinite. The location of these poles is directly related to the filtering properties of a given network. For example, for the network to be stable, the poles of its network function must be in the left-half plane, or, if on the $j\omega$ axis, they must be simple, i.e., only of first order. Networks whose network functions have right-half-plane poles, or have $j\omega$-axis poles of greater than first order, are unstable; i.e., they have an unbounded response for a bounded input, and thus they do not represent a physically useful situation.

Now let us consider how the form of a network function for a given circuit is affected by the elements of which the filter is comprised. In general for lumped, linear, finite networks, $A(s)$ and $B(s)$ will be polynomials with real coefficients. Thus the network function will have the form

$$N(s) = \frac{A(s)}{B(s)} = \frac{a_0 + a_1 s + a_2 s^2 + \cdots}{b_0 + b_1 s + b_2 s^2 + \cdots} \qquad (2)$$

In this case $N(s)$ is called a (real) *rational function*, i.e., a ratio of polynomials. For a three-terminal network containing only the passive elements resistors, capacitors, and inductors (an *RLC* network), the coefficients b_i of (2) will all be nonnegative and the poles of $N(s)$ will always be in the left-half plane and/or be simple on the $j\omega$ axis. If active elements such as controlled sources are present, however, the coefficients b_i may also be negative and the poles of $N(s)$ may be in the right half of the complex-frequency plane. In this case, of course, the network will be unstable.

A further classification of the properties of the poles of $N(s)$ may be made by considering the possible location of these poles for networks comprising various types of elements. For example, for networks with *RLC* elements, the poles may be anywhere in the left half of the complex-frequency plane and/or on the $j\omega$ axis. In the latter case they must be simple. For *LC* elements the poles will lie only on the $j\omega$ axis (and be simple). For *RC* or *RL* elements the poles will be only on the negative-real axis or at the origin. However, for networks consisting of *RC* elements and amplifiers (modeled as controlled sources), the poles may be anywhere in the complex plane. Filters realized by such networks are referred to as *RC-amplifier filters*. They are one example of the class of filters referred to as *active filters* or *active RC filters*. They basically provide all the advantages of *RLC* networks, in that they may have any desired left-half-plane pole locations, but

Table 1.2-1 Characteristics of poles for various classes of network elements

Elements				Operational amplifier or controlled source	Order and location of poles
R	C	L	GIC		
X	X				Simple on the negative-real axis and at the origin
X		X			Simple on the negative-real axis and at the origin
	X	X			Simple on the $j\omega$ axis
X	X	X			Any order in the left-half plane and simple on the $j\omega$ axis
X	X		X		Any order in the left-half plane or on the $j\omega$ axis
X	X			X	Any order anywhere in the complex plane

they do not have the disadvantages of requiring the use of inductors. Since inductors are heavy, *RC* circuits avoid weight problems, an important factor in space satellite applications. In addition, they avoid inductor nonlinearities due to saturation, and inductor nonidealness or dissipation due to wiring and core losses. Finally, they are readily manufactured in integrated form, whereas the inductor is basically nonintegrable. Thus the active *RC* filter has many advantages over the passive *RLC* one, except for those applications where extremely large voltages or currents must be filtered, as for example in radar pulsing networks. In addition to the *RC*-amplifier filter referred to above, there are several other types of active *RC* filters, for example, ones which use GICs (*generalized impedance converters*) or operational amplifiers for their active elements. The properties of network functions for circuits composed of various combinations of these and the other network elements referred to above are summarized in Table 1.2-1.

1.3 TYPES OF FILTERS

In this section we briefly review the major types of filters that we shall be concerned with in this book. The first of these is the *low-pass filter*. By definition, such a filter has the properties that low-frequency excitation signal components, down to and including direct current, are transmitted, while high-frequency components, up to and including infinite ones, are blocked. Thus the magnitude of a low-pass network function has ideally the appearance shown in Fig. 1.3-1a. The range of low frequencies which are passed is called the *passband* or the *bandwidth* of the filter. As shown in the figure, it is equal to the value of the highest frequency ω_c which is transmitted. This frequency is also called the *cutoff frequency*. In practice, the ideal magnitude characteristic shown can only be approximated. Several such approximations are shown in Fig. 1.3-1b through *d*. Note that in

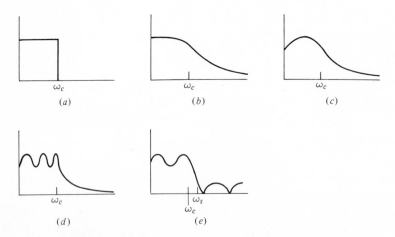

Figure 1.3-1 Some magnitude functions for low-pass filters.

every case, the magnitude of the network function reaches zero only at infinite frequency. For these characteristics, the bandwidth is defined in terms of some maximum excursion of the magnitude characteristic from its peak value. The peak value may occur at zero frequency, as shown in Fig. 1.1-3b or at some inter-mediate frequency or frequencies, as shown in Fig. 1.3-1c and d. The excursion or ripple is frequently specified as 3 dB, but other values may also be chosen. Magni-tude characteristics in which all the maximum peak and the minimum valleys in a given range of frequencies have the same value, as shown in Fig. 1.3-1c and d for the passband, are called *equal-ripple*. A characteristic in which the derivative of the magnitude does not change sign over a given range of frequencies is called *mono-tonic*. For example, the characteristic shown in Fig. 1.3-1b is completely mono-tonic. As other examples, the characteristics shown in Fig. 1.3-1c and d may be described as being equal-ripple in the passband and monotonic outside of the passband. The low-pass network functions whose magnitudes are shown in Fig. 1.3-1b through d most commonly have all their zeros located at infinity. Thus their numerator polynomials are of zero degree, i.e., a constant, independent of the degree of the denominator polynomial. Their general form is

$$N(s) = \frac{H}{B(s)} \tag{1}$$

where H is not a function of s, and where the form of the polynomial $B(s)$ depends on the elements of the network. For example, for the network shown in Fig. 1.1-2, $B(s) = s^3 + 2s^2 + 2s + 1$.

The magnitude characteristic of a low-pass filter has a nonzero positive value at zero frequency. Thus, the phase characteristic for all low-pass filters whose network functions have the form shown in (1) starts at zero (assuming $H > 0$), and decreases for increasing frequency. If the network function is rational, the maxi-mum phase is $-90n°$ at infinite frequency, where n is the order of the denominator polynomial $B(s)$. Thus, the maximum phase of the filter shown in Fig. 1.1-2 is $-270°$.

A modification of the basic low-pass magnitude characteristic occurs when equal-ripple behavior occurs in both the passband and in a stopband having a range $\omega_s \leq \omega < \infty$ as shown in Fig. 1.3-1e. Such a characteristic is called an *elliptic* one. The infinite frequency magnitude can be zero or (as shown) nonzero. To have an elliptic characteristic, the network function must have zeros on the $j\omega$ axis of the complex-frequency plane. Thus elliptic network functions have the form

$$N(s) = \frac{A(s)}{B(s)} \tag{2}$$

where the zeros of $A(s)$ are on the $j\omega$ axis and the zeros of $B(s)$ are in the left-half plane.

The second type of filter that we shall consider in this section is the *high-pass filter*. It has the property that low frequencies (the stopband) are blocked while high frequencies (the passband) are transmitted. Several magnitude characteristics

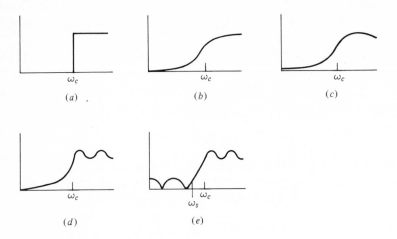

(a)   (b)   (c)

(d)   (e)

Figure 1.3-2 Some magnitude functions for high-pass filters.

for high-pass network functions are shown in Fig. 1.3-2. The terminology to describe these is similar to that used for the low-pass ones. Figure 1.3-2*a* shows an ideal (not realizable) characteristic. Figure 1.3-2*b* shows a completely monotonic characteristic. Figure 1.3-2*c* and *d* shows characteristics which are equal-ripple in the passband and monotonic outside the passband. Figure 1.3-2*e* shows a characteristic which is equal-ripple in the passband and in the stopband $0 \leq \omega \leq \omega_s$, that is, an elliptic function. Its zero-frequency magnitude may be zero or (as shown) nonzero. For high-pass filters the passband is of infinite width, since in theory it extends to infinite frequency. As a result, rather than specifying the bandwidth, we more meaningfully specify the *cutoff frequency*, which is shown as ω_c in the figures. The high-pass functions with magnitude characteristics shown in Fig. 1.3-2*a* through *d* most commonly have all their zeros located at the origin of the complex-frequency plane. Thus, for rational functions they have the form

$$N(s) = \frac{Hs^n}{B(s)} \tag{3}$$

where H is a constant and n is the degree of the denominator polynomial $B(s)$. For example, for the high-pass network shown in Fig. 1.3-3, the network function is readily shown to be

$$\frac{V_2(s)}{V_1(s)} = \frac{0.5s^3}{s^3 + 2s^2 + 2s + 1} \tag{4}$$

The phase characteristic for a rational high-pass network function having the form given in (3) starts at $+90n°$ at zero frequency and decreases to zero degrees at infinite frequency.

The third general filter type that we shall consider here is the *bandpass filter*. It has the property that one band of frequencies (the passband) is transmitted, while

Ohms, henrys, farads

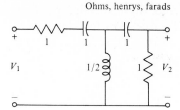

Figure 1.3-3 A high-pass filter.

two bands of frequencies, namely, those below and above the passband, are blocked (the stopbands). Several network function magnitude characteristics for bandpass filters are shown in Fig. 1.3-4. The range of frequencies which is passed is called the *bandwidth* (BW), and is defined as the difference between the frequencies which define the edges of the passband. Using ω_1 and ω_2 as shown in the figures to define the passband edges, we obtain

$$\text{BW} = \omega_2 - \omega_1 \tag{5}$$

The *center frequency* ω_0 of the passband is defined as the geometric mean of the band-edge frequencies. Thus

$$\omega_0 = \sqrt{\omega_1 \omega_2} \tag{6}$$

Figure 1.3-4a shows an ideal (not realizable) bandpass characteristic. Figure 1.3-4b shows a characteristic which is completely monotonic in the sense that on either side of the center frequency ω_0 the derivative of its magnitude characteristic does not change sign. Similarly, in Fig. 1.3-4c and d, in the passband the behavior is equal-ripple, while outside the passband the behavior is monotonic. Finally, in Fig. 1.3-4e an elliptic bandpass characteristic is shown. Its magnitude at zero and

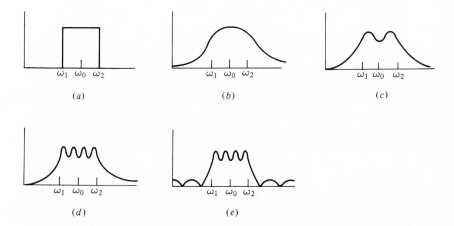

Figure 1.3-4 Some magnitude functions for bandpass filters.

infinite frequencies may be zero, or (as shown) nonzero. All the bandpass magnitude characteristics shown in Fig. 1.3-4a through d have network functions which most commonly have half their zeros at the origin and the other half at infinity. Thus, for rational functions they have the form

$$N(s) = \frac{Hs^{n/2}}{B(s)} \tag{7}$$

where H is a constant and where n is the degree of the denominator polynomial $B(s)$, and is always even. The phase characteristic for such a function starts at $(+90n/2)°$ at zero frequency and decreases to $(-90n/2)°$ at infinite frequency. It is zero at the center frequency. There are many other types of network functions such as those having band-elimination characteristics or all-pass characteristics. We will defer a treatment of these until later in the text.

1.4 FREQUENCY AND IMPEDANCE NORMALIZATIONS

The specifications which are usually given for the design of filters involve frequencies which have values of thousands of cycles per second. Synthesis calculations, however, are most easily done using frequencies of a few hertz or radians per second, since the numerical computations are simplified (and mistakes are minimized) by not having to carry along various powers of 10. Design tables for various filter characteristics also use similar convenient frequency values. Such values are usually referred to as *normalized frequency* values. To convert such normalized values to the "real world" frequencies actually required in a given filter application, we use a process called *frequency denormalization*. This involves a change of complex-frequency variable. If we consider p as the normalized complex-frequency variable and s as the denormalized one, then the frequency-denormalization process is defined by the relation

$$s = \Omega_n p \tag{1}$$

where Ω_n is called the *frequency-denormalization constant*. For example, consider an inductor with a frequency-normalized value of L henrys and a corresponding frequency-normalized impedance $Z_n(p) = pL$. The denormalized impedance $Z(s)$ is found from (1) to be $Z(s) = sL/\Omega_n$; thus, it represents an inductance of value L/Ω_n. Similarly a denormalized capacitor will have a value of C/Ω_n. Additional relations are readily developed for the way in which other network elements are affected by frequency denormalization. These are summarized in the first line of Table 1.4-1. The corresponding effects on network functions are given in the first line of Table 1.4-2. In using these relations, it is helpful to remember that elements such as resistors and controlled sources, whose network functions are not functions of s, are left invariant by a frequency denormalization; whereas reactive elements, such as inductors and capacitors, which are characterized by network functions in which s appears, change value. For such reactive elements, the change in their value is in inverse proportion to the change in frequency.

Table 1.4-1 Effect of frequency and impedance denormalization on network elements

Denormal- ization	R	C	L	VCVS gain $= \alpha$	CCCS gain $= \beta$	VCCS gain $= g$	CCVS gain $= r$	Ideal trans- former, turns ratio $= N$	FDNR*
$s = \Omega_n p$ (frequency)	R	$\dfrac{C}{\Omega_n}$	$\dfrac{L}{\Omega_n}$	α	β	g	r	N	$\dfrac{D}{\Omega_n^2}$
$Z = z_n Z_n$ (impedance)	$z_n R$	$\dfrac{C}{z_n}$	$z_n L$	α	β	$\dfrac{g}{z_n}$	$z_n r$	N	$\dfrac{D}{z_n}$
Frequency and impedance	$z_n R$	$\dfrac{C}{\Omega_n z_n}$	$\dfrac{z_n L}{\Omega_n}$	α	β	$\dfrac{g}{z_n}$	$z_n r$	N	$\dfrac{D}{\Omega_n^2 z_n}$

* This element is introduced in Chap. 5.

As an example of frequency normalization, consider the (normalized) network shown in Fig. 1.4-1a. The voltage transfer function for this, using p as the complex-frequency variable, is readily shown to be

$$\frac{V_2(p)}{V_1(p)} = N_n(p) = \frac{0.6p}{p^2 + 0.6p + 2} \tag{2}$$

The poles are at $p = -0.3 \pm j1.38203$. From the results given in Sec. 1.3 we may identify this as a bandpass characteristic. It may be shown to have a center frequency of $\sqrt{2}$ rad/s (the square root of the zero degree coefficient of the denominator). The frequency denormalization of (1) may be applied to (2) to develop a magnitude characteristic having exactly the same shape, but with a center

Table 1.4-2 Effect of frequency and impedance denormalization on network functions

Denormal- ization	Poles p_i	Zeros z_i	Voltage transfer function $V_2/V_1 = A(p)$	Current transfer function $I_2/I_1 = B(p)$	Transfer impedance $V_2/I_1 = Z(p)$	Transfer admittance $I_2/V_1 = Y(p)$
$s = \Omega_n p$ (frequency)	$\Omega_n p_i$	$\Omega_n z_i$	$A\left(\dfrac{s}{\Omega_n}\right)$	$B\left(\dfrac{s}{\Omega_n}\right)$	$Z\left(\dfrac{s}{\Omega_n}\right)$	$Y\left(\dfrac{s}{\Omega_n}\right)$
$Z = z_n Z_n$ (impedance)	p_i	z_i	$A(p)$	$B(p)$	$z_n Z(p)$	$\dfrac{Y(p)}{z_n}$
Frequency and impedance	$\Omega_n p_i$	$\Omega_n z_i$	$A\left(\dfrac{s}{\Omega_n}\right)$	$B\left(\dfrac{s}{\Omega_n}\right)$	$z_n Z\left(\dfrac{s}{\Omega_n}\right)$	$Y\left(\dfrac{s}{\Omega_n}\right)/z_n$

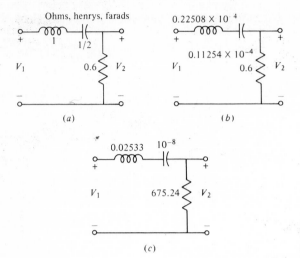

(a)

(b)

(c)

Figure 1.4-1 Frequency and impedance denormalization of a filter.

frequency of 10 kHz, by specifying the frequency normalization constant $\Omega_n = (2\pi/\sqrt{2}) \times 10^4 = 4.4429 \times 10^4$. Substituting $s/4.4429 \times 10^4$ for p in (2) we obtain the denormalized network function $N(s)$

$$\frac{V_2(s)}{V_1(s)} = N(s) = \frac{0.6 \times 4.4429 \times 10^4 s}{s^2 + 0.6 \times 4.4429 \times 10^4 s + 2 \times (4.4429 \times 10^4)^2} \qquad (3)$$

which has poles at $s = -1.33287 \times 10^4 \pm j6.14019 \times 10^4$, and which is readily shown to have a center frequency of 10^4 Hz. Applying the same frequency denormalizations to the elements of the network shown in Fig. 1.4-1a, we obtain the frequency-denormalized network shown in Fig. 1.4-1b, which realizes (3).

The role of frequency normalization in synthesis can be further visualized by reference to Fig. 1.4-2. Either of two paths may be followed to synthesize a denormalized filter (the block at the right) starting from a normalized network function (the block at the left). In the upper path the synthesis operation is applied first, followed by a frequency denormalization. In the lower path the order of the two operations is reversed. Following either path produces the same result; however,

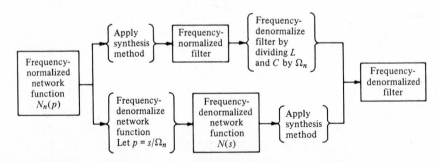

Figure 1.4-2 Flowchart for performing synthesis and frequency denormalization.

the upper one is recommended since in it the numerical computations are greatly simplified.

A second type of denormalization which is frequently used to simplify the numerical computations made during a synthesis operation is called an *impedance denormalization*. It permits the use of elements with numerical values in a range close to unity, rather than ones with real-world multipliers of 10^3, 10^6, 10^{-6}, 10^{-12}, etc. A normalized impedance $Z_n(s)$ can be denormalized to the impedance $Z(s)$, that is, converted to practical values by the relation

$$Z(s) = z_n Z_n(s) \tag{4}$$

where z_n is called the *impedance-denormalization constant*. For example, consider an inductor with an impedance-normalized value of L henrys. Its impedance-normalized impedance is $Z_n(s) = sL$, and the denormalized (practical) impedance is $Z(s) = sz_n L$, representing an inductor of value $z_n L$. On the other hand, for a capacitor with an impedance-normalized value of C farads, the impedance-normalized impedance is $Z_n(s) = 1/sC$, and the denormalized impedance is $Z(s) = z_n/sC$, representing a capacitor of value C/z_n. Thus we see that when an impedance denormalization is made, the values of inductors (in henrys) and capacitors (in farads) move in opposite directions. Similar relations are readily developed for other types of network elements and for network functions. They are summarized in the second lines of Tables 1.4-1 and 1.4-2. As an example of an impedance denormalization, let us choose z_n such that the capacitor in the network shown in Fig. 1.4-1*b* is changed to a value of 10^{-8} F. From (4) we find that $1/s10^{-8} = z_n/(s \times 0.11254 \times 10^{-4})$. Thus $z_n = 1125.4$. The denormalized values of the network elements are given in Fig. 1.4-1*c*. Since the network function of (3) for the network in Fig. 1.4-1*b* is dimensionless, it is unaffected by the impedance transformation; thus, it also applies to the network shown in Fig. 1.4-1*c*. It should be noted that, as indicated in the third lines of Tables 1.4-1 and 1.4-2, the operations of frequency and impedance denormalization are commutative: they may be applied in either order, and the results obtained will be the same.

1.5 AN EXAMPLE OF FILTER USAGE

An interesting example of how filters are used in a real-life application is the Touch-tone® dialing system now used in many telephone installations. In such a system, the telephone handset has a group of 12 pushbuttons which replace the rotating dial. When depressed, each of these pushbuttons energizes oscillators which simultaneously generate and put on the line a low-band and a high-band audio frequency signal. The frequency assignments and the arrangement of the pushbuttons are shown in Fig. 1.5-1*a*. For example, the 3 pushbutton generates 697 and 1477 Hz. Thus, each pushbutton is identified by a two-tone signal code. At the central telephone office, these signals are decoded using a set of filters arranged as shown in Fig. 1.5-1*b*. Low-pass and high-pass filters are used to

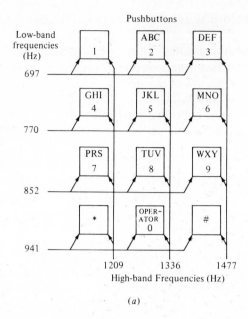

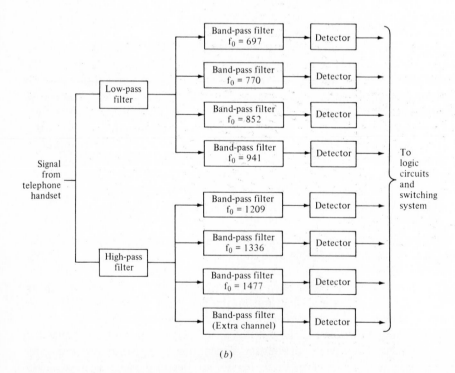

Figure 1.5-1 The Touch-tone® telephone dialing system.

separate the low-band and high-band groups of frequencies, and bandpass filters are used to identify specific frequencies in each band. Detectors monitor the output of the bandpass filters and, through the use of simple logic circuits, permit identification of the pushbutton which was pressed. This system permits far more rapid signaling than is possible with the older rotary-dial system, in which individual dc pulses are counted to identify each digit. Obviously, filters play an important role in the functioning of the Touch-tone® system.

PROBLEMS

1-1 (*Sec. 1.1*) Derive the expression given in (3) of Sec. 1.1 for the voltage transfer function of the network shown in Fig. 1.1-2.

1-2 (*Sec. 1.1*) Derive the expressions given in (10) and (11) of Sec. 1.1 for the voltage transfer function magnitude and phase for the network shown in Fig. 1.1-2.

1-3 (*Sec. 1.1*) (*a*) Find the voltage transfer function for the network shown in Fig. 1.3-3.
 (*b*) Find expressions for the magnitude and phase of the transfer function.

1-4 (*Sec. 1.2*) Find the voltage transfer functions for each of the networks shown in Fig. P1-4, and show that the pole locations agree with the cases given in Table 1.2-1.

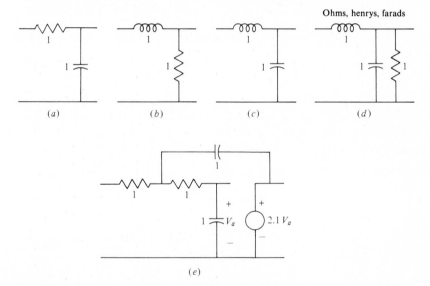

Figure P1-4

1-5 (*Sec. 1.2*) Find the driving-point admittance (at port 1) of the network shown in Fig. 1.1-2, and show that the poles are the same as those of the voltage transfer function for the network, given in (3) of Sec. 1.1.

1-6 (*Sec. 1.2*) Find the driving-point admittance (at port 1) of the network shown in Fig. 1.3-3, and show that the poles are the same as those of the voltage transfer function determined in Prob. 1-3.

1-7 (*Sec. 1.3*) Magnitude characteristics of network functions are frequently illustrated using units of attenuation rather than gain for the ordinate of the plots. Assuming such a convention, identify the ideal plots shown in Fig. P1-7 as low-pass, high-pass, bandpass, or band elimination.

(a)   (b)   (c)   (d)

Figure P1-7

1-8 (*Sec. 1.4*)   The low-pass network shown in Fig. 1.1-2 has a bandwidth of 1 rad/s for the specified element values.

(a) Find the element values that will result when a frequency denormalization is made such that the bandwidth is changed to 1 kHz.

(b) Write the voltage transfer function for the frequency-denormalized network.

(c) Perform an impedance denormalization such that the terminating resistor has a value of 10 kΩ.

(d) Write the voltage transfer function for the frequency- and impedance-denormalized network.

1-9 (*Sec. 1.4*)   (a) The frequency denormalization used in Prob. 1-8 is applied to the network shown in Fig. 1.1-2. Write the driving-point admittance function (originally derived in Prob. 1-5) for the frequency denormalized network.

(b) If the impedance denormalization specified in Prob. 1-8 is now also applied to the network, write the driving-point impedance function for the resulting circuit.

1-10 (*Sec. 1.4*)   The high-pass network shown in Fig. 1.3-3 has a cutoff frequency of 1 rad/s for the specified element values.

(a) Find the element values that will result when a frequency denormalization is made such that the cutoff frequency is changed to 1 kHz.

(b) Write the voltage transfer function for the frequency-denormalized network.

(c) Perform an impedance denormalization such that the terminating resistor has a value of 10 kΩ.

(d) Write the voltage transfer function for the frequency- and impedance-denormalized network.

1-11 (*Sec. 1.4*)   The voltage transfer function for the bandpass network shown in Fig. P1-11 has a center frequency of 1 rad/s and a bandwidth of 0.1 rad/s.

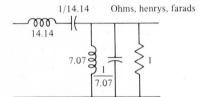

Figure P1-11

(a) Find the element values that result when a frequency denormalization is made such that the center frequency is 1 kHz.

(b) What is the bandwidth of the denormalized network?

(c) What are the element values when an additional impedance denormalization is made so that the terminating resistor has a value of 1 kΩ?

TWO

APPROXIMATION

In Chaps. 4 through 7 we will discuss methods for using active *RC* networks to synthesize transfer functions. In applying these methods we will assume that a network function—i.e., a ratio of polynomials defining a set of pole and zero locations in the *complex frequency* (or *s*) *plane*—is given as the starting point for the realization procedure. In practical filter synthesis, however, the filter designer's main concern is usually to meet some specifications on the *sinusoidal steady-state* performance of the network, i.e., its magnitude and/or phase characteristics as a function of the real frequency variable ω. In this chapter we present a discussion of how these two methods of characterizing network performance correspond to each other, i.e., how sinusoidal steady-state frequency response characteristics are related to pole and zero locations. Such relations are characterized by the general term *approximation*.

2.1 MAGNITUDE APPROXIMATION—THE MAXIMALLY FLAT CHARACTERISTIC

One of the most frequently used types of approximation is that relating the magnitude $|N(j\omega)|$, specified either by a mathematical expression, a set of data values, or a plotted waveshape, to a rational function $F(s)$, so that in some specified sense $|F(j\omega)|$ approximates $|N(j\omega)|$. Ideally, of course, we would like the two functions to be identical, and in many cases we shall find that this is true. Magnitude specifications are usually given either in linear or logarithmic measure. In the latter case decibels [$20 \log |N(j\omega)|$], abbreviated dB, are used. Table 2.1-1 gives some relations between magnitude and decibel values.

Table 2.1-1 Relations between decibels, magnitude, and ε

dB	Mag	ε^2	ε
-.010000	.998849	.002305	.048013
-.020000	.997700	.004616	.067940
-.050000	.994260	.011579	.107608
-.100000	.988553	.023293	.152620
-.200000	.977237	.047129	.217091
-.500000	.944061	.122018	.349311
-1.000000	.891251	.258925	.508847
-2.000000	.794328	.584893	.764783
-3.010300	.707107	1.000000	1.000000
-5.000000	.562341	2.162278	1.470469
-10.000000	.316228	9.000000	3.000000
-20.000000	.100000	99.000000	9.949874
-30.000000	.031623	999.000000	31.606961
-40.000000	.010000	9999.000000	99.995000
-50.000000	.003162	99999.000000	316.226185

Mag	dB	ε^2	ε
.999000	-.008690	.002003	.044755
.990000	-.087296	.020304	.142492
.980000	-.175478	.041233	.203059
.970000	-.264565	.062812	.250624
.950000	-.445528	.108033	.328684
.900000	-.915150	.234568	.484322
.800000	-1.938200	.562500	.750000
.707107	-3.010300	1.000000	1.000000
.500000	-6.020600	3.000000	1.732051
.200000	-13.979400	24.000000	4.898979
.100000	-20.000000	99.000000	9.949874
.050000	-26.020600	399.000000	19.974984
.020000	-33.979400	2499.000000	49.989999
.001000	-60.000000	999999.000000	999.999500
.000500	-66.020600	3999999.000000	1999.999750

We begin our study of magnitude approximation by considering the necessary properties that a magnitude function must have. It is more convenient to actually consider the square of the magnitude function. Thus, we may write

$$|N(j\omega)|^2 = N(j\omega)N^*(j\omega) = N(j\omega)N(-j\omega) \qquad (1)$$

where the justification for the right member of the equation is that, for rational functions with real coefficients, the conjugate of the function is found by replacing the variable by its conjugate, i.e., by replacing $j\omega$ with $-j\omega$. Now let $N(s)$ have the form

$$N(s) = \frac{b_0 + b_1 s + b_2 s^2 + b_3 s^3 + b_4 s^4 + \cdots}{a_0 + a_1 s + a_2 s^2 + a_3 s^3 + a_4 s^4 + \cdots} \tag{2}$$

Thus $N(j\omega)$ will have the form

$$N(j\omega) = \frac{b_0 - b_2 \omega^2 + b_4 \omega^4 - \cdots + j(b_1 \omega - b_3 \omega^3 + \cdots)}{a_0 - a_2 \omega^2 + a_4 \omega^4 - \cdots + j(a_1 \omega - a_3 \omega^3 + \cdots)} \tag{3}$$

Inserting this relation in the right member of (1) we readily see a first property of $|N(j\omega)|^2$, that it will be a ratio of even polynomials. If we now evaluate (1) by letting $\omega = s/j$ we may define a function $T(s^2)$ as

$$T(s^2) = |N(j\omega)|^2 \Big|_{\omega = s/j} = N(s)N(-s) \tag{4}$$

From (4) we see the poles and zeros of $T(s^2)$ must have *quadrantal symmetry* in the s plane, i.e., they must be symmetrically located in the right- and left-half planes with respect to the origin. This, of course, is necessary for $T(s^2)$ to equal the product $N(s)N(-s)$. In general the numerator and denominator polynomials of $T(s^2)$ can have only three kinds of factors: (1) $s^4 + as^2 + b$, (2) $s^2 - a$ $(a > 0)$, and (3) $s^2 + a$ $(a > 0)$. The first and second types of factors have the necessary quadrantal symmetry, but the third type does not unless it has even multiplicity, that is, unless it appears as $(s^2 + a)^2$, $(s^2 + a)^4$, etc., in which case the resulting even-multiplicity $j\omega$-axis zeros have the necessary symmetry. Considering the above discussion, we see that *for a given* $|N(j\omega)|^2$ *to be the magnitude-squared function of some rational function $N(s)$, it is necessary that* (1) *the function* $|N(j\omega)|^2$ *be a ratio of even polynomials in ω, and that* (2) *in the related function $T(s^2)$ defined in (4), any poles or zeros on the $j\omega$ axis be of even order.* The sufficiency of these two conditions is readily demonstrated by factoring $T(s^2)$ into the product $N(s)N(-s)$, taking the left-half-plane poles, and half of any even-order $j\omega$-axis pole pairs from $T(s^2)$ as the poles of $N(s)$, and similarly assigning either right- or left-half-plane zeros and half of any even-order $j\omega$-axis zeros from $T(s^2)$ as the zeros of $N(s)$. The restriction of using only the *left-half-plane* poles from $T(s^2)$ is of course simply a stability consideration. An example follows.

Example 2.1-1 *Finding a Network Function from Its Magnitude Specification* As an example of the sufficiency argument given above, consider the function

$$|N(j\omega)|^2 = \frac{\omega^2 + 1}{\omega^4 + 1} \tag{5}$$

The related function $T(s^2)$ is easily seen to be

$$T(s^2) = \frac{-s^2 + 1}{s^4 + 1} = \frac{(s + 1)(-s + 1)}{(s^2 + \sqrt{2}s + 1)(s^2 - \sqrt{2}s + 1)} \tag{6}$$

Comparing this with (4), we see that there are two network functions $N(s)$ which satisfy (5). These are

$$N_1(s) = \frac{s+1}{s^2 + \sqrt{2}\,s + 1} \qquad \text{and} \qquad N_2(s) = \frac{s-1}{s^2 + \sqrt{2}\,s + 1} \qquad (7)$$

where, in the expression for $N_2(s)$, we have reversed the signs in the numerator, since such an operation has no effect on $|N(j\omega)|^2$. □

The necessary and sufficient conditions developed above for magnitude-squared functions in general are readily applied to specific filter characteristics. As an example, consider the determination of a magnitude-squared function which, in the low-frequency range starting at zero, has as flat a characteristic as possible. One way of obtaining such a flatness is to set as many derivatives of the function as possible to zero at $\omega = 0$ rad/s. Such a function is called *maximally flat*. To see how this can be done, we may write an expression for a general magnitude-squared function $|N(j\omega)|^2$ as follows:

$$|N(j\omega)|^2 = H^2 \frac{1 + b_1\omega^2 + b_2\omega^4 + \cdots}{1 + a_1\omega^2 + a_2\omega^4 + \cdots} \qquad (8)$$

If we now divide the denominator into the numerator we obtain

$$|N(j\omega)|^2 = H^2[1 + (b_1 - a_1)\omega^2 + (b_2 - a_2 + a_1^2 - a_1 b_1)\omega^4 + \cdots] \qquad (9)$$

Now consider a general MacLaurin series, i.e., a Taylor series expansion at the origin, of an arbitrary function $F(\omega)$. This has the form

$$F(\omega) = F(0) + \frac{F^{(1)}(0)}{1!}\omega + \frac{F^{(2)}(0)}{2!}\omega^2 + \frac{F^{(3)}(0)}{3!}\omega^3 + \frac{F^{(4)}(0)}{4!}\omega^4 \cdots \qquad (10)$$

where $F^{(i)}(0)$ is the ith derivative of $F(\omega)$ evaluated at $\omega = 0$. Comparing this expression with the expansion for $|N(j\omega)|^2$ given in (9), and recalling that such an expansion must be unique, we see that due to the even nature of $|N(j\omega)|^2$, all its odd-ordered derivatives are already zero. In addition, for the second derivative to be zero we require that the coefficients a_1 and b_1 be equal. Similarly, for the fourth derivative to also be zero requires that, in addition, a_2 equal b_2, etc. Thus, the general maximally flat magnitude-squared function $|N(j\omega)|^2$ in (8) is characterized by the restriction that

$$a_i = b_i \qquad (11)$$

for as many coefficients as possible. As an example of the application of this criterion, consider the magnitude-squared function for a *low-pass network*. The plot of such a function will show a flat characteristic at low frequencies and a drop-off to some low value at high frequencies. Thus, ideally we might conceive that it would appear as shown in Fig. 2.1-1. The characteristic shown in the figure,

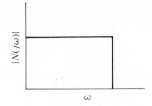

Figure 2.1-1 An ideal low-pass magnitude characteristic.

however, is not realizable.[1] As a more practical approach to finding a low-pass function, let us try to approximate it by choosing a magnitude-squared function $|N(j\omega)|^2$ which satisfies the maximally flat criteria at $\omega = 0$. This should generate the desired flatness of the curve, at least for low frequencies. Next, to provide the eventual drop-off of the characteristic at higher frequencies, we will locate all the transmission zeros of the function at infinity; thus, the numerator of $N(j\omega)$ will simply be a constant and all the coefficients b_i of (8) will be zero. For a maximally flat characteristic from (11), the coefficients a_i must also be set to zero, except of course for the highest-order one. The resulting magnitude-squared function has the form

$$|N(j\omega)|^2 = \frac{H^2}{1 + \varepsilon^2 \omega^{2n}} \tag{12}$$

where H is the value of $|N(0)|$, the maximum value that $|N(j\omega)|$ reaches, and where ε is used to adjust the rate at which the magnitude decreases. Some values of ε and ε^2 are tabulated in Table 2.1-1. The range of frequencies $0 \le \omega \le 1$ rad/s is called the *passband* of this function, and the range $\omega > 1$ rad/s is called the *stopband*. It should be noted that at $\omega = 1$ rad/s, $|N(j1)| = H/(1 + \varepsilon^2)^{1/2}$, which is independent of the value of n. The value of ε is usually chosen equal to 1, and the function is then referred to as a *Butterworth function*.[2] In this case $|N(j1)| = H/\sqrt{2} = 0.7071\,H$, and $20 \log [|N(j1)|/|N(0)|] = 20 \log 0.7071 = -3.01$ dB; thus, the frequency of 1 rad/s is usually referred to as the -3-dB *frequency* or the *3-dB-down frequency*. From (12) we see that the slope of $|N(j\omega)|^2$ at this frequency is proportional to $-n/2$. Plots of the magnitude of the Butterworth function for $n = 2$, 5, and 10 are shown in Fig. 2.1-2. Some specific values of $|N(j\omega)|$ are given in Table 2.1-2. For sufficiently large values of frequency the

---

[1] A general necessary and sufficient condition for realizability is the Paley-Wiener criterion, which requires

$$\int_{-\infty}^{\infty} \left| \frac{\ln |N(j\omega)|}{1 + \omega^2} \right| d\omega < \infty$$

where $|N(j\omega)|$ is the magnitude characteristic being tested.

[2] S. Butterworth was a British engineer who described this type of response in connection with electronic amplifiers in his paper "On the Theory of Filter Amplifiers," *Wireless Engineer*, vol. 7, 1930, pp. 536–541. Over a decade later, V. D. Landon applied the phrase *maximally flat* in his paper "Cascade Amplifiers with Maximal Flatness," *RCA Rev.*, vol. 5, 1941, pp. 347–362.

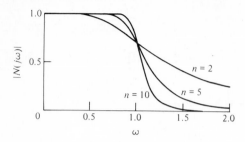

Figure 2.1-2 The magnitude of Butterworth functions of various orders.

attenuation is $20n$ dB per decade of frequency beyond $\omega = 1$ rad/s, where n is the degree of the function.

The locations of the poles of a network function $N(s)$ which has a Butterworth magnitude characteristic may be found using (4) and (12). Thus we obtain

$$N(s)N(-s) = \frac{H^2}{1 + \omega^{2n}}\bigg|_{\omega^2 = -s^2} = \frac{H^2}{1 + (-1)^n s^{2n}} \tag{13}$$

Setting the denominator polynomial of (13) to zero we find that the poles are located at the values of s which satisfy the relation

$$s = [-(-1)^n]^{1/2n} \tag{14}$$

Thus, for n even, $s = (-1)^{1/2n} = e^{j\pi k/2n}$ $(k = 1, 3, 5, \ldots, 4n - 1)$, and for n odd, $s = (1)^{1/2n} = e^{j\pi k/2n}$ $(k = 0, 2, 4, \ldots, 4n - 2)$. From these relations we see that the poles of $N(s)N(-s)$ are equiangularly spaced around the unit circle, as shown in Fig. 2.1-3. Retaining only the left-half-plane singularities, we find that the poles of $N(s)$ are given as $p_k = \sigma_k + j\omega_k$ where

$$\sigma_k = -\sin\frac{2k - 1}{2n}\pi \qquad \omega_k = \cos\frac{2k - 1}{2n}\pi \qquad k = 1, 2, 3, \ldots, n \tag{15}$$

Table 2.1-2 Values of the magnitude of maximally flat magnitude (Butterworth) functions ($\varepsilon = 1$) with normalized ($\omega = 1$ rad/s) passband

ω (rad/s)

n	0.7 Mag	dB	0.8 Mag	dB	0.9 Mag	dB	1.1 Mag	dB	1.2 Mag	dB	1.5 Mag	dB
2	.8980	-.93	.8423	-1.49	.7771	-2.19	.6370	-3.92	.5704	-4.88	.4061	-7.83
3	.9459	-.48	.8901	-1.01	.8081	-1.85	.6007	-4.43	.5009	-6.01	.2841	-10.93
4	.9724	-.24	.9254	-.67	.8361	-1.55	.5640	-4.97	.4344	-7.24	.1938	-14.25
5	.9862	-.12	.9503	-.44	.8611	-1.30	.5275	-5.56	.3729	-8.57	.1306	-17.68
6	.9932	-.06	.9673	-.29	.8830	-1.08	.4916	-6.17	.3176	-9.96	.0875	-21.16
7	.9966	-.03	.9787	-.19	.9021	-.89	.4566	-6.81	.2688	-11.41	.0584	-24.67
8	.9983	-.01	.9862	-.12	.9185	-.74	.4228	-7.48	.2265	-12.90	.0390	-28.18
9	.9992	-.01	.9911	-.08	.9325	-.61	.3904	-8.17	.1903	-14.41	.0260	-31.70
10	.9996	-.00	.9943	-.05	.9442	-.50	.3597	-8.88	.1594	-15.95	.0173	-35.22

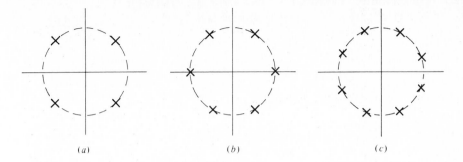

Figure 2.1-3 The roots of Butterworth polynomials.

The denominator polynomials characterized by these roots are called *Butterworth polynomials*. Values of the pole locations, coefficients, and quadratic factors of these polynomials for different values of n are given in Table 2.1-3. Passive network realizations of Butterworth functions are given in App. A. In designing a network function to meet some specific requirements, the relation given in (12) may be used to determine the value of n, that is, to determine the order of the filter that is needed.

Example 2.1-2 *Determination of the Order of a Butterworth Function* Consider the determination of a low-pass network function having a maximally flat magnitude characteristic which is 3 dB down from its maximum value (which is at zero frequency) at the edge of the passband, $\omega = 1$ rad/s. In addition, it is required that the magnitude be at least 15 dB down at all frequencies above $\omega = 2$ rad/s. Setting $\varepsilon = 1$ in (12) satisfies the specification at $\omega = 1$ rad/s. Because of the monotonic nature of the maximally flat characteristic, satisfying the second requirement at $\omega = 2$ rad/s ensures its satisfaction at all higher frequencies. We now first note that having exactly -15 dB at $\omega = 2$ rad/s corresponds to requiring $20 \log [|N(j0)|/|N(j2)|] = 15$, which requires $|N(j0)|/|N(j2)| = \log^{-1} 0.75 = 5.623$. Thus we must satisfy the inequality

$$\frac{|N(j0)|}{|N(j2)|} = (1 + 2^{2n})^{1/2} \geq 5.623 \qquad (16)$$

This may be put in the form

$$n \geq 1/2 \log_2 [(5.623)^2 - 1] = \log_2 5.53 \qquad (17)$$

Since $\log_2$ of the numbers 2, 4, 8, 16, 32, ... have the values respectively 1, 2, 3, 4, 5, ..., $\log_2 5.53$ lies between 2 and 3; thus, a third-order network function is

Table 2.1-3a Denominator coefficients of maximally flat magnitude (Butterworth) functions of the form $s^n + a_1 s^{n-1} + a_2 s^{n-2} + \cdots + a_2 s^2 + a_1 s + 1$ with passband 0 to 1 rad/s

n	a_1	a_2	a_3	a_4	a_5
2	1.414214				
3	2.000000				
4	2.613126	3.414214			
5	3.236068	5.236068			
6	3.863703	7.464102	9.141620		
7	4.493959	10.097835	14.591794		
8	5.125831	13.137071	21.846151	25.688356	
9	5.758770	16.581719	31.163437	41.986386	
10	6.392453	20.431729	42.802061	64.882396	74.233429

Table 2.1-3b Pole locations and quadratic factors $(s^2 + a_1 s + 1)$ of maximally flat magnitude (Butterworth) functions with passband 0 to 1 rad/s*

n	Poles	a_1
2	$-0.70711 \pm j0.70711$	1.41421
3	$-0.50000 \pm j0.86603$	1.00000
4	$-0.38268 \pm j0.92388$	0.76536
	$-0.92388 \pm j0.38268$	1.84776
5	$-0.30902 \pm j0.95106$	0.61804
	$-0.80902 \pm j0.58779$	1.61804
6	$-0.25882 \pm j0.96593$	0.51764
	$-0.70711 \pm j0.70711$	1.41421
	$-0.96593 \pm j0.25882$	1.93186
7	$-0.22252 \pm j0.97493$	0.44504
	$-0.62349 \pm j0.78183$	1.24698
	$-0.90097 \pm j0.43388$	1.80194
8	$-0.19509 \pm j0.98079$	0.39018
	$-0.55557 \pm j0.83147$	1.11114
	$-0.83147 \pm j0.55557$	1.66294
	$-0.98079 \pm j0.19509$	1.96158
9	$-0.17365 \pm j0.98481$	0.34730
	$-0.50000 \pm j0.86603$	1.00000
	$-0.76604 \pm j0.64279$	1.53208
	$-0.93969 \pm j0.34202$	1.87938
10	$-0.15643 \pm j0.98769$	0.31286
	$-0.45399 \pm j0.89101$	0.90798
	$-0.70711 \pm j0.70711$	1.41421
	$-0.89101 \pm j0.45399$	1.78202
	$-0.98769 \pm j0.15643$	1.97538

* *Note:* All odd-order functions also have a pole at $s = -1$.

required. From Table 2.1-3 this will have the form

$$N(s) = \frac{H}{s^3 + 2s^2 + 2s + 1} \tag{18}$$

Note that $|N(j2)| = 0.124$, and the reciprocal of this is 8.06 which, as is required by (16), is greater than 5.623. □

Example 2.1-3 *A Maximally Flat Magnitude Characteristic in Which* $\varepsilon \neq 1$ As another example of the determination of a network function with a maximally flat magnitude characteristic to satisfy a set of specifications, let us assume the attenuation at the passband edge ($\omega = 1$ rad/s) is required to be $\frac{1}{2}$ dB, while the attenuation at $\omega = 2$ rad/s must be a minimum of 18 dB. The first specification may be used to determine the value of ε in (12). We note that $20 \log [|N(j0)|/|N(j1)|] = \frac{1}{2}$, thus $|N(j0)|/|N(j1)| = (1 + \varepsilon^2)^{1/2} = \log^{-1} 0.025 = 1.0593$. The resulting value of ε^2 is 0.122. To obtain exactly 18-dB attenuation at $\omega = 2$ we must have $20 \log [|N(j0)|/|N(j2)|] = 18$, which requires $|N(j0)|/|N(j2)| = \log^{-1} 0.9 = 7.943$. Thus we must satisfy the relation

$$\frac{|N(j0)|}{|N(j2)|} = (1 + 0.122 \times 2^{2n})^{1/2} \geq 7.943 \tag{19}$$

Solving for the value of n, we find it equals $\log_2 22.56$, which lies between 4 and 5, thus a fifth-order network function is required. The denominator polynomial for this function may be found using (12). □

Computations of the type made in the two preceding examples can be simplified by the use of a nomograph giving the required degree of the filter as a function of the attenuation A_{PB} at the edge of the passband (normalized to 1 rad/s), and the attenuation A_Ω, at some (normalized) frequency Ω. Such a nomograph is shown in Fig. 2.1-4.[1] Its use is illustrated in Fig. 2.1-5. A straight line is drawn through the specified values of A_{PB} and A_Ω, shown as points 1 and 2 in Fig. 2.1-5. The intersection of this line with the ordinate of the graph determines point 3. This intersection is extended horizontally until it meets a line drawn vertically from point 4 giving the specified frequency Ω. The resulting intersection at point 5 establishes the required order of the filter. If the point is between two of the "order loci," the higher one must, of course, be used. This nomograph is readily used to verify the results of Examples 2.1-2 and 2.1-3.

[1] M. Kawakami, "Nomographs for Butterworth and Chebyshev Filters," *IEEE Trans. Circuit Theory*, vol. CT-10, June 1963, pp. 288–298. Dr. Kawakami was appointed president of Tokyo Kogyo Daigaku (Tokyo Institute of Technology) in 1973.

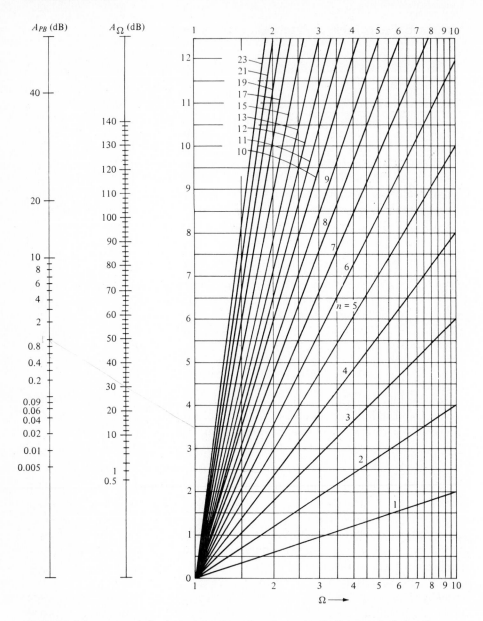

Figure 2.1-4 A nomograph for determining the order of a maximally flat magnitude function.

---

[1] These polynomials were originally used in studying the construction of steam engines by P. L. Chebyshev in the paper "Théorie des mécanismes connus sous le nom de parallelogrammes," *Oeuvres*, vol. I, St. Petersburg, 1899. The spellings *Tschebyscheff* and *Tchebysheff* for the author's name also appear frequently in the literature. All these spellings are transliterations of the Russian name Чебышёв.

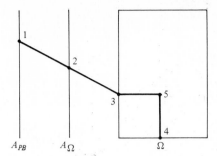

Figure 2.1-5 The method for using the nomograph of Fig. 2.1-4.

2.2 MAGNITUDE APPROXIMATION—THE EQUAL-RIPPLE CHARACTERISTIC

In the preceding section we introduced one kind of magnitude approximation, namely, the maximally flat type. This particular approximation was shown to be characterized by the fact that the derivatives of the magnitude-squared function are set to zero at zero frequency. Thus, the approximating effect is concentrated at a single frequency, i.e., zero. One result of this is that the transition from passband to stopband is not as sharp as is needed for many applications. In this section we describe a different type of approximation, one in which the approximating effect is spread over the entire passband. Such an approximation is said to have an *equal-ripple* characteristic.

The low-pass equal-ripple magnitude approximation may be developed by writing the magnitude-squared function $|N(j\omega)|^2$ in the form

$$|N(j\omega)|^2 = \frac{H^2}{1 + \varepsilon^2 C_n^2(\omega)} \tag{1}$$

where $C_n(\omega)$ is a polynomial of order n. If these polynomials have the properties $0 \le C_n^2(\omega) \le 1$ for $0 \le \omega \le 1$ rad/s and $C_n^2(\omega) > 1$ for $\omega > 1$ rad/s, then the passband will have the range $0 \le \omega \le 1$ rad/s and will be characterized by $|N(j\omega)|_{\max} = H$ and $|N(j\omega)|_{\min} = H/(1 + \varepsilon^2)^{1/2}$. Thus the value of ε will determine the limits of variation of the magnitude characteristic in the passband. For example, for $\varepsilon = 1$, the minimum value is 3.01 dB lower than the maximum value. Other pairs of values of ε and dB are given in Table 2.1-1. The stopband is defined for $\omega > 1$. In this band, $|N(j\omega)| < H/(1 + \varepsilon^2)^{1/2}$.

A set of polynomials $C_n(\omega)$ which have the properties specified above are the *Chebyshev polynomials*.[1] These are defined as follows:

$$C_1(\omega) = \omega$$

$$C_2(\omega) = 2\omega^2 - 1$$

$$C_3(\omega) = 4\omega^3 - 3\omega \tag{2}$$

$$\cdots \cdots \cdots \cdots \cdots \cdots \cdots$$

$$C_{n+1}(\omega) = 2\omega C_n(\omega) - C_{n-1}(\omega)$$

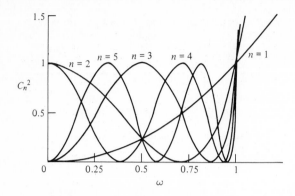

Figure 2.2-1 The magnitude of some Chebyshev polynomials $C_n^2(\omega)$.

where the last expression is valid for all $n > 1$. The Chebyshev polynomials may also be written using the trigonometric expressions

$$C_n(\omega) = \cos\left(n \cos^{-1} \omega\right) \qquad 0 \leq \omega \leq 1 \tag{3a}$$

$$C_n(\omega) = \cosh\left(n \cosh^{-1} \omega\right) \qquad \omega \geq 1 \tag{3b}$$

Plots of the values of some of the polynomials $C_n^2(\omega)$ are shown in Fig. 2.2-1. It is readily apparent that these polynomials have the desired properties. Plots of the magnitude of equal-ripple functions for $n = 2$, 5, and 10 (for $\varepsilon = 1$) are shown in Fig. 2.2-2. A comparison of these with the corresponding plots for the maximally flat magnitude case given in Fig. 2.1-2 shows that the stopband attenuation in the vicinity of the cutoff frequency is considerably higher for an equal-ripple characteristic of any given order. For sufficiently high values of frequency, of course, the attenuation will be $20n$ dB per decade of frequency past $\omega = 1$ rad/s, just as was the case for the maximally flat magnitude characteristic. Some specific values of $|N(j\omega)|$ are given in Table 2.2-1a. The -3-dB frequencies for various ripples and orders are given in Table 2.2-1b. These frequencies are determined by the relation

$$\omega_{3\,\text{dB}} = \cosh\left(\frac{1}{n}\cosh^{-1}\frac{1}{\varepsilon}\right) \qquad (\varepsilon \leq 1) \tag{4}$$

We may now use the expression for $|N(j\omega)|^2$ given in (1) to determine the pole locations for an equal-ripple network function. Following the development

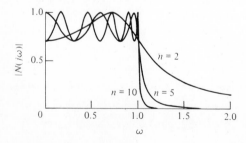

Figure 2.2-2 The magnitude of equal-ripple (3 dB) functions of various orders.

Table 2.2-1a Values of the magnitude of equal-ripple (Chebyshev) functions with normalized ($\omega = 1$ rad/s) passband

		ω (rad/s)					
		1.1		1.2		1.5	
Ripple	n	Mag	dB	Mag	dB	Mag	dB
$\frac{1}{2}$ dB	2	.8958	−.96	.8359	−1.56	.6331	−3.97
	3	.8165	−1.76	.6539	−3.69	.3031	−10.37
	4	.6864	−3.27	.4266	−7.40	.1209	−18.35
	5	.5244	−5.61	.2465	−12.16	.0465	−26.65
	6	.3698	−8.64	.1355	−17.36	.0178	−35.00
	7	.2481	−12.11	.0732	−22.71	.0068	−43.36
	8	.1624	−15.79	.0394	−28.10	.0026	−51.72
	9	.1051	−19.57	.0211	−33.50	.0010	−60.08
	10	.0677	−23.39	.0113	−38.90	.0004	−68.44
1 dB	2	.8105	−1.82	.7226	−2.82	.4896	−6.20
	3	.6966	−3.14	.5103	−5.84	.2133	−13.42
	4	.5438	−5.29	.3081	−10.23	.0833	−21.58
	5	.3894	−8.19	.1720	−15.29	.0319	−29.91
	6	.2635	−11.58	.0934	−20.59	.0122	−38.27
	7	.1732	−15.23	.0503	−25.96	.0047	−46.63
	8	.1123	−19.00	.0270	−31.36	.0018	−54.99
	9	.0723	−22.81	.0145	−36.76	.0007	−63.35
	10	.0465	−26.65	.0078	−42.17	.0003	−71.71

Table 2.2-1b Half-power (-3-dB) frequencies for equal-ripple magnitude (Chebyshev) low-pass network functions with passband 0 to 1 rad/s

Ripple (dB)	2	3	4	5	n 6	7	8	9	10
.010000	3.303615	1.877180	1.466904	1.291217	1.199412	1.145268	1.110609	1.087064	1.070331
.100000	1.943219	1.388995	1.213099	1.134718	1.092931	1.068001	1.051927	1.040955	1.033131
.200000	1.674270	1.283455	1.156346	1.099154	1.068517	1.050188	1.038351	1.030262	1.024489
.500000	1.389744	1.167485	1.093102	1.059259	1.041030	1.030090	1.023011	1.018167	1.014707
1.000000	1.217626	1.094868	1.053002	1.033815	1.023442	1.017205	1.013164	1.010396	1.008418

given for a general magnitude-squared network function in the preceding section, we may write

$$N(s)N(-s) = \left| N(j\omega) \right|^2 \Bigg|_{\omega=s/j} = \frac{H^2}{1 + \varepsilon^2 C_n^2(s/j)} \tag{5}$$

Thus, the poles of the product $N(s)N(-s)$ are the roots of $C_n^2(s/j) = -1/\varepsilon^2$ or $C_n(s/j) = \pm j/\varepsilon$. Using the trigonometric form for $C_n(\omega)$ given in (3) we may write

$$C_n\left(\frac{s}{j}\right) = \cos\left(n \cos^{-1}\frac{s}{j}\right) = \pm\frac{j}{\varepsilon} \tag{6}$$

To solve this equation we first define a complex function as

$$w = u + jv = \cos^{-1}\frac{s}{j} \tag{7}$$

Substituting this expression in (6) we obtain

$$\cos n(u + jv) = \cos nu \cosh nv - j \sin nu \sinh nv = \pm j/\varepsilon \tag{8}$$

Equating the real parts of the second and third members of this relation gives $\cos nu \cosh nv = 0$. Since $\cosh nv \geq 1$ for all values of nv, this equality requires $\cos nu = 0$. This may be written in the form

$$u_k = \frac{2k-1}{2n}\pi \qquad k = 1, 2, 3, \ldots, 2n \tag{9}$$

Equating the imaginary parts of (8), and recognizing that for the values of u defined by (9), $\sin nu = \pm 1$, we obtain

$$v = \frac{1}{n}\sinh^{-1}\frac{1}{\varepsilon} \tag{10}$$

where we have retained only the positive value for v. Equation (7) may now be put in the form

$$s = j \cos(u_k + jv) = \sin u_k \sinh v + j \cos u_k \cosh v \tag{11}$$

This relation specifies the poles of the product $N(s)N(-s)$. The left-half-plane poles are assigned to $N(s)$ to complete the determination of the network function. Thus we see that the poles of $N(s)$ will be at $p_k = \sigma_k + j\omega_k$, where

$$\left. \begin{array}{l} \sigma_k = -\sin u_k \sinh v \\[2mm] \omega_k = \cos u_k \cosh v \end{array} \right\} \qquad k = 1, 2, \ldots, n \tag{12}$$

and where u_k and v are defined in (9) and (10).

Example 2.2-1 *Determination of a Chebyshev Function* As an example of the use of (9), (10), and (11), consider the determination of the poles of a second-order equal-ripple low-pass network function having a 3-dB ripple in the passband. In this case $n = 2$, and from (9), $u_1 = \pi/4$ and $u_2 = 3\pi/4$. For a 3-dB ripple, $\varepsilon = 1$, and from (10), $v = 0.441$. Using (12) we now obtain

$$\sigma_1 = -\sin\frac{\pi}{4}\sinh 0.441 = -0.322$$

$$\omega_1 = \cos\frac{\pi}{4}\cosh 0.441 = 0.777$$

$$\sigma_2 = -\sin\frac{3\pi}{4}\sinh 0.441 = -0.322$$

$$\omega_2 = \cos\frac{3\pi}{4}\cosh 0.441 = -0.777$$

(13)

Thus the network function is

$$N(s) = \frac{H}{(s + 0.322 + j0.777)(s + 0.322 - j0.777)}$$

$$= \frac{H}{s^2 + 0.644s + 0.7074}$$

(14)   □

Passive network realizations of equal-ripple functions are given in App. A. Values of the pole locations and the coefficients of the denominator polynomials for different values of n and some common values of ε are given in Table 2.2-2. The locus on which the poles lie may be determined by starting with the basic trigonometric relation $\sin^2 u_k + \cos^2 u_k = 1$. Inserting the relations for $\sin^2 u_k$ and $\cos^2 u_k$ from (12) we obtain

$$\frac{\sigma_k^2}{\sinh^2 v} + \frac{\omega_k^2}{\cosh^2 v} = 1$$

(15)

This equation represents an ellipse centered at the origin of the p_k plane, with an ordinate semiaxis of length $\cosh v$, and an abscissa semiaxis of length $\sinh v$ as shown in Fig. 2.2-3.

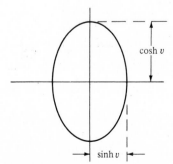

cosh v

sinh v

Figure 2.2-3 The loci of the poles of equal-ripple magnitude functions.

Table 2.2-2a Denominator coefficients of equal-ripple magnitude (Chebyshev) functions of the form $a_0 + a_1 s + a_2 s^2 + \cdots + a_{n-1} s^{n-1} + s^n$ with passband 0 to 1 rad/s

n	a_0	a_1	a_2	a_3	a_4	a_5	a_6	a_7	a_8	a_9
					½-dB ripple					
2	1.516203	1.425625								
3	.715694	1.534895	1.252913							
4	.379051	1.025455	1.716866	1.197386						
5	.178923	.752518	1.309575	1.937367	1.172491					
6	.094763	.432367	1.171861	1.589763	2.171845	1.159176				
7	.044731	.282072	.755651	1.647903	1.869408	2.412651	1.151218			
8	.023091	.152544	.573560	1.148589	2.184015	2.149217	2.656750	1.146080		
9	.011183	.094120	.340819	.983620	1.611388	2.781499	2.429330	2.902734	1.142571	
10	.005923	.049285	.237269	.626969	1.527431	2.144237	3.440927	2.709741	3.149876	1.140066
					1-dB ripple					
2	1.102510	1.097734								
3	.491307	1.238409	.988341							
4	.275628	.742619	1.453925	.952811						
5	.122827	.580534	.974396	1.688816	.936820					
6	.068907	.307081	.939346	1.202140	1.930825	.928251				
7	.030707	.213671	.548620	1.357545	1.428794	2.176078	.923123			
8	.017227	.107345	.447826	.846824	1.836902	1.655156	2.423026	.919811		
9	.007677	.070605	.244186	.786311	1.201607	2.378119	1.881480	2.670947	.917548	
10	.004307	.034497	.182451	.455389	1.244491	1.612986	2.981509	2.107852	2.919466	.915932

34

Table 2.2-2b Pole locations and quadratic factors $(a_0 + a_1s + s^2)$ of 0.5-dB equal-ripple magnitude (Chebyshev) low-pass functions with passband 0 to 1 rad/s

n	Poles	a_0	a_1
2	$-.71281 + J\ 1.00404$	1.51620	1.42562
3	$-.31323 + J\ 1.02193$	1.14245	.62646
	$-.62646$		
4	$-.17535 + J\ 1.01625$	1.06352	.35071
	$-.42334 + J\ \ .42095$	.35641	.84668
5	$-.11196 + J\ 1.01156$	1.03578	.22393
	$-.29312 + J\ \ .62518$	.47677	.58625
	$-.36232$		
6	$-.07765 + J\ 1.00846$	1.02302	.15530
	$-.21214 + J\ \ .73824$	.59001	.42429
	$-.28979 + J\ \ .27022$	.15700	.57959
7	$-.05700 + J\ 1.00641$	1.01611	.11401
	$-.15972 + J\ \ .80708$	.67688	.31944
	$-.23080 + J\ \ .44789$	.25388	.46160
	$-.25617$		
8	$-.04362 + J\ 1.00500$	1.01193	.08724
	$-.12422 + J\ \ .85200$	.74133	.24844
	$-.18591 + J\ \ .56929$	.35865	.37182
	$-.21929 + J\ \ .19991$	.08805	.43859
9	$-.03445 + J\ 1.00400$	1.00921	.06891
	$-.09920 + J\ \ .88291$	.78936	.19841
	$-.15199 + J\ \ .65532$	.45254	.30397
	$-.18644 + J\ \ .34869$	.15634	.37288
	$-.19841$		
10	$-.02790 + J\ 1.00327$	1.00734	.05580
	$-.08097 + J\ \ .90507$	.82570	.16193
	$-.12611 + J\ \ .71826$	.53181	.25222
	$-.15891 + J\ \ .46115$	.23791	.31781
	$-.17615 + J\ \ .15890$	.05628	.35230

The locus defined by (15) may be used in an alternate method for finding the pole locations of an equal-ripple function. To do this, we first define a frequency normalization such that the passband has the range $0 \leq \omega \leq \omega_c$, where $\omega_c = 1/\cosh v$. From (12), the normalized pole locations are thus defined as

$$\sigma_k = -\sin u \tanh v$$
$$\omega_k = \cos u \tag{16}$$

The elliptic locus for these poles has an ordinate semiaxis of length unity and an abscissa semiaxis of length $\tanh v$. Comparing the pole locations determined by

Table 2.2-2c Pole locations and quadratic factors $(a_0 + a_1 s + s^2)$ of 1.0-dB equal-ripple magnitude (Chebyshev) low-pass functions with passband 0 to 1 rad/s

n	Poles			a_0	a_1
2	$-.54887$	$+$ J	$.89513$	1.10251	1.09773
3	$-.24709$	$+$ J	$.96600$	.99420	.49417
	$-.49417$				
4	$-.13954$	$+$ J	$.98338$	.98650	.27907
	$-.33687$	$+$ J	$.40733$	.27940	.67374
5	$-.08946$	$+$ J	$.99011$	.98831	.17892
	$-.23421$	$+$ J	$.61192$	.42930	.46841
	$-.28949$				
6	$-.06218$	$+$ J	$.99341$	.99073	.12436
	$-.16988$	$+$ J	$.72723$	.55772	.33976
	$-.23206$	$+$ J	$.26618$	.12471	.46413
7	$-.04571$	$+$ J	$.99528$	.99268	.09142
	$-.12807$	$+$ J	$.79816$	.65346	.25615
	$-.18507$	$+$ J	$.44294$	.23045	.37014
	$-.20541$				
8	$-.03501$	$+$ J	$.99645$	.99414	.07002
	$-.09970$	$+$ J	$.84475$	.72354	.19939
	$-.14920$	$+$ J	$.56444$	.34086	.29841
	$-.17600$	$+$ J	$.19821$	.07026	.35200
9	$-.02767$	$+$ J	$.99723$	.99523	.05533
	$-.07967$	$+$ J	$.87695$	.77539	.15933
	$-.12205$	$+$ J	$.65090$	.43856	.24411
	$-.14972$	$+$ J	$.34633$	.14236	.29944
	$-.15933$				
10	$-.02241$	$+$ J	$.99778$	.99606	.04483
	$-.06505$	$+$ J	$.90011$	.81442	.13010
	$-.10132$	$+$ J	$.71433$	.52053	.20263
	$-.12767$	$+$ J	$.45863$	.22664	.25533
	$-.14152$	$+$ J	$.15803$	.04500	.28304

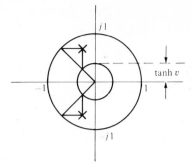

Figure 2.2-4 The relation between the pole loci for maximally flat and equal-ripple magnitude functions.

(16) with those given by (15) of Sec. 2.1 for the maximally flat magnitude case (for a unity passband) we see that, for the given frequency normalization of the equal-ripple function, the imaginary parts of the pole locations of the two types of functions are the same; however, the magnitudes of the real parts of the equal-ripple poles are decreased by a factor of tanh v. As a result, the equiangularly spaced maximally flat magnitude pole locations may be used to graphically determine the equal-ripple pole locations by drawing horizontal lines from the intercepts of a set of equiangular lines with a unit radius circle, and drawing vertical lines from the intercepts with a circle of radius tanh v. As shown in Fig. 2.2-4, for the second-order case, the intersections of the horizontal and vertical lines determine the equal-ripple pole locations. Of course, these pole locations must be frequency denormalized if the function is to have unity passband width.

Let us now consider the use of equal-ripple approximation to satisfy specific frequency-domain requirements.

Example 2.2-2 *Finding the Order of a Chebyshev Function* In Example 2.1-2 we determined a maximally flat magnitude function which was required to be 3 dB down from its maximum value at $\omega = 1$ rad/s and at least 15 dB down at all frequencies above $\omega = 2$ rad/s. Now consider using an equal-ripple function to meet the same specifications. Setting $\varepsilon = 1$ in (1) sets the ripple to 3 dB and thus satisfies the specifications at $\omega = 1$ rad/s. Since the equal-ripple function, like the maximally flat function, is monotonic in the stopband, satisfying the second requirement at $\omega = 2$ rad/s satisfies it at all higher frequencies. Following the development leading to (16) in Sec. 2.1 we now require

$$\frac{|N(j\omega)|_{\max}}{|N(j2)|} = [1 + C_n^2(2)]^{1/2} \geq 5.623 \qquad (17)$$

Note that in the numerator of the left member of (17) we use $|N(j\omega)|_{\max}$ rather than $|N(0)|$ as we did in (16) of Sec. 2.1. The reason for this is that only odd-order equal-ripple functions have their maximum value at $\omega = 0$, as is readily verified from the plots shown in Fig. 2.2-2. Substituting the expression for $C_n(\omega)$ from (3b) in (17), we obtain

$$[1 + \cosh^2 (n \cosh^{-1} 2)]^{1/2} \geq 5.623 \tag{18}$$

This may be solved for n to yield

$$n \geq \frac{1}{\cosh^{-1} 2} \cosh^{-1} [(5.623^2 - 1)^{1/2}] = 1.82 \tag{19}$$

Thus a second-order equal-ripple function satisfies the specifications, although as was shown in Example 2.1-2 a third-order maximally flat magnitude function is required due to the slower rate of falloff of the maximally flat function at the beginning of the stopband. □

Example 2.2-3 *A Chebyshev Function in Which $\varepsilon \neq 1$* As another comparison of the relative efficiency of the equal-ripple and the maximally flat characteristics, consider the specification of a network function with a $\frac{1}{2}$-dB ripple in the passband $0 \leq \omega \leq 1$ rad/s, and a minimum of 18-dB attenuation at all frequencies $\omega > 2$ rad/s. In Example 2.1-3 it was shown that a maximally flat function satisfying this characteristic would have to be of fifth order. Using the results of that example, we see that for the equal-ripple function

$$n \geq \frac{1}{\cosh^{-1} 2} \cosh^{-1} \left[\left(\frac{(7.943)^2 - 1}{0.122} \right)^{1/2} \right] = 2.89 \tag{20}$$

Thus only a third-order function is required for the equal-ripple case, and obviously a considerable saving in network complexity will result. □

Computations of the type made in the preceding two examples can also be made by the use of the nomograph shown in Fig. 2.2-5.[1] Its use is explained in Sec. 2.1 and illustrated in Fig. 2.1-5.

Considering the superior stopband attenuation provided by the equal-ripple characteristic as evidenced by the above examples, it is evident that of the functions discussed so far the equal-ripple function is the one to use if only the magnitude characteristic is of interest. In any given filter application, however, it is frequently necessary to consider not only the magnitude performance of the network function but also its phase characteristic. In this respect the maximally flat magnitude network function has some properties superior to those of the equal-ripple one. We will defer a detailed discussion of phase approximation to a later section.

---

[1] Kawakami, op. cit.

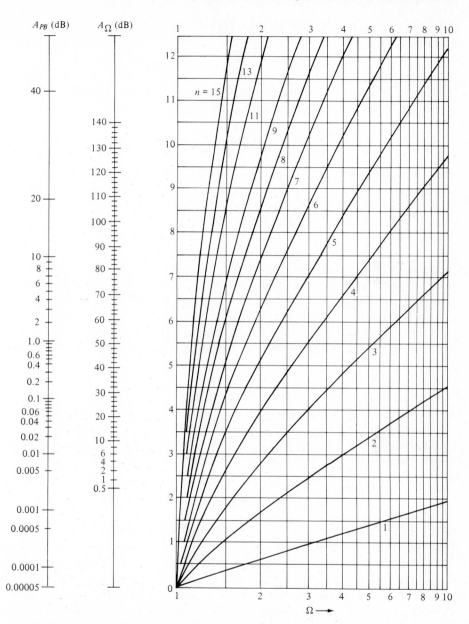

Figure 2.2-5 A nomograph for determining the order of an equal-ripple magnitude function.

2.3 MAGNITUDE APPROXIMATION—THE ELLIPTIC CHARACTERISTIC

In the preceding sections of this chapter we discussed two types of magnitude approximation, namely, the maximally flat and the equal-ripple. These two types of characteristics may both be written in the form

$$|N(j\omega)|^2 = \frac{H^2}{1 + P_n^2(\omega)} \tag{1}$$

where $P_n^2(\omega)$ is a polynomial which for the maximally flat case is ω^{2n} and for the equal-ripple case is $C_n^2(\omega)$ (a Chebyshev polynomial). In this section we consider a quite different type of low-pass magnitude characteristic, in which the *polynomial* $P_n(\omega)$ is replaced with a *rational function* $R_n(\omega)$. By choosing a specific rational function, called a *Chebyshev rational function*, it is possible to produce a magnitude characteristic which is equal-ripple in both the passband and the stopband, as shown in Fig. 2.3-1. In the figure we have assumed that the value of H in (1) has been chosen so that the maximum value of $|N(j\omega)|$ is unity. For a given-order filter, the resulting magnitude characteristic drops off even more abruptly than it does in the equal-ripple case, thus providing the sharpest cutoff of any of the three types of low-pass characteristics. The determination of the form of the rational function $R_n(\omega)$ in general requires the use of (jacobian) elliptic functions of the first kind and complete elliptic integrals. Thus, the resulting network functions are frequently referred to as *elliptic network functions*. Since the original work on such functions was done by the famous German network theorist W. Cauer, filters with an elliptic characteristic are also frequently called *Cauer filters*.[1]

Before discussing the actual rational function $R_n(\omega)$ that we shall use, let us first consider a related function $R_n'(\omega)$. This function has two forms, depending on whether n is even or odd. For the even case

$$R_n'(\omega) = \frac{(\omega_1^2 - \omega^2)(\omega_2^2 - \omega^2) \cdots (\omega_{n/2}^2 - \omega^2)}{(1 - \omega_1^2\omega^2)(1 - \omega_2^2\omega^2) \cdots (1 - \omega_{n/2}^2\omega^2)} \tag{2}$$

where the critical frequencies ω_i are chosen so as to give the desired equal-ripple properties. For the odd case

$$R_n'(\omega) = \frac{\omega(\omega_1^2 - \omega^2)(\omega_2^2 - \omega^2) \cdots (\omega_{(n-1)/2}^2 - \omega^2)}{(1 - \omega_1^2\omega^2)(1 - \omega_2^2\omega^2) \cdots (1 - \omega_{(n-1)/2}^2\omega^2)} \tag{3}$$

If we examine the functions given in (2) and (3) we see that the pole locations are the reciprocals of the zero locations, and that $R_n'(1/\omega) = 1/R_n'(\omega)$. Because of this, the value of $R_n'(\omega)$ at any frequency ω_0 in the passband $0 \le \omega < \omega_{\text{cutoff}}$ is the reciprocal of its value at the inverse frequency $1/\omega_0$ in the stopband $\omega_{\text{stopband}} \le \omega < \infty$. Thus, finding values for the frequencies ω_i such that $R_n(\omega)$ has

---

[1] W. Cauer, *Synthesis of Linear Communication Networks* (translated from the German edition), McGraw-Hill Book Company, New York, 1958.

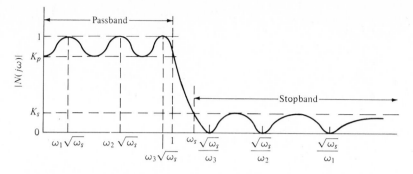

Figure 2.3-1 A magnitude function which is equal ripple in both the pass- and stopbands.

an equal-ripple characteristic in the passband automatically produces an equal ripple in the stopband. Although the functions $R'_n(\omega)$ given in (2) and (3) have the desired ripple characteristics (for a properly chosen set of values ω_i), the form of these functions is inconvenient, since the passband goes from zero to a frequency less than unity, and the exact value of this frequency varies as a function of n. Similarly, the stopband starts at a frequency greater than unity whose value also varies with n. A more useful rational function is obtained by choosing a pair of functions $R_n(\omega)$ which are related to the ones given in (2) and (3) by a frequency transformation. The even function is

$$R_n(\omega) = M \frac{(\omega_1^2 - \omega^2/\omega_s)(\omega_2^2 - \omega^2/\omega_s) \cdots (\omega_{n/2}^2 - \omega^2/\omega_s)}{(1 - \omega_1^2 \omega^2/\omega_s)(1 - \omega_2^2 \omega^2/\omega_s) \cdots (1 - \omega_{n/2}^2 \omega^2/\omega_s)} \tag{4}$$

where M is a multiplicative constant and ω_s is the frequency at which the stopband begins. The odd function is

$$R_n(\omega) = N \frac{\omega(\omega_1^2 - \omega^2/\omega_s)(\omega_2^2 - \omega^2/\omega_s) \cdots (\omega_{(n-1)/2}^2 - \omega^2/\omega_s)}{(1 - \omega_1^2/\omega_s)(1 - \omega_2^2/\omega_s) \cdots (1 - \omega_{(n-1)/2}^2 \omega^2/\omega_s)} \tag{5}$$

The normalization used on the functions given in (4) and (5) produces a passband of $0 \leq \omega \leq 1$ rad/s independent of the value of n and a stopband $\omega \geq \omega_s$, as shown in Fig. 2.3-1. The magnitude characteristic of the elliptic network function may now be written in the form

$$|N(j\omega)|^2 = \frac{H^2}{1 + R_n^2(\omega)} \tag{6}$$

where $R_n(\omega)$ is given in (4) and (5). Let us now separately consider the odd and even cases. If the odd form of $R_n(\omega)$ given in (5) is substituted in (6), we may then obtain the network function by replacing ω with s/j, and selecting the left-half-plane poles and half of the $j\omega$-axis zeros (see the discussion in Sec. 2.1). Thus we obtain the following general form for the odd-order elliptic network function:

$$N_o(s) = \frac{H_o \prod_{i=1}^{(n-1)/2} (s^2 + \Omega_i^2)}{a_0 + a_1 s + \cdots + a_{n-1} s^{n-1} + a_n s^n} \tag{7}$$

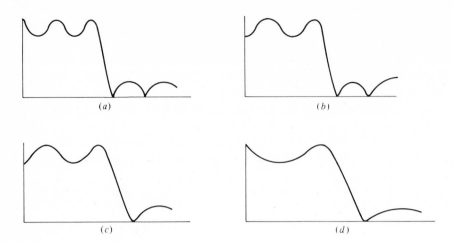

(a)

(b)

(c)

(d)

Figure 2.3-2 Various forms of elliptic magnitude functions.

where the $j\omega$-axis zeros are located at $s = \pm j\Omega_i$. The degree of the denominator polynomial of $N_o(s)$ will be n, while the degree of the numerator polynomial will be $n - 1$. $|N_o(j\omega)|$ will have $(n - 1)/2$ peaks in the passband (plus a peak which occurs at $\omega = 0$), $(n - 1)/2$ transmission zeros in the stopband, and a zero value at $\omega = \infty$. An example for $n = 5$ is shown in Fig. 2.3-2a.

Now let us consider the case where n is even. Using the even form of $R_n(\omega)$ given in (4), and proceeding as in the above, we obtain a first general form for the even-order elliptic network function:

$$N_a(s) = \frac{H_a \prod\limits_{i=1}^{n/2} (s^2 + \Omega_i^2)}{a_0 + a_1 s + \cdots + a_{n-1} s^{n-1} + a_n s^n} \tag{8}$$

We call this *case A for n even*. The degrees of both the numerator and denominator polynomials are equal to n. $|N_a(j\omega)|$ will have $n/2$ peaks in the passband, $n/2$ transmission zeros in the stopband, and a nonzero value at $\omega = \infty$. An example is shown for $n = 4$ in Fig. 2.3-2b. A consequence of the fact that the function is finite at $\omega = \infty$ is that a passive RLC ladder network realization of the function is only possible if coupled coils, i.e., transformers, are used in the circuit. This is usually undesirable from the standpoint both of cost and of accurate determination of element values. Thus, this function is normally considered only for realization by active RC filters of the type which will be introduced in Chaps. 4 through 7. If passive realizations are required, however, the need for transformers may be eliminated by modifying the form of the function $R_n(\omega)$ given in (4). Such a modification effectively consists of a frequency transformation such that the location of the highest frequency pole is shifted to infinity. In this case, the function will have the form

$$R_b(\omega) = M_b \frac{(\omega_1^2 - \omega^2/\omega_s)(\omega_2 - \omega^2/\omega_s) \cdots (\omega_{n/2}^2 - \omega^2/\omega_s)}{(1 - \omega_2'^2\omega^2/\omega_s) \cdots (1 - \omega_{n/2}'^2 \omega^2/\omega_s)} \tag{9}$$

where the values of ω_i and ω_s are different from those given in (4), and where the values of ω_i' are different from the values ω_i.

Inserting this in (6), and performing operations as before, we obtain the second general form for the even-order elliptic network function:

$$N_b(s) = \frac{H_b \prod\limits_{i=2}^{n/2} (s^2 + \Omega_i^2)}{a_0 + a_1 s + \cdots + a_{n-1}s^{n-1} + a_n s^n} \tag{10}$$

We call this *case B for n even*. The quantities Ω_i and a_i, of course, have different values from the ones in (8). The degree of the numerator is $n - 2$ (note that the product index i starts at 2, not 1), while that of the denominator is n. This function may be realized without coupled coils as long as the source and load resistance terminations do not have the same value. A price is required, however, for the simplification of the realization, in that for a given order and a given stopband attenuation, the stopband frequency ω_s lies somewhat higher than it does in case A. An example is shown for $n = 4$ in Fig. 2.3-2c. The requirement for unequal resistance values is a result of the fact that the function $R_b(\omega)$ of (9) is finite at $\omega = 0$. If we make another frequency transformation such that the first zero location is effectively shifted to the origin, a function is produced having the general form

$$R_c(\omega) = M_c \frac{\omega^2(\omega_2^2 - \omega^2/\omega_s) \cdots (\omega_{n/2}^2 - \omega^2/\omega_s)}{(1 - \omega_2^2\omega^2/\omega_s) \cdots [\omega_{n/2}^2(1 - \omega_{n/2}^2\omega^2/\omega_s)]} \tag{11}$$

where the values of ω_i and ω_s are different from those given in (4) or (9). Inserting this in (6) we obtain a network function having the general form

$$N_c(s) = \frac{H_c \prod\limits_{i=2}^{n/2} (s^2 + \Omega_i^2)}{a_0 + a_1 s + \cdots + a_{n-1}s^{n-1} + a_n s^n} \tag{12}$$

This is *case C for n even*. An example of the form of this network function for $n = 4$ is shown in Fig. 2.3-2d. Note that the magnitude at zero frequency is the maximum passband value. The stopband frequency ω_s will be even higher than that of case B, as a result of the additional requirement on the form of the realization; however, this function can be realized with equal-resistance terminations.

Table 2.3-1 shows selected examples of odd- and even-order normalized elliptic functions, their pole and zero locations, and their quadratic denominator

Table 2.3-1 Elliptic functions with poles at p_i and having the form*

$$N(s) = H \prod_i \frac{s^2 + c_i}{s^2 + a_i s + b_i}$$

(a) Odd and case A even: 0.1-dB passband ripple:

n	ω_s	K_s (dB)	c_i	p_i	a_i	b_i
2	1.05	.343	1.438664	$-.075407 + J1.180400$	.150814	1.399030
	1.10	.559	1.714083	$-.129483 + J1.268507$	.258966	1.625877
	1.20	1.075	2.235990	$-.236268 + J1.393844$	.472537	1.998624
	1.50	3.210	3.927051	$-.534107 + J1.568367$	1.068213	2.745046
	2.00	7.418	7.464102	$-.843443 + J1.581991$	1.686887	3.214092
3	1.05	1.748	1.205410	$-.044853 + J1.079332$ -2.812966	.089707	1.166969
	1.10	3.374	1.370314	$-.085421 + J1.121848$ -2.240832	.170843	1.265840
	1.20	6.691	1.699617	$-.156766 + J1.170259$ -1.744102	.313532	1.394082
	1.50	14.848	2.806014	$-.289646 + J1.212428$ -1.298182	.579292	1.553876
	2.00	24.010	5.153209	$-.381858 + J1.217905$ -1.116765	.763717	1.629108
4	1.05	6.397	1.153634 3.312518	$-.618576 + J1.143244$ $-.037598 + J1.045948$	1.237152 .075196	1.689644 1.095422
	1.10	10.721	1.290925 4.349930	$-.703816 + J .976495$ $-.066734 + J1.066126$	1.407633 .133467	1.448899 1.141079
	1.20	17.051	1.572430 6.224402	$-.108448 + J1.086869$ $-.726853 + J .798154$	.216897 1.453706	1.193044 1.165365
	1.50	29.064	2.535553 12.099310	$-.698734 + J .616949$ $-.173627 + J1.108114$	1.397469 .347253	.868855 1.258062
	2.00	41.447	4.593261 24.227201	$-.670443 + J .535639$ $-.216254 + J1.116820$	1.340886 .432509	.736403 1.294053

* This table was computed using a program from David Jose Miguel Baezlopez, "Determination of Elliptic Network Functions," M.S. thesis, University of Arizona, Tucson, 1977.

Table 2.3-1—*Continued*

(*a*) **Odd and case *A* even: 0.1-dB passband ripple:**

n	ω_s	K_s (dB)	c_i	p_i	a_i	b_i
5	1.05	13.641	1.133422	$-.266902 + J1.015887$	.533804	1.103263
			1.773739	$-.030115 + J1.028040$	.060229	1.057772
				-1.128858		
	1.10	20.050	1.259320	$-.329692 + J .953290$	.659383	1.017475
			2.193003	$-.049533 + J1.039346$	.099067	1.082694
				$-.932112$		
	1.20	28.303	1.521127	$-.379155 + J .875398$	.758311	.910681
			2.968367	$-.075430 + J1.051645$	.150860	1.111647
				$-.782858$		
	1.50	43.415	2.425515	$-.417037 + J .775766$	.834075	.775733
			5.437645	$-.114129 + J1.066151$	.228259	1.149703
				$-.649753$		
	2.00	58.901	4.364951	$-.429092 + J .721329$	.858183	.704436
			10.567732	$-.138913 + J1.073567$	.277825	1.171844
				$-.590933$		
6	1.05	22.088	1.123326	$-.647026 + J .628506$	1.294052	.813662
			1.438664	$-.151511 + J .985417$	.303023	.994002
			6.528768	$-.023386 + J1.018380$	.046771	1.037644
	1.10	29.686	1.243362	$-.599771 + J .517581$	1.199542	.627615
			1.714083	$-.194450 + J .951604$	.388900	.943361
			8.826455	$-.036964 + J1.025840$	.073927	1.053714
	1.20	39.630	1.495035	$-.547628 + J .429686$	1.095257	.484527
			2.235990	$-.235429 + J .907696$	.470859	.879339
			12.952671	$-.054595 + J1.034294$	.109189	1.072744
	1.50	57.772	2.369289	$-.487832 + J .349732$	.975663	.360292
			3.927051	$-.277388 + J .846778$	.554775	.793976
			25.827242	$-.080385 + J1.044897$	.160770	1.098271
	2.00	76.355	4.248155	$-.457235 + J .314305$	.914470	.307851
			7.464102	$-.296650 + J .810843$	.593299	.745467
			52.356841	$-.096676 + J1.050734$	.193352	1.113387
7	1.05	30.470	1.117521	$-.362386 + J .791219$	.724773	.757351
			1.308341	$-.097930 + J .979496$	.195860	.969003
			2.714372	$-.018274 + J1.012906$	.036549	1.026313
				$-.697913$		
	1.10	39.357	1.234128	$-.372606 + J .706869$	.745212	.638500
			1.523943	$-.129118 + J .957427$	.258235	.933330
			3.514769	$-.028279 + J1.018274$	.056558	1.037683
				$-.599630$		
	1.20	50.963	1.479872	$-.371195 + J .627191$	.742389	.531154
			1.941341	$-.161448 + J .928552$	.322896	.888274
			4.966697	$-.041080 + J1.024498$	.082161	1.051284
				$-.519720$		

(continued)

Table 2.3-1—*Continued*

(*a*) **Odd and case *A* even: 0.1-dB passband ripple:**

n	ω_s	K_s (dB)	c_i	p_i	a_i	b_i
	1.50	72.129	2.336522	$-.359474 + J .543128$	.718047	.424210
			3.313990	$-.198305 + J .887345$	.396611	.826705
			2.530074	$-.059550 + J1.032553$	.119119	1.069713
				$-.443710$		
	2.00	93.809	4.189043	$-.350169 + J .501993$	.700339	.374616
			5.201776	$-.217125 + J .862267$	.434251	.790648
			18.961005	$-.071124 + J1.037140$	.142248	1.080719
				$-.408602$		
8	1.05	39.872	1.113864	$-.518427 + J .408220$	1.036554	.435410
			1.243118	$-.223630 + J .856767$	.447259	.784060
			1.906139	$-.068627 + J .979963$	.137255	.965037
			11.046606	$-.014531 + J1.009559$	.029062	1.019421
	1.10	49.032	1.228286	$-.466763 + J .344817$	.933526	.336767
			1.427103	$-.246416 + J .795957$	.492832	.694269
			2.380411	$-.092340 + J .964023$	.184680	.937868
			15.106224	$-.022205 + J1.013628$	.044410	1.027534
	1.20	62.296	1.470253	$-.419007 + J .293877$	.838014	.261931
			1.789509	$-.261711 + J .732751$	.523423	.605416
			3.252831	$-.118065 + J .943252$	.236131	.903664
			22.383599	$-.031932 + J1.018411$	.063864	1.038180
	1.50	86.485	2.315697	$-.369042 + J .246297$	.738083	.196854
			2.995660	$-.270992 + J .659775$	.541983	.508739
			6.021824	$-.149084 + J .913365$	.298167	.856462
			45.059978	$-.045849 + J1.024717$	.091699	1.052148
	2.00	111.263	4.136734	$-.344561 + J .224683$	.689123	.169205
			5.545063	$-.272775 + J .621654$	.545549	.460860
			11.766674	$-.165771 + J .894945$	.331543	.828407
			91.761533	$-.054503 + J1.028377$	.109007	1.060529
9	1.05	47.276	1.111406	$-.355257 + J .616991$	.710513	.506886
			1.205410	$-.149551 + J .893074$	.299102	.819946
			1.594271	$-.050863 + J .982010$	.101725	.966931
			3.993674	$-.011772 + J1.007371$	.023544	1.014935
				$-.514179$		
	1.10	58.707	1.224347	$-.341731 + J .544813$	.683461	.413601
			1.370314	$-.173149 + J .847267$	.346297	.747841
			1.937719	$-.069513 + J .969793$	.139026	.945331
			5.299248	$-.017844 + J1.010567$	.035688	1.021564
				$-.448275$		
	1.20	73.629	1.463756	$-.323598 + J .480740$	.647197	.335827
			1.699617	$-.192808 + J .797069$	.385616	.672494
			2.579104	$-.090331 + J .953984$	.180663	.918245
			7.652393	$-.025490 + J1.014360$	.050980	1.029577
				$-.392972$		

Table 2.3-1—*Continued*

(a) Odd and case A even: 0.1-dB passband ripple:

n	ω_s	K_s (dB)	c_i	p_i	a_i	b_i
	1.50	100.842	2.301616	$-.299692 + J .415879$	.599383	.262770
			2.806014	$-.210182 + J .735885$	.420364	.585703
			4.636336	$-.116302 + J .931223$	.232604	.880703
			15.014341	$-.036362 + J1.019421$	.072724	1.040542
				$-.339006$		
	2.00	128.717	4.107442	$-.286338 + J .384821$	.572676	.230077
			5.153209	$-.217131 + J .702561$	.434263	.540739
			8.922191	$-.130710 + J .917130$	.261421	.858212
			30.201059	$-.043092 + J1.022392$	.086183	1.047142
				$-.313657$		
10	1.05	55.681	1.109672	$-.422209 + J .302646$	.844417	.269855
			1.181462	$-.246341 + J .730422$	.492682	.594200
			1.438664	$-.106451 + J .916451$	.212634	.851186
			2.533648	$-.039276 + J .984256$	.078553	.970303
			16.859610	$-.009704 + J1.005863$	.019407	1.011854
	1.10	68.382	1.221564	$-.377369 + J .259863$	.754739	.209936
			1.333834	$-.251250 + J .665084$	.502500	.505463
			1.714083	$-.127784 + J .880488$	.255569	.791588
			3.261942	$-.054334 + J .974525$	.108668	.952652
			23.183803	$-.014626 + J1.008442$	.029253	1.017169
	1.20	84.962	1.459158	$-.337237 + J .224734$	.674475	.164234
			1.641431	$-.250236 + J .602564$	.500472	.425702
			2.235990	$-.147516 + J .839832$	.295032	.727079
			4.585492	$-.071471 + J .962020$	.142943	.930591
			34.512249	$-.020797 + J1.011524$	.041594	1.023613
	1.50	115.199	2.291641	$-.296075 + J .151193$	.592150	.124215
			2.682641	$-.243660 + J .535025$	.487320	.345621
			3.927051	$-.167330 + J .788570$	.334660	.649842
			8.750764	$-.093337 + J .944057$	.186675	.899055
			69.791499	$-.029532 + J1.015670$	.059064	1.032457
	2.00	146.171	4.086685	$-.276117 + J .175698$	.552234	.107110
			4.897971	$-.238417 + J .501208$	.476833	.308052
			7.464102	$-.176419 + J .759877$	.352837	.608537
			17.363454	$-.105713 + J .932921$	.211426	.881517
			142.430967	$-.034920 + J1.018121$	.069840	1.037791

(*continued*)

Table 2.3-1—*Continued*

(*b*) **Odd and case *A* even: 1.0-dB passband ripple:**

n	ω_s	K_s(dB)	c_i	p_i	a_i	b_i
2	1.05	2.816	1.438664	$-.157083 + J1.068900$	.314166	1.167222
	1.10	4.025	1.714083	$-.229129 + J1.075841$	.458258	1.209934
	1.20	6.150	2.235990	$-.320565 + J1.064452$	.641131	1.235820
	1.50	11.194	3.927051	$-.439709 + J1.010488$	.879418	1.214431
	2.00	17.095	7.464102	$-.499471 + J .959482$	.998942	1.170077
3	1.05	8.134	1.205410	$-.065504 + J1.017106$ $-.947805$	.131007	1.038796
	1.10	11.480	1.370314	$-.097651 + J1.016303$ $-.816161$	.195302	1.042407
	1.20	16.209	1.699617	$-.136461 + J1.010059$ $-.701999$	.272923	1.038841
	1.50	25.176	2.806014	$-.187698 + J .994225$ $-.591015$	.375396	1.023714
	2.00	34.454	5.153209	$-.217034 + J .981575$ $-.530958$	.434067	1.010594
4	1.05	15.840	1.153634 3.312518	$-.400926 + J .723958$ $-.036963 + J1.004642$	.801852 .073925	.684857 1.010671
	1.10	20.832	1.290925 4.349930	$-.399229 + J .638481$ $-.054484 + J1.003351$	.798458 .108969	.567042 1.009681
	1.20	27.432	1.572430 6.224402	$-.386971 + J .560447$ $-.075673 + J1.000256$	.773942 .151346	.463847 1.006238

Table 2.3-1—*Continued*

(*b*) **Odd and case *A* even: 1.0-dB passband ripple:**

n	ω_s	K_s (dB)	c_i	p_i	a_i	b_i
	1.50	39.518	2.535553	$-.364988 + J\ .480692$	.729977	.364281
			12.039310	$-.104409 + J\ .993937$	.208819	.998811
	2.00	51.906	4.593261	$-.351273 + J\ .442498$	.702546	.319197
			24.227201	$-.121478 + J\ .989176$	.242957	.993226
5	1.05	24.135	1.133422	$-.181185 + J\ .858432$	.362371	.769820
			1.773739	$-.023559 + J1.001164$	.047118	1.002885
				$-.511794$		
	1.10	30.471	1.259320	$-.202145 + J\ .804785$	.404289	.688541
			2.193093	$-.034621 + J1.000221$	.069241	1.001640
				$-.446562$		
	1.20	38.757	1.521127	$-.217568 + J\ .748167$	.435136	.607089
			2.968367	$-.048084 + J\ .998478$	.096167	.999271
				$-.391579$		
	1.50	53.875	2.425515	$-.228875 + J\ .681678$	.457749	.517069
			5.437645	$-.066541 + J\ .995254$	.133081	.994957
				$-.337846$		
	2.00	69.360	4.364951	$-.232338 + J\ .646440$	.464676	.471866
			10.567732	$-.077625 + J\ .992914$	.155249	.991903
				$-.312599$		
6	1.05	32.523	1.123326	$-.340554 + J\ .466561$	.681109	.333656
			1.438664	$-.099253 + J\ .910440$	.198505	.838752
			6.528768	$-.016283 + J1.000095$	.032567	1.000456
	1.10	40.142	1.243362	$-.315089 + J\ .409244$	.630179	.266762
			1.714083	$-.118730 + J\ .874514$	.237461	.778873
			8.826455	$-.023927 + J\ .999416$	.047854	.999404
	1.20	50.089	1.495035	$-.289467 + J\ .359828$	.578933	.213267
			2.235990	$-.136580 + J\ .834256$	.273161	.714637
			12.952671	$-.033261 + J\ .998304$	.066522	.997718
	1.50	68.231	2.369289	$-.260804 + J\ .310775$	.521608	.164600
			3.927051	$-.154480 + J\ .783831$	.308960	.638255
			25.827242	$-.046116 + J\ .996374$	.092233	.994887
	2.00	86.814	4.248155	$-.246136 + J\ .287533$	.492272	.143258
			7.464102	$-.162691 + J\ .755715$	.325383	.597574
			52.356841	$-.053871 + J\ .995015$	.107742	.992957

(*continued*)

Table 2.3-1—*Continued*

(b) **Odd and case** *A* **even: 1.0-dB passband ripple:**

n	ω_s	K_s (dB)	c_i	p_i		a_i	b_i
7	1.05	40.926	1.117521	$-.206293 +$ J	$.681553$	.412587	.507071
			1.338341	$-.061953 +$ J	$.937640$	.123905	.883009
			2.714372	$-.011920 +$ J	$.999752$	.023840	.999646
				$-.352248$			
	1.10	49.816	1.234128	$-.206797 +$ J	$.621264$	.413594	.428734
			1.523943	$-.077646 +$ J	$.911762$	.155292	.837339
			3.514769	$-.017524 +$ J	$.999244$	.035048	.998795
				$-.310175$			
	1.20	61.422	1.479872	$-.203316 +$ J	$.564153$	.406632	.359606
			1.941341	$-.093371 +$ J	$.881886$	.186742	.786440
			4.966697	$-.024380 +$ J	$.998471$	.048760	.997530
				$-.274069$			
	1.50	82.588	2.336522	$-.195790 +$ J	$.502754$	.391580	.291095
			3.313990	$-.110878 +$ J	$.843206$	.221757	.723291
			9.530078	$-.033844 +$ J	$.997189$	.067688	.995531
				$-.238188$			
	2.00	104.268	4.180043	$-.190684 +$ J	$.472047$	.381369	.259189
			6.201776	$-.119725 +$ J	$.821044$	.239450	.688447
			18.961095	$-.039566 +$ J	$.996309$	.079132	.994196
				$-.221131$			
8	1.05	49.331	1.113864	$-.274078 +$ J	$.341142$	.548157	.191497
			1.243118	$-.130371 +$ J	$.789532$	.260743	.640357
			1.906139	$-.042262 +$ J	$.953923$	.084524	.911755
			11.046606	$-.009102 +$ J	$.999652$	.018205	.999387
	1.10	59.491	1.228286	$-.248664 +$ J	$.300352$	.497328	.152045
			1.427103	$-.139688 +$ J	$.737745$	.279377	.563780
			2.380411	$-.054736 +$ J	$.934315$	.109472	.875940
			15.106224	$-.013388 +$ J	$.999259$	.026777	.998698
	1.20	72.755	1.470253	$-.225049 +$ J	$.265193$	.450098	.120974
			1.789509	$-.145710 +$ J	$.685429$	.291419	.491044
			3.252831	$-.067891 +$ J	$.911293$	.135782	.835065
			22.383599	$-.018638 +$ J	$.998689$	.037276	.997727
	1.50	96.945	2.315697	$-.200053 +$ J	$.230188$	.400106	.093008
			2.995660	$-.149085 +$ J	$.625789$	.298170	.413938
			6.021824	$-.083400 +$ J	$.880916$	.166800	.782968
			45.059978	$-.025894 +$ J	$.997774$	.051789	.996224
	2.00	121.722	4.136734	$-.187672 +$ J	$.213541$	.375344	.080821
			5.545063	$-.149543 +$ J	$.594686$	.299085	.376615
			11.766674	$-.091635 +$ J	$.863229$	.183270	.753561
			91.761533	$-.030287 +$ J	$.997159$	.060574	.995244

Table 2.3-1—*Continued*

(b) Odd and case *A* even: 1.0-dB passband ripple:

n	ω_s	K_s (dB)	c_i	p_i	a_i	b_i
9	1.05	57.736	1.111406	-.195247 + J .551838	.390495	.342647
			1.205410	-.087514 + J .850482	.175028	.730978
			1.594271	-.030697 + J .964496	.061394	.931195
			3.993674	-.007178 + J .999640	.014357	.999331
				-.269878		
	1.10	69.167	1.224347	-.186735 + J .497328	.373473	.282206
			1.370314	-.098792 + J .807558	.197584	.661910
			1.937719	-.040723 + J .949094	.081447	.902438
			5.299248	-.010563 + J .999327	.021126	.998766
				-.238541		
	1.20	84.089	1.463756	-.176578 + J .447596	.353156	.231522
			1.699617	-.108069 + J .762286	.216138	.592758
			2.579104	-.051655 + J .930819	.103311	.869092
			7.652393	-.014711 + J .998888	.029422	.997993
				-.211419		
	1.50	111.302	2.301616	-.163781 + J .395695	.327563	.183399
			2.806014	-.116231 + J .708498	.232461	.515479
			4.636336	-.065021 + J .906413	.130043	.825813
			15.014341	-.020451 + J .998201	.040901	.996824
				-.184275		
	2.00	139.176	4.107442	-.156746 + J .370225	.313491	.161635
			5.153209	-.119505 + J .679573	.239009	.476101
			8.922191	-.072341 + J .892058	.144681	.801001
			30.201059	-.023928 + J .997747	.047856	.996072
				-.171309		
10	1.05	66.141	1.109672	-.225977 + J .269047	.451954	.123452
			1.181462	-.138223 + J .680726	.276445	.482493
			1.438664	-.062031 + J .888118	.124062	.792601
			2.533648	-.023347 + J .971768	.046694	.944878
			16.859610	-.005806 + J .999650	.011612	.999352
	1.10	78.842	1.221564	-.203387 + J .237538	.406775	.097791
			1.333834	-.139376 + J .625733	.278752	.410967
			1.714083	-.072952 + J .852542	.145905	.732149
			3.261942	-.031537 + J .959337	.063074	.921321
			23.183803	-.008547 + J .999404	.017093	.998881
	1.20	95.422	1.459158	-.182921 + J .210279	.365841	.077677
			1.641431	-.137940 + J .572984	.275879	.347338
			2.235990	-.082871 + J .813874	.165743	.669259
			4.585492	-.040676 + J .944491	.081352	.893698
			34.512249	-.011907 + J .999055	.023815	.998253
	1.50	125.659	2.291641	-.161666 + J .183031	.323333	.059636
			2.682641	-.133903 + J .515483	.267806	.283652
			3.927051	-.092777 + J .766550	.185554	.596207
			8.750764	-.052134 + J .924481	.104268	.857383
			69.791499	-.016560 + J .998520	.033120	.997316
	2.00	156.630	4.086686	-.151264 + J .170031	.302528	.051791
			4.897971	-.130976 + J .486399	.261951	.253739
			7.464102	-.097323 + J .740520	.194647	.557842
			17.363454	-.058541 + J .912636	.117082	.836331
			142.430967	-.019380 + J .998171	.038760	.996720

(*continued*)

Table 2.3-1—*Continued*

(*c*) **Case *B* even: 0.1-dB passband ripple:**

n	ω_s	K_s(dB)	c_i	p_i	a_i	b_i
4	1.05	4.485	1.166586	$-.852205 + J\ .918084$	1.704411	1.569133
				$-.040627 + J1.053933$	.681255	1.112426
	1.10	8.308	1.309737	$-.073101 + J1.076009$	.146201	1.163130
				$-.796684 + J\ .809361$	1.593368	1.289771
	1.20	14.387	1.601406	$-.118529 + J1.095968$	.237059	1.215196
				$-.746825 + J\ .700487$	1.493650	1.048420
	1.50	26.320	2.595517	$-.183989 + J1.112642$	.367978	1.271824
				$-.693645 + J\ .586090$	1.387209	.818602
	2.00	38.697	4.716540	$-.223115 + J1.118520$	.446230	1.300868
				$-.655767 + J\ .521238$	1.331534	.714935
6	1.05	20.307	1.125244	$-.639897 + J\ .518952$	1.279794	.721895
			1.500649	$-.166247 + J\ .984394$	.332494	.996669
				$-.024644 + J1.019788$	.049288	1.040575
	1.10	27.899	1.246215	$-.206675 + J\ .947127$	.413349	.939764
			1.800145	$-.038741 + J1.027332$	.077482	1.056912
				$-.588055 + J\ .479414$	1.176111	.575647
	1.20	37.827	1.499501	$-.243668 + J\ .901847$	.487336	.872702
			2.364951	$-.056665 + J1.035618$	.113330	1.075715
				$-.538160 + J\ .410727$	1.075320	.458312
	1.50	55.966	2.378628	$-.280868 + J\ .642354$	.561736	.788449
			4.189497	$-.082131 + J1.045701$	.164262	1.100276
				$-.483133 + J\ .343213$	.966267	.351213
	2.00	74.548	4.267406	$-.298091 + J\ .808340$	.596182	.742272
			8.001486	$-.097774 + J1.051149$	.195549	1.114475
				$-.455028 + J\ .311687$	.910055	.304199
8	1.05	37.529	1.114416	$-.506414 + J\ .386312$	1.012827	.405692
			1.252994	$-.232472 + J\ .847640$	.464944	.772537
			2.034525	$-.015010 + J1.009939$	.030021	1.020203
				$-.071703 + J\ .979388$	.143406	.964341
	1.10	47.685	1.229114	$-.251503 + J\ .787764$	.503006	.653826
			1.441163	$-.022814 + J1.014037$	.045627	1.028791
			2.557144	$-.095240 + J\ .962010$	.190490	.936269
				$-.458373 + J\ .332941$	.916746	.320955

Table 2.3-1—*Continued*

(c) **Case B even: 0.1-dB passband ripple:**

n	ω_s	K_s (dB)	c_i	p_i	a_i	b_i
	1.20	60.949	1.471557	$-.254032 + J\ .726645$	.528064	.597725
			1.810911	$-.032591 + J1.018790$	.065183	1.038996
			3.516389	$-.120409 + J\ .941863$	.240619	.901605
				$-.413887 + J\ .287903$	.827774	.254190
	1.50	85.138	2.318435	$-.271506 + J\ .656662$	.543012	.504921
			3.039648	$-.046374 + J1.024966$	.092748	1.052713
			6.536721	$-.150389 + J\ .912223$	.300778	.854768
				$-.366858 + J\ .244169$	.733715	.194203
	2.00	109.915	4.142383	$-.272874 + J\ .620149$	.545748	.459045
			5.635349	$-.054826 + J1.028517$	.109652	1.060853
			12.851234	$-.166427 + J\ .894253$	.332854	.827387
				$-.343582 + J\ .223806$	.687165	.168139
10	1.05	54.608	1.109888	$-.414731 + J\ .292587$	.829462	.258195
			1.184476	$-.249669 + J\ .721758$	.499237	.583270
			1.459121	$-.109291 + J\ .913096$	.218582	.847333
			2.745052	$-.040314 + J\ .983945$	.090628	.969774
				$-.009914 + J1.006605$	.019828	1.012145
	1.10	67.307	1.221890	$-.252592 + J\ .658710$	.505184	.497701
			1.338203	$-.130002 + J\ .877950$	.260004	.787698
			1.742587	$-.014883 + J1.008599$	.029767	1.017493
			3.552883	$-.055379 + J\ .974066$	.110756	.951871
				$-.372589 + J\ .254860$	.745178	.203776
	1.20	83.887	1.459673	$-.250516 + J\ .598450$	.501033	.420901
			1.648160	$-.148915 + J\ .827636$	.297830	.723809
			2.278803	$-.072379 + J\ .961490$	.144758	.929701
			5.019460	$-.021068 + J1.011673$	.042135	1.023927
				$-.334463 + J\ .222157$	.668925	.161219
	1.50	114.123	2.292725	$-.243524 + J\ .533167$	.487049	.343572
			2.696572	$-.167901 + J\ .787240$	.335803	.647937
			4.014312	$-.093903 + J\ .943626$	.187801	.899247
			9.631781	$-.029742 + J1.015773$	.059483	1.032670
				$-.294931 + J\ .190246$	.589862	.123178
	2.00	145.095	4.085922	$-.238296 + J\ .500354$	.476591	.307139
			4.926617	$-.176655 + J\ .759172$	.353311	.607551
			7.642845	$-.106015 + J\ .932657$	.212031	.881089
			19.166456	$-.035048 + J1.018190$	.070095	1.037920
				$-.275610 + J\ .175300$	.551220	.106601

(continued)

Table 2.3-1—*Continued*

(d) Case B even: 1.0-dB passband ripple:

n	ω_s	K_s(dB)	c_i	p_i	a_i	b_i
4	1.05	13.243	1.156566	$-.422751 + J\ .627439$	.645502	.572396
				$-.043154 + J1.006305$	.046309	1.014512
	1.10	18.140	1.309737	$-.403039 + J\ .574970$	.606078	.493031
				$-.061668 + J1.004223$	.123336	1.012268
	1.20	24.700	1.601406	$-.383555 + J\ .523216$	.767109	.420870
				$-.082727 + J1.000117$	.165455	1.007078
	1.50	36.771	2.595517	$-.361063 + J\ .465487$	.722126	.347045
				$-.109356 + J\ .993062$	.218712	.998131
	2.00	49.156	4.716543	$-.348975 + J\ .435853$	.697949	.311751
				$-.124305 + J\ .988442$	.248611	.992470
6	1.05	30.730	1.125244	$-.332474 + J\ .436556$	.664948	.301121
			1.500649	$-.107214 + J\ .902829$	.214429	.826595
				$-.017380 + J1.000183$	.034760	1.000669
	1.10	38.342	1.246215	$-.308336 + J\ .391152$	.616673	.248071
			1.800145	$-.124563 + J\ .866975$	.249126	.767162
				$-.025185 + J\ .999425$	.050371	.999485
	1.20	48.285	1.499501	$-.284782 + J\ .349826$	.569565	.203480
			2.364951	$-.140216 + J\ .827910$	.280431	.705096
				$-.034509 + J\ .998231$	.069019	.997656
	1.50	66.425	2.378628	$-.258566 + J\ .306851$	.517132	.161014
			4.189497	$-.155955 + J\ .780065$	.311909	.632824
				$-.047022 + J\ .996256$	.094044	.994738
	2.00	85.008	4.267406	$-.245081 + J\ .285837$	.490161	.141767
			8.001485	$-.163303 + J\ .753736$	.326605	.594786
				$-.054402 + J\ .994927$	.108805	.992839
8	1.05	47.987	1.114416	$-.268114 + J\ .329175$	.536228	.180241
			1.252994	$-.134219 + J\ .780319$	.268437	.626912
			2.034525	$-.044114 + J\ .952060$	.088228	.908365
				$-.009435 + J\ .999653$	.018870	.999396
	1.10	58.146	1.229114	$-.244589 + J\ .293222$	.489178	.145803
			1.441163	$-.141813 + J\ .730432$	.283626	.553642
			2.557144	$-.056360 + J\ .932329$	.112720	.872413
				$-.013771 + J\ .999245$	.027542	.998679

Table 2.3-1—*Continued*

(*d*) **Case *B* even: 1.0-dB passband ripple:**

n	ω_s	K_s (dB)	c_i	p_i		a_i	b_i
	1.20	71.408	1.471557	$-.222550 + J$	$.261260$	.445090	.117780
			1.810911	$-.146643 + J$	$.680312$	.292298	.484330
			3.516369	$-.059126 + J$	$.909468$	.138252	.631011
				$-.019020 + J$	$.998650$	.038040	.997684
	1.50	95.597	2.318435	$-.198966 + J$	$.228644$	.397931	.091866
			3.039648	$-.149279 + J$	$.623251$	.298559	.410726
			6.556721	$-.084056 + J$	$.879706$	.168112	.780940
				$-.026174 + J$	$.997742$	.052349	.996174
	2.00	120.374	4.142383	$-.187178 + J$	$.212872$	.374357	.080350
			5.635349	$-.149573 + J$	$.593460$	.299147	.374567
			12.861234	$-.091959 + J$	$.862553$	.183918	.752454
				$-.030453 + J$	$.907137$	.060905	.995209
10	1.05	65.067	1.109888	$-.222316 + J$	$.263196$	.444633	.118697
			1.184476	$-.139553 + J$	$.673297$	.279106	.472804
			1.459121	$-.063624 + J$	$.885077$	.127249	.787410
			2.745052	$-.023970 + J$	$.971078$	.047940	.943567
				$-.005940 + J$	$.999655$	.011880	.999346
	1.10	77.767	1.221890	$-.201018 + J$	$.234049$	.402036	.095187
			1.338203	$-.139872 + J$	$.620398$	.279743	.404458
			1.742587	$-.074098 + J$	$.849790$	.148196	.727633
			3.552883	$-.032133 + J$	$.958580$	.064266	.919909
				$-.008701 + J$	$.999395$	.017402	.998866
	1.20	94.346	1.459673	$-.181522 + J$	$.208352$	.363045	.076361
			1.648160	$-.138003 + J$	$.569550$	.276006	.343432
			2.278803	$-.083577 + J$	$.811692$	.167155	.665829
			5.019460	$-.041174 + J$	$.943762$	.082348	.892382
				$-.012062 + J$	$.999042$	.024123	.998231
	1.50	124.583	2.292725	$-.161078 + J$	$.182269$	.322155	.059168
			2.696572	$-.133813 + J$	$.513911$	.267627	.282010
			4.014312	$-.093062 + J$	$.765311$	.186124	.594362
			9.631781	$-.052431 + J$	$.923983$	.104862	.856494
				$-.016673 + J$	$.998508$	.033347	.997296
	2.00	155.554	4.088922	$-.151000 + J$	$.169700$	.302000	.051599
			4.926617	$-.130907 + J$	$.485667$	.261815	.253009
			7.642845	$-.097442 + J$	$.739879$	.194883	.556916
			19.166456	$-.058698 + J$	$.912351$	.117396	.835830
				$-.019447 + J$	$.998163$	.038895	.996707

(*continued*)

Table 2.3-1—*Continued*

(e) **Case *C* even: 0.1-dB passband ripple:**

n	ω_s	K_s(dB)	c_i	p_i	a_i	b_i
4	1.05	3.284	1.176045	$-.041450 + J1.062080$	.082900	1.129732
				$-1.142752 + J1.056906$	2.285503	2.422931
	1.10	6.478	1.321589	$-.076408 + J1.089646$	.152816	1.193166
				$-1.041973 + J .905418$	2.083946	1.905490
	1.20	12.085	1.615455	$-.128382 + J1.115527$	.256764	1.260882
				$-.953405 + J .756606$	1.906811	1.481435
	1.50	23.736	2.611679	$-.206296 + J1.136431$	.412592	1.334033
				$-.863022 + J .597833$	1.726044	1.102212
	2.00	36.023	4.733595	$-.817435 + J .520713$	1.634871	.939343
				$-.253437 + J1.142940$	.506873	1.370542
6	1.05	18.727	1.126696	$-.789645 + J .570579$	1.579289	.949099
			1.535284	$-.185878 + J .993057$	.371755	1.020713
				$-.025910 + J1.021602$	.051819	1.044342
	1.10	26.230	1.248053	$-.715339 + J .476877$	1.430677	.739121
			1.837658	$-.230107 + J .952094$	.460214	.959433
				$-.040931 + J1.029755$	.081861	1.062071
	1.20	36.113	1.501690	$-.646187 + J .398581$	1.292374	.576425
			2.404505	$-.269710 + J .902237$	.539420	.886775
				$-.060077 + J1.038657$	.120154	1.082417
	1.50	54.202	2.381154	$-.572096 + J .324068$	1.144191	.432313
			4.230449	$-.308434 + J .836973$	.616868	.795656
				$-.087368 + J1.049425$	.174736	1.108925
	2.00	72.761	4.270072	$-.534998 + J .290136$	1.069995	.370401
			8.042806	$-.325870 + J .799822$	.651740	.745907
				$-.104187 + J1.055209$	.208374	1.124320
8	1.05	36.268	1.114839	$-.604065 + J .371382$	1.208130	.502820
			1.259509	$-.257536 + J .843660$	.515072	.778086
			2.035094	$-.076089 + J .979591$	.152178	.965388
				$-.015568 + J1.010441$	.031137	1.021234
	1.10	46.399	1.229651	$-.541763 + J .314616$	1.083525	.392490
			1.448643	$-.276093 + J .780152$	.552186	.684864
			2.608880	$-.023659 + J1.014703$	.047318	1.030181
				$-.100633 + J .962426$	.201267	.936390

Table 2.3-1—*Continued*

(e) **Case C even: 0.1-dB passband ripple:**

n	ω_s	K_s (dB)	c_i	p_i	a_i	b_i
	1.20	59.639	1.472198	$-.485227 + J\ .267901$	.070453	.307216
			1.819178	$-.287400 + J\ .715925$	.574801	.595147
			3.568474	$-.126736 + J\ .940478$	.253471	.900562
				$-.033798 + J1.019637$	.067597	1.040801
	1.50	83.807	2.319176	$-.426488 + J\ .223590$	.852977	.231885
			3.048578	$-.292845 + J\ .643036$	.585691	.499254
			6.608405	$-.157688 + J\ .909548$	.315377	.852143
				$-.048107 + J1.026042$	.096214	1.055076
	2.00	108.575	4.143165	$-.397764 + J\ .203316$	.795529	.199554
			5.644525	$-.292952 + J\ .605259$	.585904	.452160
			12.912424	$-.174180 + J\ .890785$	.348360	.823836
				$-.056895 + J1.029716$	.113790	1.063553
10	1.05	53.576	1.110055	$-.486272 + J\ .273776$	.972543	.311413
			1.186597	$-.272544 + J\ .710688$	.545089	.579358
			1.470226	$-.115389 + J\ .911607$	.230777	.844342
			2.804853	$-.041773 + J\ .983714$	.083546	.969439
				$-.010169 + J1.006194$	.020339	1.012529
	1.10	66.262	1.222102	$-.434146 + J\ .234898$	.868292	.243660
			1.340720	$-.273555 + J\ .645766$	.547110	.491847
			1.754581	$-.136496 + J\ .874512$	.272992	.783403
			3.612427	$-.057220 + J\ .973594$	.114440	.951160
				$-.015256 + J1.008851$	.030511	1.018013
	1.20	82.830	1.459927	$-.387617 + J\ .202683$	.775235	.191328
			1.651014	$-.269401 + J\ .584235$	.538801	.413911
			2.291423	$-.155607 + J\ .833128$	.311213	.718316
			5.078099	$-.074620 + J\ .960707$	.149239	.928526
				$-.021584 + J1.011998$	.043168	1.024605
	1.50	113.056	2.293019	$-.339922 + J\ .171783$	.679845	.145057
			2.699727	$-.259970 + J\ .518096$	.519940	.336007
			4.027356	$-.174590 + J\ .781514$	.349180	.641246
			9.688795	$-.096612 + J\ .942388$	.193225	.897429
				$-.030465 + J1.016193$	.060930	1.033575
	2.00	144.023	4.089232	$-.316795 + J\ .157490$	.633589	.125162
			4.929889	$-.253464 + J\ .485036$	.506929	.299505
			7.655998	$-.183250 + J\ .752817$	.366500	.600315
			19.222417	$-.108980 + J\ .931131$	.217960	.878881
				$-.035902 + J1.018656$	.071804	1.038948

(continued)

Table 2.3-1—*Continued*

(*f*) **Case *C* even: 1.0-dB passband ripple:**

n	ω_s	K_s(dB)	c_i	p_i	a_i	b_i
4	1.05	11.322	1.176045	$-.050129 + J1.009723$	.100257	1.022053
				$-.664397 + J .601548$	1.328793	.803283
	1.10	15.942	1.321589	$-.617627 + J .535240$	1.235255	.667946
				$-.072023 + J1.007836$	.144045	1.020920
	1.20	22.293	1.615455	$-.573215 + J .472186$	1.146429	.551534
				$-.096950 + J1.003197$	.193901	1.015804
	1.50	34.179	2.611679	$-.524307 + J .404475$	1.048614	.438498
				$-.128327 + J .994626$	.256654	1.005748
	2.00	46.481	4.733595	$-.499020 + J .370743$	.998040	.386471
				$-.145844 + J .988871$	.291689	.999136
6	1.05	29.133	1.126696	$-.479520 + J .376458$	.959040	.371660
			1.535284	$-.124739 + J .894415$	.249477	.815537
				$-.018753 + J1.000393$	.037507	1.001138
	1.10	36.680	1.248053	$-.437212 + J .330308$	.874424	.300258
			1.837653	$-.142909 + J .855552$	.285819	.752392
				$-.027110 + J .999598$	.054220	.999932
	1.20	46.571	1.501690	$-.397576 + J .289708$	.795151	.241997
			2.404505	$-.158828 + J .813561$	.317656	.687107
				$-.037077 + J .998324$	.074154	.998026
	1.50	64.661	2.381154	$-.174302 + J .762525$	.348604	.611825
			4.230449	$-.050450 + J .996191$	.100900	.994942
				$-.355021 + J .248804$	.710041	.187943
	2.00	83.221	4.270072	$-.333676 + J .229258$	.667352	.163904
			9.042806	$-.181299 + J .734590$	.362598	.572492
				$-.058342 + J .994745$	.116684	.992921
8	1.05	46.727	1.114839	$-.371900 + J .270505$	.743801	.211483
			1.259509	$-.151848 + J .763118$	.303695	.605407
			2.085094	$-.047269 + J .949664$	.094537	.904097
				$-.009859 + J .999670$	.019717	.999437
	1.10	56.498	1.229651	$-.335499 + J .237676$	.670998	.169050
			1.448643	$-.158468 + J .711532$	.316936	.531390
			2.608880	$-.059982 + J .929136$	.119964	.866892
				$-.014360 + J .999248$	.028719	.998703

Table 2.3-1—*Continued*

(*f*) **Case *C* even: 1.0-dB passband ripple:**

n	ω_s	K_s (dB)	c_i	p_i	a_i	b_i
	1.20	70.098	1.472198	$-.302267 + J\ .209205$	.604534	.135132
			1.819178	$-.162095 + J\ .660152$	.324190	.462075
			3.568474	$-.073157 + J\ .905406$	.146315	.825112
				$-.019803 + J\ .998642$	.039607	.997679
	1.50	94.266	2.319176	$-.267488 + J\ .180775$	.534975	.104229
			3.048578	$-.163159 + J\ .662103$	.326319	.389149
			6.608405	$-.088478 + J\ .874558$	.176956	.772679
				$-.027224 + J\ .997687$	.054449	.996120
	2.00	119.034	4.143165	$-.250369 + J\ .167247$	.500739	.090656
			5.644525	$-.162577 + J\ .571959$	.325154	.353568
			12.912424	$-.096552 + J\ .856791$	.193104	.743414
				$-.031665 + J\ .997054$	.063330	.995120
10	1.05	64.036	1.110055	$-.302228 + J\ .211052$	.604457	.135885
			1.186597	$-.154774 + J\ .652742$	.309547	.450027
			1.470226	$-.047537 + J\ .880131$	.135073	.779193
			2.804853	$-.024923 + J\ .970175$	.049846	.941860
				$-.006111 + J\ .999653$	.012221	.999343
	1.10	76.722	1.222102	$-.271183 + J\ .185925$	.542366	.108108
			1.340720	$-.153648 + J\ .599498$	.307296	.383006
			1.754581	$-.078131 + J\ .843990$	.156263	.718423
			3.612427	$-.033289 + J\ .957369$	.066578	.917664
				$-.008938 + J\ .999388$	.017875	.998855
	1.20	93.289	1.459927	$-.243254 + J\ .164164$	.486508	.086122
			1.651014	$-.150336 + J\ .548673$	.300673	.323643
			2.291423	$-.087630 + J\ .805089$	.175260	.655848
			5.078099	$-.042532 + J\ .942201$	.085063	.889552
				$-.012377 + J\ .999027$	.024754	.998208
	1.50	123.515	2.293018	$-.214394 + J\ .142421$	.428788	.066248
			2.699727	$-.144539 + J\ .493399$	.289078	.264334
			4.027356	$-.097026 + J\ .757852$	.194052	.583753
			9.688795	$-.054019 + J\ .921965$	.108037	.852937
				$-.017097 + J\ .998479$	.034194	.997252
	2.00	154.482	4.089232	$-.200312 + J\ .132059$	.400624	.057564
			4.929889	$-.140813 + J\ .465461$	.281625	.236482
			7.655998	$-.101316 + J\ .731995$	.202633	.546082
			19.222417	$-.060405 + J\ .910063$	.120810	.831863
				$-.019938 + J\ .998124$	.039876	.996650

factors. Passive network elliptic realizations are given in App. A. For the even-order functions, cases *B* and *C* have been used. Additional realizations may be found in the references listed in the Bibliography.[1]

In using the tables it should be noted that a given elliptic network function and its realizations are characterized by four items of information: (1) the order *n* of the function; (2) the passband ripple constant K_p (see Fig. 2.3-1), expressed either as a magnitude or in decibels; (3) the stopband tolerance constant K_s similarly expressed; and (4) the stopband frequency ω_s. From data given in the table, we see that when any two of the three items 2 through 4 are specified, a network function of a given order is completely defined.

Example 2.3-1 *Use of Table 2.3-1*   It is desired to find an elliptic network function with the following specifications: (1) a passband of 0 to 1 rad/s with a ripple of 1.0 dB; (2) a stopband of $\omega \geq 2$ rad/s with a minimum attenuation of 34 dB; and (3) a maximum magnitude of unity in the passband. Using the 1.0-dB ripple entries of Table 2.3-1, we see that specifications 1 and 2 are satisfied by a third-order function. Using the tabulated poles and zeros, the function has the form

$$N(s) = \frac{H(s^2 + 5.153209)}{(s + 0.539958)(s^2 + 0.434067s + 1.010594)}$$

To satisfy specification 3, we evaluate $N(0) = H \times 5.153209/(0.539958 \times 1.010594) = 9.443675H = 1$, from which we obtain $H = 0.105891$. It should be noted that, if a type *A* or *B* even-order function had been required to meet the specifications, then, to determine *H*, $N(0)$ would be set to 0.89125 (the magnitude equivalent of -1.0 dB) rather than to unity. This, of course, is because such even-order functions have their *minimum* passband magnitude at zero frequency, while odd-order functions have their *maximum* magnitude there.   □

The order of an elliptic network function required to meet a given set of specifications can be readily determined from the nomograph given in Fig. 2.3-3.[2]

---

[1] The tables given in the references are usually catalogued using values of the reflection coefficient ρ, rather than of the passband ripple K_p. The values may be related by using $K_p = -10 \log [1 - (\rho/100)^2]$, where ρ is in percent. Some examples are given in the following table:

ρ (%)	1	2	5	10	15	20	25	50
K_p (dB)	0.00043	0.0017	0.011	0.044	0.098	0.18	0.28	1.25

In addition, the selection of the values of ω_s is made by the use of integer values of a modular angle θ, where $\omega_s = 1/\sin \theta$ ($\theta < 90°$) for case *A*. Unfortunately, the use of these parameters does not permit the specification of convenient numerical values of K_p and ω_s.

[2] Kawakami, op. cit.

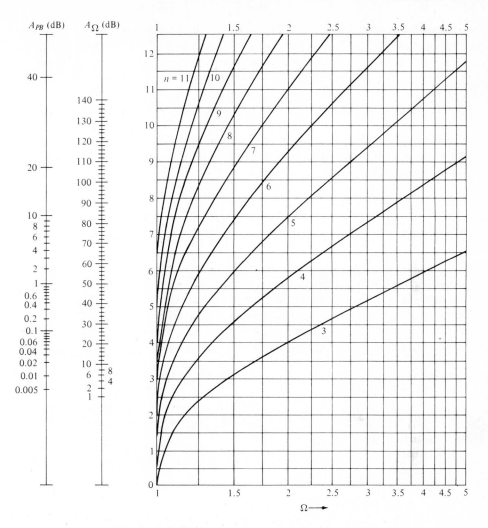

Figure 2.3-3 A nomograph for determining the order of an elliptic magnitude function.

Its use is explained in Sec. 2.1 and illustrated in Fig. 2.1-5. It should be noted that for even-order functions, the nomograph *gives the required order of a case A function*. If case *B* or *C* functions are to be used, a higher-order one than indicated by the nomograph may be required to meet the given specifications.

Example 2.3-2 *Use of the Nomograph for Elliptic Functions* It is desired to find the necessary order of an elliptic network function which has a ripple of 0.1 dB in the passband and a minimum attenuation of 40 dB at all frequencies greater than 1.5 times the cutoff frequency. From Fig. 2.3-3 we find that the required order is 5. Comparing this with the required orders of Chebyshev and Butterworth filters to meet the same specifications, using Figs. 2.2-5 and 2.1-4, we find that orders of 8 and approximately 17, respectively, would be required. This readily demonstrates the superiority of the elliptic characteristic in providing sharp cutoff characteristics. □

In this section we have introduced one of the most useful network functions for sharp cutoff filters, the elliptic one. Like the previously defined Butterworth and Chebyshev functions, the elliptic one has been defined for a low-pass characteristic. In the next section we shall see how all three of these characteristics can be used in high-pass and bandpass filters.

2.4 TRANSFORMATIONS OF THE COMPLEX-FREQUENCY VARIABLE

In the preceding sections of this chapter we have considered three methods for approximating magnitude characteristics. The techniques which were developed all apply to low-pass network functions. In this section we shall show how these approximations may be extended to other types of network functions. The types to be considered are the high-pass, bandpass, and band-elimination ones. The extension is done through the use of transformations made on the complex-frequency variable. We shall discuss the use of these transformations from three different viewpoints: (1) their effect on the magnitude characteristic, (2) their effect on the network function, and (3) their effect on the elements of a given network realization.

The first transformation of the complex-frequency variable which we shall describe is called the *low-pass to high-pass transformation*. If we let $s = \sigma + j\omega$ be the original complex-frequency variable, and $p = u + jv$ be the resulting transformed complex-frequency variable, then the transformation is defined as

$$s = \sigma + j\omega = \frac{\omega_0^2}{p} = \frac{\omega_0^2}{u + jv} = \omega_0^2 \left(\frac{u - jv}{u^2 + v^2} \right) \tag{1}$$

where ω_0 is a constant. In this relation, if we confine our range of interest to the sinusoidal steady-state case by letting $\sigma = 0$, then equating real and imaginary parts in (1) we see that $j\omega = -j\omega_0^2/v$. Thus the positive and negative imaginary axes in the (original) s plane become respectively the negative and positive imaginary axes in the (transformed) p plane. In addition, the points at the origin and at infinity are interchanged by the transformation. As a result of this interchange, a low-pass magnitude characteristic on the $j\omega$ axis is transformed to a high-pass one on the jv axis.[1] From (1) we also see that, under this transformation, corresponding points of the magnitude characteristics on the $j\omega$ and jv axes are geometrically

---

[1] Since magnitude characteristics are symmetrical about the origin, the reversal of sign has no effect.

centered about the frequency ω_0. For example, if the 3-dB-down (from its zero frequency value) frequency of a low-pass function $N_{LP}(s)$ is $\omega_{3\,dB}$, then the 3-dB-down (from its infinite frequency value) frequency $v_{3\,dB}$ of the high-pass function $N_{HP}(p)$ produced by the transformation of (1) satisfies the relation $|\omega_{3\,dB}||v_{3\,dB}| = \omega_0^2$. Frequently, for convenience, we set $\omega_0 = 1$ rad/s. Thus we define a *normalized low-pass to high-pass transformation* as

$$s = \frac{1}{p} \tag{2}$$

For this transformation $|\omega_{3\,dB}||v_{3\,dB}| = 1$. An example of applying such a transformation to an equal-ripple low-pass characteristic is shown in Fig. 2.4-1. Corresponding values on the original and the transformed characteristics are indicated by the letters $A, B, \ldots, A', B', \ldots$.

Now let us consider what happens when the normalized transformation of (2) is applied to a general low-pass network function defined as

$$N_{LP}(s) = \frac{H}{a_0 + a_1 s + a_2 s^2 + \cdots + a_{n-1} s^{n-1} + s^n} \tag{3}$$

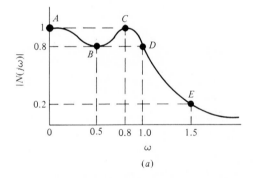

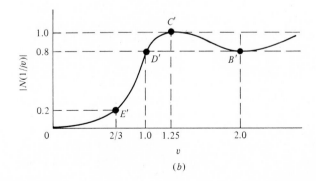

Figure 2.4-1 The normalized low-pass to high-pass transformation.

Applying the transformation and multiplying numerator and denominator by p^n we obtain

$$N_{HP}(p) = N_{LP}\left(\frac{1}{p}\right) = \frac{Hp^n}{a_0 p^n + a_1 p^{n-1} + a_2 p^{n-2} + \cdots + a_{n-1} p + 1} \qquad (4)$$

Obviously, the n zeros at infinity of $N_{LP}(s)$ have been transformed into n zeros at the origin in $N_{HP}(p)$. As an example of this transformation, consider the two functions

$$N_{LP}(s) = \frac{H}{s^3 + 2s^2 + 2s + 1} \qquad N_{HP}(p) = \frac{Hp^3}{p^3 + 2p^2 + 2p + 1} \qquad (5)$$

The function $N_{LP}(s)$ is readily identified (see Table 2.1-3) as a third-order maximally flat magnitude low-pass function with a 3-dB-down frequency of 1 rad/s. Since $N_{HP}(p)$ is derived from $N_{LP}(s)$ by the transformation $s = 1/p$, it is a third-order maximally flat magnitude high-pass function and its 3-dB-down frequency is also 1 rad/s. For this function, the term *maximally flat magnitude* of course refers to its behavior at infinite frequency.

From the derivation given above we see that, for normalized Butterworth low-pass functions with 3-dB-down frequencies of 1 rad/s, the pole locations for the high-pass functions are the same as those of the low-pass ones. For other magnitude characteristics, however, this is not the case. To see this, consider a general low-pass function written in the factored form

$$N_{LP}(s) = \frac{H}{(s + \sigma) \prod\limits_{i=1}^{n/2} (s^2 + a_{2i}s + a_{1i})} \qquad (6)$$

Applying the transformation of (2) we obtain the general high-pass factored function

$$N_{HP}(p) = \frac{Hp^{n+1} \Big/ \left(\sigma \prod\limits_{i=1}^{n/2} a_{1i}\right)}{(p + 1/\sigma) \prod\limits_{i=1}^{n/2} [p^2 + (a_{2i}/a_{1i})p + 1/a_{1i}]} \qquad (7)$$

Thus, the quadratic factors and poles are readily found from the low-pass ones. A tabulation of values for the $\frac{1}{2}$- and 1-dB ripple high-pass Chebyshev filter is given in Table 2.4-1. The entries correspond to the ones given for the low-pass case in Table 2.2-2.

The transformation defined by (2) can also be applied directly to the elements of a network realization. Thus, an impedance $Z_{LP}(s) = Ks$, defining an inductor of K henrys, becomes, as a result of the transformation, an impedance $Z_{HP}(p) = K/p$ which defines a capacitor of $1/K$ farads. Similarly, an admittance $Y_{LP}(s) = Ks$ defining a capacitor of K farads becomes an admittance $Y_{HP}(p) = K/p$ which defines an inductor of $1/K$ henrys. As an example of applying the transformation of (2) directly to the elements of a network realization, consider the third-order

Table 2.4-1a Pole locations and quadratic factors $(a_0 + a_1 s + s^2)$ of 0.5-dB equal-ripple magnitude (Chebyshev) high-pass functions with normalized passband 1 to ∞ rad/s

n	Poles	a_0	a_1
2	$-.47013 \pm J \ .66221$	.65954	.94026
3	$-.27417 \pm J \ .89451$ -1.59628	.87531	.54835
4	$-.16488 \pm J \ .95556$ $-1.18778 \pm J \ 1.18107$	.94028 2.80574	.32976 2.37557
5	$-.10809 \pm J \ .97661$ $-.61481 \pm J \ 1.31128$ -2.75999	.96545 2.09746	.21619 1.22963
6	$-.07590 \pm J \ .98577$ $-.35956 \pm J \ 1.25124$ $-1.84585 \pm J \ 1.72115$	.97750 1.69489 6.36953	.15181 .71912 3.69170
7	$-.05610 \pm J \ .99045$ $-.23596 \pm J \ 1.19234$ $-.90910 \pm J \ 1.76421$ -3.90366	.98415 1.47736 3.93890	.11220 .47193 1.81820
8	$-.04311 \pm J \ .99315$ $-.16756 \pm J \ 1.14928$ $-.51835 \pm J \ 1.58731$ $-2.49048 \pm J \ 2.27032$	.98821 1.34892 2.78823 11.35688	.08621 .33512 1.03671 4.98097
9	$-.03414 \pm J \ .99484$ $-.12567 \pm J \ 1.11850$ $-.33585 \pm J \ 1.44808$ $-1.19251 \pm J \ 2.23028$ -5.04019	.99087 1.26684 2.20975 6.39622	.06828 .25135 .67171 2.38502
10	$-.02770 \pm J \ .99597$ $-.09806 \pm J \ 1.09612$ $-.23713 \pm J \ 1.35061$ $-.66792 \pm J \ 1.93832$ $-3.12995 \pm J \ 2.82349$	.99272 1.21109 1.88038 4.20319 17.76864	.05539 .19612 .47427 1.33583 6.25989

Table 2.4-1b Pole locations and quadratic factors $(a_0 + a_1 s + s^2)$ of 1.0-dB equal-ripple magnitude (Chebyshev) high-pass functions with normalized passband 1 to ∞ rad/s

n	Poles	a_0	a_1
2	$-.49783 + J \quad .81190$	.90702	.99567
3	$-.24853 + J \quad .97163$ -2.02359	1.00583	.49705
4	$-.14144 + J \quad .99683$ $-1.20570 + J \; 1.45788$	1.01368 3.57912	.28289 2.41140
5	$-.09052 + J \; 1.00181$ $-.54555 + J \; 1.42540$ -3.45431	1.01182 2.32938	.18103 1.09111
6	$-.06276 + J \; 1.00270$ $-.30460 + J \; 1.30393$ $-1.86087 + J \; 2.13447$	1.00935 1.79302 8.01880	.12553 .60920 3.72173
7	$-.04605 + J \; 1.00262$ $-.19599 + J \; 1.22144$ $-.80309 + J \; 1.92208$ -4.86821	1.00737 1.53033 4.33933	.09209 .39199 1.60618
8	$-.03521 + J \; 1.00232$ $-.13779 + J \; 1.16752$ $-.43773 + J \; 1.65595$ $-2.50491 + J \; 2.82099$	1.00589 1.38209 2.93376 14.23261	.07043 .27557 .87546 5.00983
9	$-.02780 + J \; 1.00201$ $-.10274 + J \; 1.13098$ $-.27831 + J \; 1.48416$ $-1.05168 + J \; 2.43274$ -6.27626	1.00479 1.28968 2.28018 7.02425	.05560 .20549 .55661 2.10336
10	$-.02250 + J \; 1.00172$ $-.07987 + J \; 1.10521$ $-.19464 + J \; 1.37231$ $-.56331 + J \; 2.02362$ $-3.14474 + J \; 3.51168$	1.00396 1.22786 1.92112 4.41233 22.22131	.04501 .15974 .38928 1.12661 6.28949

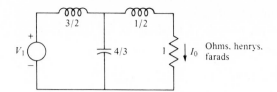

Figure 2.4-2 A third-order maximally flat magnitude low-pass network.

maximally flat magnitude low-pass network realization shown in Fig. 2.4-2. This network is shown in App. A to have the transfer admittance given by $N_{LP}(s)$ in (5) (with $H = 1$). Applying the transformation $s = 1/p$ to the network elements we obtain the realization shown in Fig. 2.4-3. Since applying the transformation to the network elements produces the same result as applying the transformation to the network function, we conclude that the transfer admittance of this network is given by $N_{HP}(p)$ in (5) (with $H = 1$).

The second transformation of the complex-frequency variable that we shall consider in this section is the *low-pass to bandpass transformation*. Like the low-pass to high-pass transformation described above, this transformation may of course be applied to transform any type of magnitude characteristic; however, when it is applied to a low-pass characteristic, it produces a bandpass one. The general form of the transformation is

$$s = \frac{1}{BW}\left(\frac{p^2 + \omega_0^2}{p}\right) \tag{8}$$

where s is the low-pass variable, p is the bandpass variable, BW is the passband bandwidth (assuming that the low-pass function has a normalized bandwidth of 1 rad/s), and ω_0 is the center frequency of the resulting passband. For convenience we shall define a *normalized low-pass to bandpass transformation* as

$$s = \frac{p^2 + 1}{p} \tag{9}$$

in which the center frequency is unity and the bandwidth of the bandpass function is equal to the bandwidth of the low-pass one. If we solve for p in (9), we obtain

$$p = \frac{s}{2} \pm \sqrt{\left(\frac{s}{2}\right)^2 - 1} \tag{10}$$

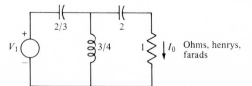

Figure 2.4-3 A third-order maximally flat magnitude high-pass network obtained by transforming the network of Fig. 2.4-2.

If we now confine our attention to the sinusoidal steady-state case by letting $s = \sigma + j\omega$ and setting $\sigma = 0$, then for $p = u + jv$ we find from (10) that $u = 0$ and v is given by

$$v = \frac{\omega}{2} \pm \sqrt{\left(\frac{\omega}{2}\right)^2 + 1} \tag{11}$$

Thus, the imaginary axis in the s plane transforms to the imaginary axis in the p plane. The nature of the transformation may be further defined by noting that, from (10), the point $s = 0$ transforms into the two points $p = \pm j1$. Similarly, the point $s = \infty$ transforms into the two points $p = 0$ and $p = \infty$. Finally, using (11) we see that any arbitrary point on the positive imaginary axis of the s plane, defined as $s = j\omega_1$, transforms into two points jv_2 and $-jv_1$ on the p plane:

$$-v_1 = \frac{\omega_1}{2} - \sqrt{\left(\frac{\omega_1}{2}\right)^2 + 1} \qquad v_2 = \frac{\omega_1}{2} + \sqrt{\left(\frac{\omega_1}{2}\right)^2 + 1} \tag{12}$$

where v_1 and v_2 are both positive and $v_2 > v_1$. The point $-j\omega_1$ on the s plane is similarly transformed into the points jv_1 and $-jv_2$ on the p plane. Now let us consider the points v_1 and v_2 in more detail. From (12) we find that

$$v_1 v_2 = 1 \qquad v_2 - v_1 = \omega_1 \tag{13}$$

From (12) and (13) we conclude that *any frequency in a low-pass magnitude characteristic is transformed into two frequencies in the bandpass characteristic; and that these frequencies have the following properties: (1) their geometric mean is the bandpass center frequency, and (2) their difference equals the low-pass frequency. Thus the bandwidth of the bandpass characteristic equals the bandwidth of the low-pass characteristic.* As an example of these conclusions, consider the transformation of a low-pass magnitude characteristic with a 3-dB-down bandwidth of 1 rad/s. The frequencies v_1 and v_2 defining the 3-dB-down bandwidth of the bandpass characteristic are found from (12) to be 0.618 and 1.618 rad/s. Obviously these values satisfy the relations of (13). As a more detailed example of the operation of the low-pass to bandpass transformation, Fig. 2.4-4 shows the transformation of an equal-ripple magnitude characteristic. The low-pass characteristic has been extended to the negative frequency axis to better illustrate the transformation. Corresponding values on the original and the transformed characteristic are indicated by the letters A, B, ..., A', B', ...

Now let us consider what happens when the transformation of (9) is applied to the general low-pass network function given in (3). After multiplying the numerator and denominator by p^n we obtain

$$N_{\text{BP}}(p) = \frac{Hp^n}{a_0 p^n + a_1 p^{n-1}(p^2 + 1) + a_2 p^{n-2}(p^2 + 1)^2 + \cdots + a_{n-1} p(p^2 + 1)^{n-1} + (p^2 + 1)^n}$$

$$\tag{14}$$

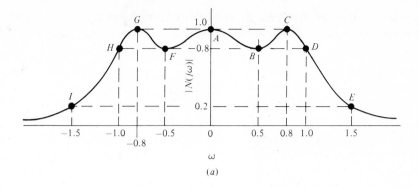

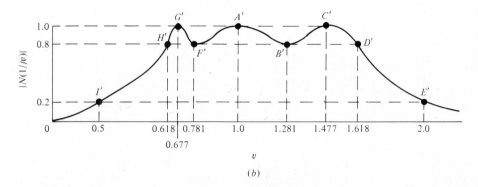

Figure 2.4-4 The normalized low-pass to bandpass transformation.

The general resulting denominator polynomials for typical values of n are given in Table 2.4-2.[1] Each resulting bandpass function has an nth-order zero at the origin and an nth-order zero at infinity. Some specific values of denominator coefficients, roots, and quadratic factors for various bandpass characteristics are given in Tables 2.4-3 through 2.4-5.

Since the degree of the denominator polynomial of a bandpass network function produced by the low-pass to bandpass transformation of (9) is twice that of the original low-pass denominator polynomial, factoring it is in general possible only through the use of digital-computer root-solving programs.

An exception to this requirement, however, occurs when the bandwidth of the resulting bandpass function is much less than its center frequency. In this case, as an alternate procedure to the application of (9) we may directly transform the pole positions of the low-pass function, thus producing a bandpass function with its

---

[1] A general algorithm for determining the coefficients of bandpass network functions from low-pass network functions may be found in L. P. Huelsman, "An Algorithm for the Low-Pass to Bandpass Transformation," *IEEE Trans. Education*, vol. E-11, March 1968, p. 72.

Table 2.4-2 Low-pass to bandpass transformation with normalized bandpass center frequency of 1 rad/s

$$N_{\text{LP}}(s) = \frac{H}{a_0 + a_1 s + \cdots + a_{n-1} s^{n-1} + s^n}$$

$$N_{\text{BP}}(p) = \frac{Hp^n}{1 + b_1 p + b_2 p^2 + \cdots + b_2 p^{2n-2} + b_1 p^{2n-1} + p^{2n}}$$

n	b_1	b_2	b_3	b_4	b_5
2	a_1	$a_0 + 2$			
3	a_2	$a_1 + 3$	$a_0 + 2a_2$		
4	a_3	$a_2 + 4$	$a_1 + 3a_3$	$a_0 + 2a_2 + 6$	
5	a_4	$a_3 + 5$	$a_2 + 4a_4$	$a_1 + 3a_3 + 10$	$a_0 + 2a_2 + 6a_4$

denominator polynomial in factored form. The procedure is referred to as the *narrow-band approximation*. To see how it operates, let us assume that the center frequency of the bandpass function is to be 1 rad/s. We now define an intermediate frequency variable p' by the relation

$$p = p' + j1 \tag{15}$$

where p is the complex frequency variable of the bandpass function. This transformation puts the origin of the p' plane at the point $j1$ in the p plane. Substituting this transformation in (9) and using the binomial expansion we obtain

$$s = (p' + j1) + \frac{1}{p' + j1} = (p' + j1) + \frac{-j}{1 + p'/j}$$

$$= (p' + j1) - j\left(1 - \frac{p'}{j} + \cdots\right) \approx 2p' \tag{16}$$

where the last relation given in the right member of (16) is a first-order approximation valid for $p' \approx 0$, that is, valid in the vicinity of the center-frequency location $j1$ in the p plane. This relation together with (15) defines the narrow-band approximation, which may be directly applied to pole locations in the (low-pass) s plane to determine locations in the (bandpass) p plane. Specifically, comparing the relation of the p-plane poles to the point $p = j1$, with the relation of the s-plane poles to the origin, we see that the angles are identical but that the p-plane poles are only half as far away from the referenced point. As an example of this conclusion, consider the use of the narrow-band approximation to determine the pole locations for a four-pole maximally flat magnitude bandpass function with a 3-dB bandwidth of 0.05 rad/s and a center frequency of 1 rad/s. The s-plane pole locations for a two-pole low-pass maximally flat magnitude function with a 3-dB bandwidth of 0.05 rad/s are $-0.05/\sqrt{2} \pm j0.05/\sqrt{2}$, as shown in Fig. 2.4-5a. The p-plane pole locations determined by (15) and (16) are $-0.025/\sqrt{2} \pm$

Table 2.4-3a Denominator coefficients of maximally flat magnitude (Butterworth) bandpass functions of the form $1 + a_1 s + a_2 s^2 + \cdots + a_2 s^{n-2} + a_1 s^{n-1} + s^n$ with normalized center frequency of 1 rad/s

n	a_1	a_2	a_3	a_4	a_5
4	.070711	2.002500			
6	.100000	3.005000	.200125	BW = 0.05	
8	.130656	4.008536	.392296	6.017077	
10	.161803	5.013090	.647868	10.039291	.972130
4	.141421	2.010000			
6	.200000	3.020000	.401000	BW = 0.1	
8	.261313	4.034142	.786551	6.068384	
10	.323607	5.052361	1.299663	10.157406	1.952123
4	.282843	2.040000			
6	.400000	3.080000	.808000	BW = 0.2	
8	.522625	4.136569	1.588781	6.274737	
10	.647214	5.209443	2.630743	10.633506	3.967379
4	.707107	2.250000			
6	1.000000	3.500000	2.125000	BW = 0.5	
8	1.306563	4.853553	4.246330	7.769607	
10	1.618034	6.309017	7.126644	14.129305	11.048471

Table 2.4-3b **Pole locations and quadratic factors** $(a_0 + a_1 s + s^2)$ **of maximally flat magnitude (Butterworth) bandpass functions with normalized center frequency of 1 rad/s**

Band width	n	Poles		a_0	a_1
0.05	4	-.01799	+J 1.01768	1.03599	.03598
		-.01737	+J .98232	.96526	.03473
	6	-.01277	+J 1.02181	1.04425	.02554
		-.01223	+J .97851	.95762	.02446
		-.02500	+J .99969	1.00000	.05000
	8	-.02332	+J 1.00935	1.01932	.04664
		-.02288	+J .99021	.98104	.04575
		-.00979	+J 1.02332	1.04728	.01958
		-.00935	+J .97712	.95486	.01869
	10	-.02052	+J 1.01460	1.02983	.04105
		-.01993	+J .98521	.97103	.03986
		-.00791	+J 1.02403	1.04870	.01582
		-.00754	+J .97648	.95356	.01508
		-.02500	+J .99969	1.00000	.05000
0.1	4	-.03661	+J 1.03536	1.07330	.07321
		-.03411	+J .96465	.93170	.06821
	6	-.02608	+J 1.04393	1.09046	.05216
		-.02392	+J .95732	.91704	.04784
		-.05000	+J .99875	1.00000	.10000
	8	-.04708	+J 1.01825	1.03905	.09416
		-.04531	+J .97998	.96242	.09062
		-.02002	+J 1.04708	1.09677	.04003
		-.01825	+J .95469	.91177	.03650
	10	-.04164	+J 1.02900	1.06058	.08328
		-.03926	+J .97023	.94288	.07852
		-.01618	+J 1.04856	1.09975	.03237
		-.01472	+J .95346	.90930	.02943
		-.05000	+J .99875	1.00000	.10000

Table 2.4-3b Pole locations and quadratic factors $(a_0 + a_1 s + s^2)$ of maximally flat magnitude (Butterworth) bandpass functions with normalized center frequency of 1 rad/s

Band width	n	Poles	a_0	a_1
0.2	4	$-.07571 \ +J \ 1.07072$	1.15218	.15142
		$-.06571 \ +J \ \ .92930$	.86792	.13142
	6	$-.05432 \ +J \ 1.08911$	1.18911	.10864
		$-.04568 \ +J \ \ .91590$	.84097	.09136
		$-.10000 \ +J \ \ .99499$	1.00000	.20000
	8	$-.09594 \ +J \ 1.03473$	1.07988	.19187
		$-.08884 \ +J \ \ .95820$	.92603	.17768
		$-.04179 \ +J \ 1.09592$	1.20279	.08358
		$-.03475 \ +J \ \ .91115$	.83140	.06949
	10	$-.08566 \ +J \ 1.05724$	1.12510	.17133
		$-.07614 \ +J \ \ .93969$	.88881	.15228
		$-.03383 \ +J \ 1.09915$	1.20927	.06766
		$-.02797 \ +J \ \ .90894$	.82695	.05595
		$-.10000 \ +J \ \ .99499$	1.00000	.20000
0.5	4	$-.20801 \ +J \ 1.17726$	1.42922	.41602
		$-.14554 \ +J \ \ .82371$	.69968	.29108
	6	$-.15164 \ +J \ 1.23236$	1.54171	.30328
		$-.09836 \ +J \ \ .79935$	.64863	.19672
		$-.25000 \ +J \ \ .96825$	1.00000	.50000
	8	$-.25357 \ +J \ 1.07359$	1.21688	.50713
		$-.20837 \ +J \ \ .88224$	.82177	.41675
		$-.11729 \ +J \ 1.25306$	1.58391	.23458
		$-.07405 \ +J \ \ .79112$	.63135	.14810
	10	$-.23225 \ +J \ 1.13770$	1.34829	.46450
		$-.17226 \ +J \ \ .84380$	.74168	.34451
		$-.09517 \ +J \ 1.26289$	1.60395	.19034
		$-.05934 \ +J \ \ .78736$	.62346	.11867
		$-.25000 \ +J \ \ .96825$	1.00000	.50000

Table 2.4-4a Denominator coefficients of 0.5-dB equal-ripple magnitude (Chebyshev) bandpass functions of the form $1 + a_1 s + a_2 s^2 + \cdots + a_2 s^{n-2} + a_1 s^{n-1} + s^n$ **with normalized center frequency of 1 rad/s**

n	a_1	a_2	a_3	a_4	a_5
4	.071281	2.003791			
6	.062646	3.003837	.125381	BW = 0.05	
8	.059869	4.004292	.179736	6.008587	
10	.058625	5.004843	.234662	10.014535	.352075
4	.142563	2.015162			
6	.125291	3.015349	.251299	BW = 0.1	
8	.119739	4.017169	.360242	6.034375	
10	.117249	5.019374	.470306	10.058196	.706116
4	.285125	2.060648			
6	.250583	3.061396	.506891	BW = 0.2	
8	.239477	4.068675	.726636	6.137956	
10	.234498	5.077495	.948470	10.233688	1.428001
4	.712813	2.379051			
6	.626457	3.383724	1.347376	BW = 0.5	
8	.598693	4.429217	1.924263	6.882125	
10	.586246	5.484342	2.508681	11.500059	3.850462

Table 2.4-4b Pole locations and quadratic factors $(a_0 + a_1 s + s^2)$ of 0.5-dB equal-ripple magnitude (Chebyshev) bandpass functions with normalized center frequency of 1 rad/s

Band width	n	Poles		a_0	a_1
0.05	4	-.01827	+J 1.02526	1.05149	.03654
		-.01737	+J .97506	.95103	.03475
	6	-.00803	+J 1.02584	1.05242	.01606
		-.00763	+J .97475	.95019	.01526
		-.01566	+J .99988	1.00000	.03132
	8	-.01069	+J 1.01052	1.02127	.02139
		-.01047	+J .98948	.97917	.02094
		-.00450	+J 1.02572	1.05212	.00899
		-.00427	+J .97491	.95046	.00854
	10	-.00744	+J 1.01572	1.03175	.01489
		-.00721	+J .98447	.96923	.01443
		-.00287	+J 1.02560	1.05187	.00574
		-.00273	+J .97503	.95068	.00546
		-.00906	+J .99996	1.00000	.01812
0.1	4	-.03743	+J 1.05083	1.10564	.07486
		-.03385	+J .95042	.90445	.06771
	6	-.01646	+J 1.05228	1.10756	.03292
		-.01486	+J .95009	.90268	.02972
		-.03132	+J .99951	1.00000	.06265
	8	-.02161	+J 1.02104	1.04300	.04323
		-.02072	+J .97895	.95877	.04144
		-.00921	+J 1.05206	1.10692	.01843
		-.00832	+J .95044	.90340	.01665
	10	-.01511	+J 1.03164	1.06451	.03023
		-.01420	+J .96912	.93940	.02840
		-.00588	+J 1.05184	1.10640	.01176
		-.00532	+J .95058	.90383	.01063
		-.01812	+J .99984	1.00000	.03623

(*continued*)

Table 2.4-4b Pole locations and quadratic factors $(a_0 + a_1 s + s^2)$ of 0.5-dB equal-ripple magnitude (Chebyshev) bandpass functions with normalized center frequency of 1 rad/s

Band width	n	Poles		a_0	a_1
0.2	4	-.07842	+J 1.10293	1.22260	.15684
		-.06414	+J .90212	.81793	.12828
	6	-.03451	+J 1.10592	1.22646	.06902
		-.02814	+J .90253	.81536	.05627
		-.06265	+J .99804	1.00000	.12529
	8	-.04412	+J 1.04209	1.08789	.08823
		-.04055	+J .95790	.91921	.08110
		-.01931	+J 1.10662	1.22499	.03862
		-.01576	+J .90337	.81633	.03152
	10	-.03114	+J 1.06404	1.13316	.06228
		-.02748	+J .93901	.88249	.05497
		-.01232	+J 1.10620	1.22382	.02465
		-.01007	+J .90389	.81711	.02014
		-.03623	+J .99934	1.00000	.07246
0.5	4	-.22221	+J 1.26747	1.65585	.44442
		-.13420	+J .76545	.60392	.26839
	6	-.09774	+J 1.28481	1.66029	.19549
		-.05887	+J .77385	.60230	.11774
		-.15661	+J .98766	1.00000	.31323
	8	-.11697	+J 1.10524	1.23523	.23395
		-.09470	+J .89476	.80957	.18939
		-.05464	+J 1.28496	1.65410	.10928
		-.03303	+J .77683	.60456	.06607
	10	-.08463	+J 1.16584	1.36635	.16925
		-.06194	+J .85325	.73188	.12387
		-.03486	+J 1.28401	1.64991	.06971
		-.02113	+J .77823	.60610	.04225
		-.09058	+J .99589	1.00000	.18116

Table 2.4-5a Denominator coefficients of 1.0-dB equal-ripple magnitude (Chebyshev) bandpass functions of the form $1 + a_1 s + a_2 s^2 + \cdots + a_2 s^{n-2} + a_1 s^{n-1} + s^n$ with normalized center frequency of 1 rad/s

n	a_1	a_2	a_3	a_4	a_5
4	.054887	2.002756			
6	.049417	3.003096	.098896	BW = 0.05	
8	.047641	4.003635	.143015	6.007271	
10	.046841	5.004222	.187486	10.012670	.281290
4	.109773	2.011025			
6	.098834	3.012394	.198160	BW = 0.1	
8	.095281	4.014539	.286586	6.029106	
10	.093682	5.016888	.375703	10.050723	.564042
4	.219547	2.044100			
6	.197668	3.049536	.399267	BW = 0.2	
8	.190562	4.058157	.577628	6.116755	
10	.187364	5.067553	.757252	10.203587	1.139814
4	.548867	2.275628			
6	.494171	3.309602	1.049755	BW = 0.5	
8	.476406	4.363481	1.522045	6.744189	
10	.468410	5.422204	1.995441	11.302896	3.057899

Table 2.4-5*b* **Pole locations and quadratic factors $(a_0 + a_1 s + s^2)$ of 1.0-dB equal-ripple magnitude (Chebyshev) bandpass functions with normalized center frequency of 1 rad/s**

Band width	n	Poles		a_0	a_1
0.05	4	-.01403	+J 1.02253	1.04577	.02806
		-.01341	+J .97778	.95623	.02683
	6	-.00633	+J 1.02442	1.04948	.01265
		-.00603	+J .97612	.95285	.01206
		-.01235	+J .99992	1.00000	.02471
	8	-.00851	+J 1.01020	1.02058	.01702
		-.00834	+J .98983	.97984	.01667
		-.00357	+J 1.02488	1.05039	.00715
		-.00340	+J .97571	.95202	.00681
	10	-.00594	+J 1.01540	1.03107	.01189
		-.00577	+J .98480	.96987	.01153
		-.00229	+J 1.02506	1.05075	.00458
		-.00218	+J .97555	.95170	.00436
		-.00724	+J .99997	1.00000	.01447
0.1	4	-.02867	+J 1.04538	1.09365	.05734
		-.02622	+J .95587	.91437	.05243
	6	-.01295	+J 1.04939	1.10139	.02590
		-.01176	+J .95279	.90795	.02352
		-.02471	+J .99969	1.00000	.04942
	8	-.01719	+J 1.02043	1.04158	.03437
		-.01650	+J .97970	.96008	.03300
		-.00732	+J 1.05035	1.10329	.01464
		-.00663	+J .95201	.90638	.01327
	10	-.01207	+J 1.03100	1.06310	.02414
		-.01135	+J .96980	.94065	.02270
		-.00469	+J 1.05072	1.10403	.00939
		-.00425	+J .95171	.90577	.00850
		-.01447	+J .99990	1.00000	.02895

Table 2.4-5b Pole locations and quadratic factors $(a_0 + a_1 s + s^2)$ of 1.0-dB equal-ripple magnitude (Chebyshev) bandpass functions with normalized center frequency of 1 rad/s

Band width	n	Poles		a_0	a_1
0.2	4	-.05979 +J	1.09202	1.19609	.11958
		-.04999 +J	.91300	.83606	.09997
	6	-.02709 +J	1.10095	1.21283	.05417
		-.02233 +J	.90775	.82452	.04466
		-.04942 +J	.99878	1.00000	.09883
	8	-.03506 +J	1.04100	1.08490	.07012
		-.03232 +J	.95953	.92174	.06463
		-.01532 +J	1.10307	1.21699	.03064
		-.01259 +J	.90639	.82170	.02518
	10	-.02485 +J	1.06279	1.13014	.04970
		-.02199 +J	.94041	.88485	.04398
		-.00983 +J	1.10386	1.21861	.01965
		-.00806 +J	.90584	.82061	.01613
		-.02895 +J	.99958	1.00000	.05790
0.5	4	-.16744 +J	1.23974	1.56498	.33488
		-.10699 +J	.79217	.63898	.21398
	6	-.07630 +J	1.26849	1.61490	.15259
		-.04725 +J	.78549	.61923	.09449
		-.12354 +J	.99234	1.00000	.24709
	8	-.09278 +J	1.10351	1.22633	.18556
		-.07566 +J	.89984	.81544	.15131
		-.04322 +J	1.27506	1.62766	.08643
		-.02655 +J	.78337	.61438	.05310
	10	-.06742 +J	1.16296	1.35701	.13484
		-.04968 +J	.85700	.73691	.09937
		-.02774 +J	1.27748	1.63272	.05548
		-.01699 +J	.78242	.61248	.03398
		-.07237 +J	.99738	1.00000	.14475

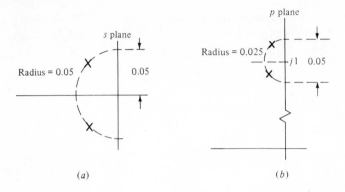

Figure 2.4-5 An example of the use of the narrow-band approximation.

$j(1 \pm 0.025/\sqrt{2})$, as shown in Fig. 2.4-5b. Adding the obvious requirements of a second-order zero at the origin and the conjugate poles, the resulting network function $N_{\text{NB}}(p)$ is

$$N_{\text{NB}}(p) = \cfrac{Hp^2}{\left[p + \dfrac{0.025}{\sqrt{2}} + j\left(1 + \dfrac{0.025}{\sqrt{2}}\right)\right]\left[p + \dfrac{0.025}{\sqrt{2}} - j\left(1 + \dfrac{0.025}{\sqrt{2}}\right)\right]}$$
$$\times \left[p + \dfrac{0.025}{\sqrt{2}} + j\left(1 - \dfrac{0.025}{\sqrt{2}}\right)\right]\left[p + \dfrac{0.025}{\sqrt{2}} - j\left(1 - \dfrac{0.025}{\sqrt{2}}\right)\right]$$

$$= \frac{Hp^2}{p^4 + 0.0707p^3 + 2.0025p^2 + 0.0707p + 1.0000} \tag{17}$$

The actual network function $N_{\text{BP}}(p)$ obtained by directly applying (9) is identical with the unfactored form of (17) within the given number of significant figures. In general, as a guideline, the narrow-band approximation gives good accuracy as long as the bandwidth is equal to or less than one-tenth of the center frequency, i.e., the Q is 10 or greater. As a somewhat extreme example of the distortion in the pole positions that occurs when the guideline given above is violated, Fig. 2.4-6 shows the loci of the actual bandpass pole locations and the pole locations determined by the narrow-band approximation for the case where the bandwidth is 1.5 rad/s and the center frequency is 1 rad/s. The actual positions on the loci of the poles for the eighteenth-order bandpass case (ninth-order low-pass) are also shown. Obviously, considerable error is produced by using the poles determined by the narrow-band approximation in this case.

The transformation defined by (9) can also be applied directly to the elements of a given network realization. Thus an inductor of K henrys, with an impedance $Z_{\text{LP}}(s) = Ks$, becomes $Z_{\text{BP}}(p) = Kp + K/p$, a *series connection* of an inductor of K henrys and a capacitor of $1/K$ farads; while a capacitor of K farads with an

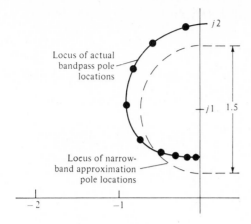

Figure 2.4-6 A case where the narrow-band approximation does not apply.

admittance $Y_{LP}(s) = Ks$ becomes $Y_{BP}(p) = Kp + K/p$, a *parallel connection* of a capacitor of K farads and an inductor of $1/K$ henrys. As an example of such an application, the network realization for a third-order maximally flat magnitude low-pass transfer admittance shown in Fig. 2.4-2 may be directly transformed to realize a sixth-order maximally flat magnitude transfer admittance with a bandwidth of 1 rad/s, and with 3-dB-down frequencies of 0.618 and 1.618 rad/s. The resulting network is shown in Fig. 2.4-7. The transfer admittance of this network is readily found by applying Table 2.4-2 to the first equation given in (5). Thus we obtain

$$\frac{I_0(p)}{V_1(p)} = \frac{Hp^3}{p^6 + 2p^5 + 5p^4 + 5p^3 + 5p^2 + 2p + 1} \tag{18}$$

where $H = 1$. A summary of the transformations of network elements is given in Table 2.4-6.

The normalized low-pass to high-pass and low-pass to bandpass transformations defined in (2) and (9), coupled with appropriate frequency normalizations, may be applied in various sequences to obtain any desired combination of center frequency and bandwidth. In addition, if a low-pass to bandpass transformation is applied to a high-pass network, a *band-elimination* characteristic results. An example of this follows.

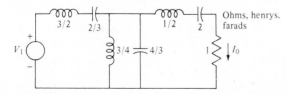

Figure 2.4-7 A sixth-order maximally flat magnitude bandpass network obtained by transforming the network of Fig. 2.4-2.

Table 2.4-6 Changes of Network Elements Under Frequency Transformations

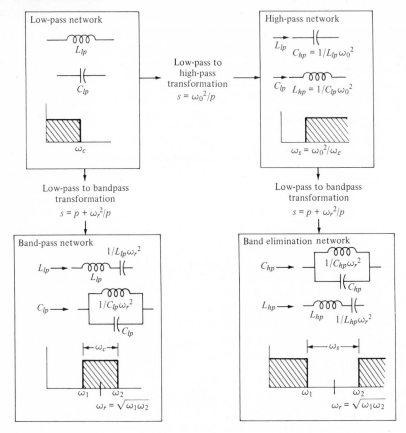

Example 2.4-1 *Determination of a Band-Elimination Filter* A sixth-order band-elimination admittance transfer function with a resistive termination of 1000 Ω, a bandwidth of 1 kHz, and a geometric center frequency of 5 kHz is to be realized. The specifications also require that there be at least 15 dB of attenuation at the band edges and that the magnitude characteristic be monotonic in the passband. Anticipating the use of (9), we first normalize the center-frequency and bandwidth specifications to 1 rad/s and 0.2 rad/s respectively. From Example 2.1-2 we know that the network shown in Fig. 2.4-2 has a minimum of 15 dB of attenuation at 2 rad/s and is monotonic in the stopband. Frequency denormalizing this network by 2.5, we obtain the network shown in Fig. 2.4-8a, which has the required attenuation at all frequencies *over* 5 rad/s. Applying the low-pass to high-pass transformation of (2) to this network, we obtain the high-pass one shown in Fig. 2.4-8b, which has the required attenuation (from its high-frequency value) at all frequencies *less than* 0.2 rad/s. Applying the transformation of (9) to this high-pass

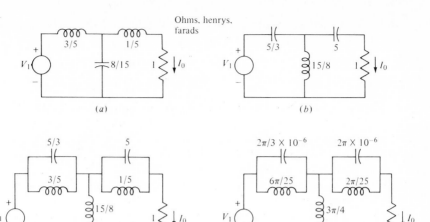

Figure 2.4-8 The band-elimination network realized in Example 2.4-1.

network, we obtain the one shown in Fig. 2.4-8c. This network has a band-elimination characteristic with a bandwidth of 0.2 rad/s and a geometric center frequency (for the eliminated frequency band) of 1 rad/s. Frequency denormalizing the elements of this network by $2\pi/5000$ and impedance denormalizing them by 1000, we obtain the configuration shown in Fig. 2.4-8d, which meets the specifications. □

The low-pass to bandpass transformation described in this section can also be applied directly to low-pass elliptic functions to produce a bandpass function with an elliptic characteristic. An example follows:

Example 2.4-2 *An Elliptic Bandpass Function*   It is desired to find a fourth-order bandpass function with a center frequency of 1 rad/s in which the magnitude has a 1-dB ripple over a bandwidth of 0.1 rad/s, and in which the magnitude is at least 17 dB down from its maximum passband value of unity for any value of frequency outside a bandwidth of 0.2 rad/s. From Table 2.3-1, we find that the second-order low-pass elliptic function

$$N(s) = \frac{0.139713(s^2 + 0.07464102)}{s^2 + 0.0998942s + 0.01170077} \tag{19}$$

has a maximum magnitude of unity, has 1-dB ripple in a passband of 0 to 1 rad/s, and has 17-dB attenuation for all frequencies greater than a cutoff frequency of 2.0 rad/s. Frequency denormalizing this function by a factor of

0.1 we change the bandwidth to 0 to 0.1 rad/s and the cutoff frequency to 0.2 rad/s. The denormalized function is

$$N(s) = \frac{0.139713(s^2 + 0.07464102)}{s^2 + 0.0998942s + 0.01170077} \tag{20}$$

The pole locations for this function are at $s = -0.0499471 \pm j0.0959482$. The poles of the bandpass function may now be found directly by applying the narrow-band approximation. They are at $s = -0.0249735 \pm j(1 \pm 0.0479741)$. These pole locations correspond to the denominator quadratic factors $s^2 + 0.0499471s + 0.906977$ and $s^2 + 0.0499471s + 1.0988733$.

The numerator is readily transformed using (9) to the form $s^4 + 2.074641s^2 + 1$, which may be factored as $(s^2 + 1.313063) \times (s^2 + 0.761578)$. Thus the bandpass network function has the form

$$N_{BP}(s) = \frac{H_1(s^2 + 1.313063)}{s^2 + 0.0499471s + 0.906977} \frac{H_2(s^2 + 0.761578)}{s^2 + 0.0499471s + 1.0988733} \tag{21}$$

where the product $H_1 \times H_2$ equals 0.139713. Note that in the above, we have grouped the poles and zeros in separate biquadratic factors, choosing them so that the maximum pole and zero separation occurs in each factor. Such a grouping reduces the sensitivity (see Chap. 3) of the network function to changes in the element values of any network realization, and also helps to minimize excessive spreads in element values in the realization. □

2.5 PHASE APPROXIMATION

In the preceding sections of this chapter we have discussed the approximation of a magnitude function. In this section we turn our attention to a consideration of the approximation of a phase function. Thus we desire to find the complex-frequency-plane pole and zero locations for a network function which has some specified sinusoidal steady-state phase characteristic. To begin our study let us consider a general rational network function $N(s)$ which has been put in the form

$$N(s) = \frac{A(s)}{B(s)} = \frac{m_1(s) + n_1(s)}{m_2(s) + n_2(s)} \tag{1}$$

where $m_1(s)$ and $n_1(s)$ are the even and odd parts respectively of the numerator polynomial $A(s)$, and $m_2(s)$ and $n_2(s)$ are the even and odd parts of the denominator polynomial $B(s)$. If we now consider $N(s)$ under sinusoidal steady-state conditions, i.e., by letting $s = j\omega$, then $m_1(j\omega)$ and $m_2(j\omega)$ being even are real, while $n_1(j\omega)$ and $n_2(j\omega)$ being odd are imaginary. Thus, the function $N(j\omega)$ may be written

$$N(j\omega) = \frac{\text{Re } A(j\omega) + j \text{ Im } A(j\omega)}{\text{Re } B(j\omega) + j \text{ Im } B(j\omega)} = \frac{m_1(j\omega) + n_1(j\omega)}{m_2(j\omega) + n_2(j\omega)} \tag{2}$$

Rationalizing this by multiplying the numerator and denominator by the conjugate of the denominator, $m_2(j\omega) - n_2(j\omega)$, we obtain

$$N(j\omega) = \text{Re } N(j\omega) + j \text{ Im } N(j\omega)$$

$$= \frac{m_1 m_2 - n_1 n_2}{m_2^2 - n_2^2} + \frac{m_2 n_1 - m_1 n_2}{m_2^2 - n_2^2} \tag{3}$$

where, for convenience, we have deleted the functional notation on the quantities m_i and n_i. The phase or argument of $N(j\omega)$ is now defined as

$$\arg N(j\omega) = \tan^{-1} \frac{\text{Im } N(j\omega)}{\text{Re } N(j\omega)} = \tan^{-1} \frac{1}{j} \frac{m_2 n_1 - m_1 n_2}{m_1 m_2 - n_1 n_2} \tag{4}$$

It is now convenient to define the function $A(\omega)$ as

$$A(\omega) = \tan \left[\arg N(j\omega)\right] = \frac{1}{j} \frac{m_2 n_1 - m_1 n_2}{m_1 m_2 - n_1 n_2} \tag{5}$$

From (5) we conclude that a necessary condition of the function $A(\omega)$ is that it be an odd rational function, i.e., the ratio of an odd polynomial to an even one. This requirement is also sufficient for the determination of a rational network function $N(s)$ from $A(\omega)$. To show the sufficiency, we first form the function $1 + jA(\omega)$. Using (5) we obtain

$$1 + jA(\omega) = \frac{m_1 m_2 - n_1 n_2 + m_2 n_1 - m_1 n_2}{m_1 m_2 - n_1 n_2} = \frac{(m_1 + n_1)(m_2 - n_2)}{m_1 m_2 - n_1 n_2} \tag{6}$$

From the form of the right member of the above relation we conclude that the numerator of the function $1 + jA(s/j)$ must contain, as factors, the zeros and the right-half-plane reflection of the poles of the network function $N(s)$, which is related to the function $A(\omega)$ by (5). Any desired assignment of these factors which is consistent with stability requirements thus satisfies the sufficiency condition. An example follows.

Example 2.5-1 *Determination of a Network Function from its Argument* As an example of the procedure given above consider the function

$$A(\omega) = \frac{-\omega^3 + 11\omega}{5\omega^2 - 15} \tag{7}$$

Thus we see that

$$1 + jA(\omega)\bigg|_{\omega = s/j} = \frac{-j\omega^3 + 5\omega^2 + j11\omega - 15}{5\omega^2 - 15}\bigg|_{\omega = s/j}$$

$$= \frac{s^3 - 5s^2 + 11s - 15}{-5s^2 - 15} \tag{8}$$

The numerator of the right member of the above is readily factored as $(s - 3)(s^2 - 2s + 5)$. Choosing $s - 3$ as $m_1 + n_1$, and $s^2 - 2s + 5$ as $m_2 - n_2$, we find the resulting network function $N(s)$ to be

$$N(s) = \frac{s - 3}{s^2 + 2s + 5} \tag{9}$$

For this network function, tan $[\arg N(j\omega)]$ is given as the function $A(\omega)$ in (7). Alternately, we could choose the factor $(s - 3)$ as $m_2 - n_2$ and $s^2 - 2s + 5$ as $m_1 + n_1$ and obtain

$$N(s) = -\frac{s^2 - 2s + 5}{s + 3} \tag{10}$$

as a second network function with the same phase characteristic. Finally, we could choose $(s - 3)(s^2 - 2s + 5)$ as $m_2 - n_2$ and 1 as $m_1 + n_1$ and obtain

$$N(s) = -\frac{1}{(s + 3)(s^2 + 2s + 5)} \tag{11}$$

as a third network function with the same phase characteristic. For all three of the network functions given above, tan $[\arg N(j\omega)]$ is given as the function $A(\omega)$ in (7). □

When we discussed the approximation of magnitude characteristics in the preceding sections, our goal was usually to keep this quantity constant within some specified tolerance in the passband. The goal for phase approximation, however, is usually quite different. We can define this latter goal for many cases by first investigating the concept of *ideal transmission*. For a network to provide ideal transmission of some arbitrary input excitation signal $e(t)$, we would like the output response signal $r(t)$ to have the same information content, i.e., the same waveform. In such a case there are only two operations which the network may be permitted to perform on $e(t)$, namely, a magnitude scaling and a time shifting (a delay). Thus we may define ideal transmission by the relation

$$r(t) = Ke(t - t_0) \qquad t_0 > 0 \tag{12}$$

In other words, the output waveform will be identical to the input waveform except that its magnitude is multiplied by K and it is delayed by t_0 seconds. Assuming $e(t)$ is zero for $t < 0$, the Laplace transformation of (12) gives $R(s) = KE(s)e^{-t_0 s}$ where $R(s)$ and $E(s)$ are the Laplace transforms of $r(t)$ and $e(t)$. Letting $s = j\omega$ we now obtain

$$\frac{\mathcal{R}}{\mathcal{E}} = N(j\omega) = Ke^{-j\omega t_0} \tag{13}$$

where $\mathcal{R}$ and $\mathcal{E}$ are output and input phasors and $N(j\omega)$ is the network function. From (13) we see that ideal transmission requires that $|N(j\omega)| = K$ (that is, the magnitude is a constant independent of frequency) and arg $N(j\omega) = -\omega t_0$ (that is, *the phase is linearly proportional to frequency*).

Let us now consider how we may approximate the linear phase which, as shown above, produces ideal transmission of a given band of frequencies. For the low-pass case, one way of obtaining the desired linearity is to make the first derivative of the phase function be nonzero but to set as many as possible of the higher-order derivatives to zero at $\omega = 0$. This approach is similar to that used to obtain maximally flat magnitude performance in Sec. 2.1. Assuming that the general low-pass network function $N(s)$ has the form

$$N(s) = \frac{1}{a_0 + a_1 s + a_2 s^2 + a_3 s^3 + a_4 s^4 + \cdots} \tag{14}$$

the corresponding phase function is

$$\arg N(j\omega) = \tan^{-1} \frac{-a_1 \omega + a_3 \omega^3 - \cdots}{a_0 - a_2 \omega^2 + a_4 \omega^4 - \cdots} \tag{15}$$

To expand this we may use the series

$$\tan^{-1} x = x - \tfrac{1}{3}x^3 + \tfrac{1}{5}x^5 - \cdots \tag{16}$$

An example of the procedure follows:

Example 2.5-2 *Determination of a Linear Phase Function* Consider the second-order low-pass function

$$N(s) = \frac{1}{s^2 + a_1 s + a_0} \tag{17}$$

The argument function may be put in the form

$$\arg N(j\omega) = -\tan^{-1} \frac{1}{a_0} \frac{a_1 \omega}{1 - \omega^2/a_0} \tag{18}$$

Using (16) this may be written

$$\arg N(j\omega) = \frac{-1}{a_0} \frac{a_1 \omega}{1 - \omega^2/a_0} + \frac{1}{3a_0^3} \left(\frac{a_1 \omega}{1 - \omega^2/a_0} \right)^3 - \cdots$$

Using the binomial expansion and simplifying we obtain

$$\arg N(j\omega) = -\frac{a_1}{a_0} \omega + \left(-\frac{a_1}{a_0^2} + \frac{a_1^3}{3a_0^3} \right) \omega^3 - \cdots \tag{19}$$

Defining $a_1/a_0 = t_0$ and setting the coefficient of ω^3 to zero in the above will provide an approximation to the linear phase $\arg N(j\omega) = -\omega t_0$. Doing this we obtain the following relations:

$$a_0 = \frac{3}{t_0^2} \qquad a_1 = t_0 a_0 \tag{20}$$

For example, for $t_0 = 1$ we require $a_0 = 3$ and $a_1 = 3$. Thus, the second-order low-pass network function having a maximally flat linear phase with a slope of -1 (at direct current) is

$$N(s) = \frac{1}{s^2 + 3s + 3} \tag{21}$$

$\square$

Extending the procedure given in the above example to higher-order low-pass functions, we obtain a series of polynomials which are related to *Bessel polynomials*. Filters whose network functions use these polynomials are called *Thomson filters*.[1] In general for such filters, the coefficients of the denominator of (14) corresponding to a linear phase characteristic with a slope of -1 (a delay of 1 s at direct current) may be found from the relation

$$a_k = \frac{(2n - k)!}{2^{n-k}k!\,(n - k)!} \qquad k = 0, 1, \ldots, n - 1 \tag{22}$$

where n is the degree of the denominator. The highest-degree coefficient is unity. A listing of polynomial coefficients, roots, and quadratic factors is given in Table 2.5-1. The recursion formula

$$B_n(s) = (2n - 1)B_{n-1}(s) + s^2 B_{n-2}(s) \tag{23}$$

may also be used to derive the denominator polynomials starting with $B_1(s) = s + 1$ and $B_2(s) = s^2 + 3s + 3$.

The actual characteristics provided by the functions tabulated in Table 2.5-1 deviate from the ideal in both delay and magnitude as indicated in Fig. 2.5-1. The actual characteristics may be related to the ideal ones by specifying the deviation from the ideal time delay in percent as shown in Fig. 2.5-2, and the magnitude error in dB as shown in Fig. 2.5-3.[2] Both of these figures use ωT, where T is the

[1] W. E. Thomson, "Delay Networks Having Maximally Flat Frequency Characteristics," *Proc. IEEE*, part 3, vol. 96, November 1949, pp. 487–490.

[2] M. E. Van Valkenburg, *Introduction to Modern Network Synthesis*, John Wiley & Sons, Inc., New York, 1960, chap. 13.

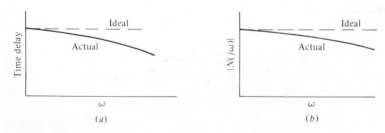

Figure 2.5-1 Deviation from the ideal of characteristics of low-pass functions.

Table 2.5-1a The denominator coefficients of linear phase (Thomson) low-pass functions of the form $a_0 + a_1 s + a_2 s^2 + \cdots$ $+ a_{n-1} s^{n-1} + s^n$ with normalized delay (at dc) of 1 s

n	a_0	a_1	a_2	a_3	a_4	a_5	a_6	a_7	a_8	a_9
2	3	3								
3	15	15	6							
4	105	105	45	10						
5	945	945	420	105	15					
6	10395	10395	4725	1260	210	21				
7	135135	135135	62370	17325	3150	378	28			
8	2027025	2027025	945945	270270	51975	6930	630	36		
9	34459425	34459425	16216200	4729725	945945	135135	13860	990	45	
10	654729075	654729075	310134825	91891800	18918900	2837835	315315	25740	1485	55

Table 2.5-1b Pole locations and quadratic factors $(a_0 + a_1 s + s^2)$ of linear phase (Thomson) low-pass functions with normalized delay (at dc) of 1 s

n	Poles	a_0	a_1
2	$-1.50000 + J \ .86603$	3.00000	3.00000
3	$-1.83891 + J \ 1.75438$ -2.32219	6.45943	3.67781
4	$-2.10379 + J \ 2.65742$ $-2.89621 + J \ .86723$	11.48780 9.14013	4.20758 5.79242
5	$-2.32467 + J \ 3.57102$ $-3.35196 + J \ 1.74266$ -3.64674	18.15632 14.27248	4.64935 6.70391
6	$-2.51593 + J \ 4.49267$ $-3.73571 + J \ 2.62627$ $-4.24836 + J \ .86751$	26.51403 20.85282 18.80113	5.03186 7.47142 8.49672
7	$-2.68568 + J \ 5.42069$ $-4.07014 + J \ 3.51717$ $-4.75829 + J \ 1.73929$ -4.97179	36.59679 28.93655 25.66644	5.37135 8.14028 9.51658
8	$-2.83898 + J \ 6.35391$ $-4.36829 + J \ 4.41444$ $-5.20484 + J \ 2.61618$ $-5.58789 + J \ .86761$	48.43202 38.56925 33.93474 31.97723	5.67797 8.73658 10.40968 11.17577
9	$-2.97926 + J \ 7.29146$ $-4.63844 + J \ 5.31727$ $-5.60442 + J \ 3.49816$ $-6.12937 + J \ 1.73785$ -6.29702	62.04144 49.78850 43.64665 40.58927	5.95852 9.27688 11.20884 12.25874
10	$-3.10892 + J \ 8.23270$ $-4.88622 + J \ 6.22499$ $-5.96753 + J \ 4.38495$ $-6.92204 + J \ .86767$ $-6.61529 + J \ 2.61157$	77.44270 62.62559 54.83916 48.66755 50.58236	6.21783 9.77244 11.93506 13.84409 13.23058

ideal delay, as abscissas. Since this product is dimensionless, it is not affected by any frequency normalization. In applying the figures to determine the order required for a given filter, if both delay and magnitude tolerances are to be met, the figure giving the higher order must of course be used. The values of the network elements for resistance-terminated lossless-ladder realizations for various orders of Thomson filters may be found in App. A.

Another set of network functions having a linear phase characteristic may be derived from the functions defined above. To see this we note that a network function having the form

$$N(s) = \frac{H}{m(s) + n(s)} \tag{24}$$

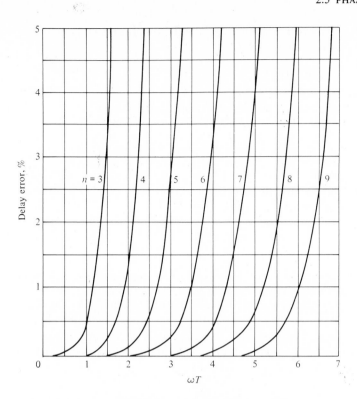

Figure 2.5-2 Delay error produced by the functions of Table 2.5-1.

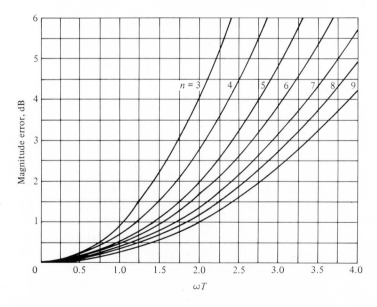

Figure 2.5-3 Magnitude error produced by the functions of Table 2.5-1.

where $m(s)$ is the even part of the denominator polynomial and $n(s)$ is the odd part, has a phase characteristic defined as

$$\arg N(j\omega) = -\tan^{-1} \frac{n(j\omega)/j}{m(j\omega)} \tag{25}$$

Now consider what happens if we form the network function

$$N_{AP}(s) = H \frac{m(s) - n(s)}{m(s) + n(s)} \tag{26}$$

For this function, it is readily shown that $|N(j\omega)| = H$, i.e., that the magnitude is constant for all frequencies. Such a function is called an *all-pass function*. The phase characteristic of this function is readily found to be

$$\arg N_{AP}(j\omega) = -2 \tan^{-1} \frac{n(j\omega)/j}{m(j\omega)} \tag{27}$$

Thus, the shape of the phase characteristic of the all-pass function is identical to that of the original function except for scaling by a factor of 2.

The treatment of ideal transmission given above can be extended by considering what happens when an impulse is applied to an arbitrary network characterized by a network function $N(j\omega)$. Such an excitation is convenient since its peak value is readily identified in the output waveform. The response $h(t)$ to the impulse excitation is found by taking the inverse Fourier transformation. Thus we obtain

$$h(t) = \frac{1}{2\pi} \int_{-\infty}^{\infty} N(j\omega)e^{j\omega t}\, d\omega = \frac{1}{\pi} \int_{0}^{\infty} |N(j\omega)|\, \cos\,[\omega t + \arg N(j\omega)]\, d\omega \tag{28}$$

In this expression, the principal contribution to the integral occurs when the argument of the cosine function in the integrand is constant. Thus, the peak value of $h(t)$ occurs when

$$\frac{d}{d\omega}[\omega t + \arg N(j\omega)] = 0 \tag{29}$$

Solving this expression we may define a *delay function* $D(\omega)$[1] for an arbitrary network as

$$D(\omega) = -\frac{d}{d\omega}\arg N(j\omega) \tag{30}$$

Obviously, for the ideal transmission case discussed above, in which $\arg N(j\omega) = -\omega t_0$, $D(\omega) = t_0$, and a delay of t_0 seconds occurs for all frequency components of the input signal. Considering the above we see that the maximally linear (at the origin) phase characteristic of Thomson filters produces a delay characteristic which is maximally flat (at the origin). Thus these filters are frequently referred to as *maximally flat delay* (MFD) filters.

---

[1] This is also referred to as the *group delay function*.

2.6 TIME-DOMAIN CONSIDERATIONS

In the preceding sections of this chapter we have considered some of the *frequency-domain* properties of network functions. In this section we present a brief discussion of some of their *time-domain* properties. Unlike the frequency-domain situation in which completely general expressions relating the ratio of response and excitation may be developed, in the time-domain situation the response must usually be separately determined for each form of excitation. For simplicity of presentation, we shall restrict our attention to the case which has the most practical interest, namely, the step response of low-pass functions. This case is of considerable importance in studying pulse and other digital transmission systems.

We now consider the situation where the excitation function $e(t)$ is a unit step occurring at $t = 0$. This may be written as

$$e(t) = u(t)\begin{vmatrix} = 0 & t < 0 \\ = 1 & t > 0 \end{vmatrix} \tag{1}$$

where $u(t)$ is the unit step function. It is shown in Fig. 2.6-1. The response $r(t)$ of a filter function to such an excitation is called its *step response*. For an ideal network function, this response would be determined by (12) of Sec. 2.5 as

$$r(t) = u(t - t_0)\begin{vmatrix} = 0 & t < t_0 \\ = 1 & t > t_0 \end{vmatrix} \tag{2}$$

This is shown in Fig. 2.6-2. Actually, for the nonideal low-pass function, the step response will usually have a form more like the one shown in Fig. 2.6-3; that is, it will demonstrate an oscillatory overshoot about its final value. This is commonly referred to as *ringing*. Also in Fig. 2.6-3, several figures of merit commonly used to evaluate a step response are defined. These are: the delay time t_d, the time required for the response to reach 50 percent of its final or steady-state value r_{final}; $t_{10\%}$, the time for the response to reach 10 percent of its final value; and $t_{90\%}$, the time for it to first reach 90 percent of its final value. In terms of these latter two quantities, we define the rise time t_r as

$$t_r = t_{90\%} - t_{10\%} \tag{3}$$

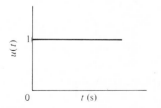

Figure 2.6-1 A unit step.

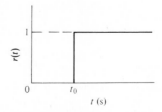

Figure 2.6-2 An ideal response to a unit step.

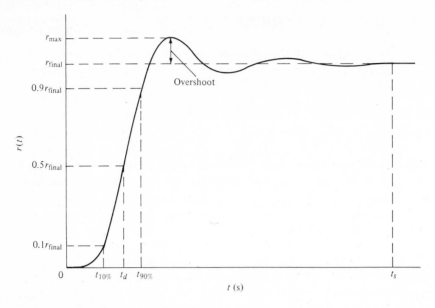

Figure 2.6-3 An actual response to a unit step.

Other step-response quantities of interest illustrated in Fig. 2.6-3 are: t_s, the settling time, meaning the time after which the response remains within some specified range, usually ± 2 percent of its final value; and the overshoot, usually defined in terms of a *peak-overshoot ratio* (**POR**) given (in percentage) as

$$\text{POR} = \frac{r_{\max} - r_{\text{final}}}{r_{\text{final}}} \times 100 \tag{4}$$

In general, good time-domain performance means small values of t_r, t_s, and POR.

Some of the step-response parameters defined above may be approximately related to the frequency-domain parameters defined in Sec. 1.3. As an example of this, for low-pass network functions in which the POR is small (less than 5 percent), the rise time t_r and the -3-dB bandwidth ω_c are approximately related as

$$t_r \omega_c \approx 2.2 \tag{5}$$

As an illustration of this relation, consider the network shown in Fig. 2.6-4. Its voltage transfer function is

$$\frac{V_2(s)}{V_1(s)} = \frac{1}{s + 1} \tag{6}$$

The bandwidth ω_c is readily shown to be 1 rad/s. The step response may be determined as

$$v_2(t) = (1 - e^{-t})u(t) \tag{7}$$

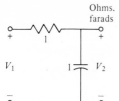

Ohms. farads

V_1

V_2

1

1

Figure 2.6-4 An example network.

From this relation we readily find $t_{10\%} = 0.1054$ s and $t_{90\%} = 2.3026$ s. Thus $t_r = 2.1972$ s and $t_r \times \omega_c$ equals 2.1972, in good agreement with the approximate value of 2.2 predicted in (5).

Some examples of step responses for various orders of the Butterworth and Chebyshev (1-dB ripple) functions introduced earlier in this chapter are shown in Figs. 2.6-5 and 2.6-6. From these we may observe another general correlation between the time and frequency domains: overshoot and settling time increase as the cutoff of the network function's magnitude response is made sharper. As a result, higher-order functions show more ringing in their step response than lower-order ones do, and Chebyshev functions have more than Butterworth ones. Another interesting property of these types of functions is that (for the constant bandwidth normalization chosen) their delay times increase with increasing order, while their rise times are very nearly constant. The properties of the Bessel (or Thomson) functions discussed in Sec. 2.5 are quite different. Their step responses

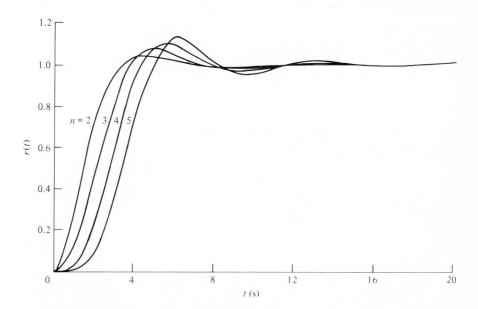

Figure 2.6-5 Step responses for Butterworth functions.

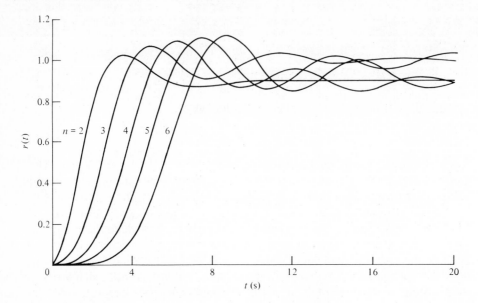

Figure 2.6-6 Step responses for Chebyshev (1-dB-ripple) functions.

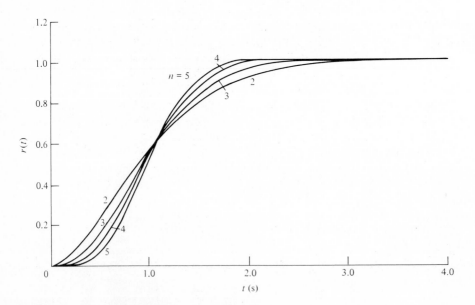

Figure 2.6-7 Step responses for Thomson functions.

are illustrated in Fig. 2.6-7. In this figure, we see some dramatic results of the linear phase properties that these functions have and the constant delay time normalization chosen for them. First of all, there is no overshoot. Second, the rise time decreases with increasing order. Obviously the Bessel functions have far better time-domain performance than do the Butterworth, which in turn are better than the Chebyshev.

2.7 SUMMARY

In this chapter we have presented several topics in the general subject area of approximation, which is the study of how the sinusoidal steady-state frequency response characteristics of network functions are related to their pole and zero positions. The first three of these topics were concerned with methods for approximating the magnitude characteristic of a low-pass filter function. Thus, in Sec. 2.1 we discussed the Butterworth or maximally flat magnitude characteristic, in which all the approximation is done at a single frequency ($\omega = 0$). Next, in Sec. 2.2 we described the Chebyshev or equal-ripple magnitude characteristic, in which the approximation is spread over the entire passband range of frequencies. Finally, in Sec. 2.3 we discussed the Cauer or elliptic magnitude characteristic, in which the approximation is made over both the passband and the stopband. Of the three approximations, for a given order of filter, the elliptic characteristic provides the sharpest cutoff between passband and stopband, the Chebyshev the next sharpest, and the Butterworth the poorest. Conversely, for the three approximations, as shown in Sec. 2.6, the Butterworth approximation provides the best transient response, the Chebyshev the next best, and the elliptic the poorest. Although all of these magnitude approximations are originally defined for low-pass functions, they are readily transformed to high-pass and bandpass ones using the methods of Sec. 2.4. The Thomson (or Bessel) linear phase (maximally flat delay) approximation provides a transient response which is considerably improved over that provided by the Butterworth function. It is discussed in Sec. 2.5.

The literature covering the theory of approximation is a broad one, encompassing the research efforts of many individuals over many years. The techniques described in this section were chosen to provide an introduction to the most useful and well-known results of their efforts. Space limitations prevent a discussion of many other valuable but less used methods, such as the inverse Chebyshev, the gaussian, and the transitional. Details of these, however, may be found in the references listed in the Bibliography.

PROBLEMS

2-1 (*Sec. 2.1*) For each of the following network functions $N_i(s)$, find the magnitude function $|N_i(j\omega)|^2$ and show that it is a ratio of even polynomials.

(a) $N_a(s) = \dfrac{H}{s^2 + as + b}$

(b) $N_b(s) = \dfrac{H(s + c)}{s^2 + as + b}$

(c) $N_c(s) = \dfrac{H}{s^3 + 2s^2 + 2s + 1}$

2-2 (*Sec. 2.1*) Show that for each of the following network functions $N_i(s)$, the magnitude functions $|N_i(j\omega)|^2$ are constants, independent of frequency. Functions such as these are called *all-pass functions*.

(a) $N_a(s) = H\dfrac{s - a}{s + a}$

(b) $N_b(s) = H\dfrac{s^2 - as + b}{s^2 + as + b}$

2-3 (*Sec. 2.1*) Find $N(s)$ for any of the following for which it exists:

(a) $|N(j\omega)|^2 = \dfrac{1 - \omega^2}{\omega^4 - 4\omega^2 + 8}$

(b) $|N(j\omega)|^2 = \dfrac{1 + \omega^2}{\omega(\omega^4 + 1)}$

(c) $|N(j\omega)|^2 = \dfrac{\omega^4 - 2\omega^2 + 1}{(1 + \omega^2)(\omega^4 - 3\omega^2/2 + 25/16)}$

2-4 (*Sec. 2.1*) Determine the value of the constant a in the following network function $N(s)$ so that $|N(j\omega)|$ satisfies a maximally flat magnitude criterion:

$$N(s) = \dfrac{s + 1}{s^2 + as + 1}$$

2-5 (*Sec. 2.1*) (a) Show that the constant ε in (12) in Sec. 2.1 may be considered as a frequency-normalization constant, and determine an expression for the frequency normalization performed by it.

(b) Rework Example 2.1-3 by first finding the frequency at which $|N(j\omega)|$ in (12) (for $\varepsilon = 1$ and $n = 5$) is down $\frac{1}{2}$ dB, and then showing that at two times this frequency the denominator of (12) has the same value as the expression given in (19) (with $n = 5$).

2-6 (*Sec. 2.1*) (a) Determine the transfer function for a low-pass filter having a maximally flat magnitude characteristic which is 1 dB down at 2 rad/s and 30 dB down at 6 rad/s. Use (12) in Sec. 2.1 to determine the order and (15) to find the pole locations.

(b) Verify the results of part a by using Fig. 2.1-4 and Table 2.1-3.

(c) Assume that the desired function is a transfer admittance and use App. A to find a resistance-terminated ($R = 1$) lossless-ladder realization.

2-7 (*Sec. 2.1*) (a) Find a low-pass maximally flat magnitude network function which is down 3 dB at 1 kHz and down 20 dB at all frequencies greater than 2.5 kHz.

(b) Using App. A, find a realization for this function as a lossless ladder with a single 1000-Ω terminating resistor.

2-8 (*Sec. 2.2*) Determine the Chebyshev polynomial $C_7(\omega)$ using the relations given in (2) of Sec. 2.2.

2-9 (*Sec. 2.2*)   Prove that the quantities ε, n, and v are related by the equation

$$e^v = \left(\sqrt{\frac{1}{\varepsilon^2} + 1} + \frac{1}{\varepsilon} \right)^{1/n}$$

2-10 (*Sec. 2.2*)   (*a*) Determine the transfer function for a low-pass filter having a 1-dB equal-ripple characteristic from 0 to 2 rad/s, and a minimum of 30-dB attenuation at 4.0 rad/s. Use (1) and (3) of Sec. 2.2 to determine the order, and use (12) to find the pole locations.

(*b*) Verify the results of part *a* by using Fig. 2.2-5 and Table 2.2-2.

(*c*) Assume that the desired function is a transfer admittance and use App. A to find a resistance-terminated ($R = 1$) lossless-ladder realization.

2-11 (*Sec. 2.2*)   Determine the minimum value of n required for an equal-ripple magnitude function which has a ripple of $\frac{1}{2}$ dB in a passband of 0 to 1 rad/s but which has a -3.01-dB-down frequency no greater than 1.1 rad/s.

2-12 (*Sec. 2.2*)   Find the pole and zero locations for a second-order network function which has equal-ripple behavior between the limits of 0 and 0.29289 in its stopband from 1 rad/s to infinity, and monotonic behavior in its passband from 0 to 1 rad/s.

2-13 (*Sec. 2.3*)   (*a*) Find the required order for a maximally flat magnitude function which is down 1 dB at 1 rad/s and 38 dB at 1.5 rad/s.

(*b*) Repeat for an equal-ripple function.

(*c*) Repeat for an elliptic function.

(*d*) If even-order case A elliptic functions are excluded, what order of elliptic function is required?

2-14 (*Sec. 2.3*)   (*a*) Find a low-pass elliptic network function which has 1.0-dB maximum ripple in a passband of 0 to 1 rad/s, and a minimum of 50-dB attenuation at all frequencies greater than 2.0 rad/s.

(*b*) For such a network function, determine the value of the multiplicative constant such that the peak magnitude value is unity.

2-15 (*Sec. 2.3*)   Find a lossless-ladder network realization for an elliptic low-pass voltage transfer function with 1.0-dB ripple in the passband 0 to 1 kHz, and a minimum of 34-dB attenuation at all frequencies greater than 2 kHz. The filter should have equal 1000-Ω terminations at the input and the output.

2-16 (*Sec. 2.4*)   (*a*) Use the narrow-band approximation to find the pole locations for a bandpass network function which has a sixth-order equal-ripple characteristic with a 1-dB tolerance in the passband. Assume a bandwidth of 0.1 rad/s and a center frequency of 1 rad/s.

(*b*) Compare the results with those given in Table 2.4-5.

2-17 (*Sec. 2.4*)   The network shown in Fig. P2-17 has a maximally flat magnitude low-pass character-istic with a -3-dB frequency of 1 rad/s. It is desired to use appropriate frequency transformations on this network to obtain a band-elimination network with a (geometric) center frequency of 1 rad/s and a bandwidth of $\frac{1}{3}$ rad/s. The following procedures are to be tried: (*a*) First transform the given network to a high-pass one (-3 dB at 1 rad/s), then apply the low-pass to bandpass transformation to the frequency-normalized high-pass network; and (*b*) first frequency normalize the given low-pass network, then apply the low-pass to bandpass transformation, and then the low-pass to high-pass trans-formation. If the resulting networks are not the same, explain the reason for any difference.

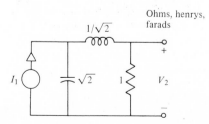

Ohms, henrys, farads

Figure P2-17

2-18 (*Sec. 2.4*) (*a*) Use the low-pass to bandpass transformation to find the network function for a fourth-order maximally flat magnitude band-elimination filter with a center frequency of 1 rad/s and bandwidth of 1.0 rad/s, defined by the frequencies which are 3 dB down from the passband transmission at zero and infinity. Express the result as a ratio of polynomials.

(*b*) Repeat the above using the narrow-band approximation, and compare the two results.

2-19 (*Sec. 2.4*) The band-elimination network shown in Fig. P2-19 has a maximally flat magnitude characteristic in its passband. The stopband has a center frequency of 1 rad/s and a bandwidth (-3 dB from maximum passband values) of $\frac{1}{3}$ rad/s. Find appropriate frequency transformations which may be made on this network so as to derive a maximally flat magnitude second-order low-pass network with a bandwidth of 1 rad/s.

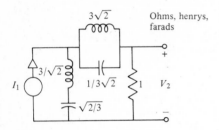

Figure P2-19

2-20 (*Sec. 2.4*) Find a fourth-order network function which has a magnitude varying within 1 dB in a passband from 0.618 to 1.618 rad/s, and has nulls at 0.3269 and 3.059 rad/s.

2-21 (*Sec. 2.5*) Determine whether the function

$$A(\omega) = \tan\left[\arg T(j\omega)\right] = \frac{\omega^5 - 2\omega^3 - 2\omega}{-2\omega^4 - \omega^2 + 2}$$

is the phase function of any of the network functions $T_i(s)$ given below:

(*a*) $T_a(s) = \dfrac{s+1}{(s^2 + s + 1)(s^2 + 2s + 2)}$

(*b*) $T_b(s) = \dfrac{s^2 - s + 1}{(s^2 + 2s + 2)(s + 1)}$

(*c*) $T_c(s) = \dfrac{s-1}{(s^2 + s + 1)(s^2 + 2s + 2)}$

(*d*) $T_d(s) = \dfrac{(s+1)(s^2 - 2s + 2)}{s^2 + s + 1}$

2-22 (*Sec. 2.5*) Apply the procedure given in Example 2.5-2 to the determination of the denominator coefficients of a third-order network function which has a maximally flat phase characteristic with a -1 slope. Verify the results using Table 2.5-1.

2-23 (*Sec. 2.5*) Use the expression given in (22) of Sec. 2.5 to determine the denominator coefficients of the network function for a fourth-order Thomson filter. Verify the results using Table 2.5-1.

2-24 (*Sec. 2.5*) Use the expression given in (23) of Sec. 2.5 to determine the denominator coefficients of a fifth-order Thomson filter. Verify the results using Table 2.5-1.

2-25 (*Sec. 2.5*) (*a*) Determine the required order for a 1-ms-delay network function with a delay error no greater than 2.5 percent and a loss no greater than 3 dB for all frequencies up to 1500 rad/s.

(*b*) Realize the network function as a transfer admittance using a lossless-ladder network with a single 1000-Ω resistance termination.

2-26 (*Sec. 2.6*)   For the network function

$$N(s) = \frac{2s + 3}{(s + 1)(s + 2)}$$

find the rise time t_r and the -3-dB bandwidth ω_c, and compare their product with the value given in (5) of Sec. 2.6.

2-27 (*Sec. 2.6*)   Repeat Prob. 2-26 for the network function

$$N(s) = \frac{1}{s^2 + s + 1}$$

THREE

SENSITIVITY

One of the problems that continually face a network designer is the evaluation of his or her design, especially in comparison with other possible realizations which meet the same specifications. To do this the designer must be concerned with the sensitivity of the filter. By *sensitivity*, we mean a measure of the change in some performance characteristic of the network (or the network function) resulting from some change in the nominal value of one or more of the elements of the network. Even if a network realization is attractive from theoretical considerations, high sensitivities may make it useless in practice. Thus, in the design of filters we are interested both in choosing realizations which have low sensitivities and in minimizing the sensitivities of realizations which we desire to use. We shall see examples of both processes in the chapters which follow.

3.1 RELATIVE SENSITIVITY RELATIONS

The symbol S is used to denote sensitivity. In addition, a superscript character is used to indicate the performance characteristic which is being evaluated, and a subscript character to indicate the network element which is causing the change. If we let y be the performance characteristic and x be the element, we define *relative sensitivity* as follows

$$S_x^y = \frac{\partial y}{\partial x}\frac{x}{y} = \frac{\partial y/y}{\partial x/x} = \frac{\partial(\ln y)}{\partial(\ln x)} \tag{1}$$

There are many choices which are common for the performance characteristic y. For example, it may be the actual network function (as a function of s or ω). Other choices are pole (and zero) locations, specific coefficients of the network function, properties such as the Q of the network function, etc. A treatment of the various types of sensitivity will be given in the following sections of this chapter. Here we shall develop some properties which apply to any sensitivity which is defined by (1). The numbers used with the properties refer to Table 3.1-1.

Table 3.1-1 Properties of the relative sensitivity function of (1)

Property no.	Relation
1	$S_x^{ky} = S_{kx}^y = S_x^y$
2	$S_x^x = S_x^{kx} = S_{kx}^{kx} = 1$
3	$S_{1/x}^y = S_x^{1/y} = -S_x^y$
4	$S_x^{y_1 y_2} = S_x^{y_1} + S_x^{y_2}$
5	$S_x^{\prod\limits_{i=1}^{n} y_i} = \sum\limits_{i=1}^{n} S_x^{y_i}$
6	$S_x^{y^n} = nS_x^y$
7	$S_x^{x^n} = S_x^{kx^n} = n$
8	$S_{x^n}^y = \dfrac{1}{n} S_x^y$
9	$S_{x^n}^x = S_{kx^n}^x = \dfrac{1}{n}$
10	$S_x^{y_1/y_2} = S_x^{y_1} - S_x^{y_2}$
11	$S_{x_1}^y = S_{x_2}^y S_{x_1}^{x_2}$
12*	$S_x^y = S_x^{|y|} + j \arg y S_x^{\arg y}$
13*	$S_x^{\arg y} = \dfrac{1}{\arg y} \operatorname{Im} S_x^y$
14*	$S_x^{|y|} = \operatorname{Re} S_x^y$
15	$S_x^{y+z} = \dfrac{1}{y+z} (yS_x^y + zS_x^z)$
16	$S_x^{\sum\limits_{i=1}^{n} y_i} = \dfrac{\sum\limits_{i=1}^{n} y_i S_x^{y_i}}{\sum\limits_{i=1}^{n} y_i}$
17	$S_x^{\ln y} = \dfrac{1}{\ln y} S_x^y$

* In this relation y is a complex quantity, and x is a real quantity.

Properties 1 and 2 *The sensitivity of any characteristic multiplied by a constant is the same as the original sensitivity.* To see this we write

$$S_x^{ky} = \frac{\partial(ky)}{\partial x} \frac{x}{ky} = \frac{\partial(ky)}{\partial y} \frac{\partial y}{\partial x} \frac{x}{ky} = \frac{\partial y}{\partial x} \frac{x}{y} = S_x^y \tag{2}$$

where k is not a function of x. Similarly, $S_{kx}^y = S_x^y$.

Property 3 *The sensitivity to a reciprocal quantity is the negative of the sensitivity to the original quantity.* To see this we may write

$$S_{1/x}^y = \frac{\partial y}{\partial(1/x)} \frac{1/x}{y} = \frac{\partial y}{\partial x} \frac{\partial x}{\partial(1/x)} \frac{1/x}{y} = -\frac{\partial y}{\partial x} \frac{x}{y} = -S_x^y \tag{3}$$

Similarly $S_x^{1/y} = -S_x^y$.

Properties 4 through 9 *The sensitivity of a product of characteristics is the sum of the sensitivities of the individual characteristics.* To see this we write

$$S_x^{y_1 y_2} = \frac{\partial(y_1 y_2)}{\partial x} \frac{x}{y_1 y_2} = \frac{y_2}{\partial x} \frac{\partial y_1}{\partial x} \frac{x}{y_1 y_2} + \frac{y_1}{\partial x} \frac{\partial y_2}{\partial x} \frac{x}{y_1 y_2}$$

$$= \frac{\partial y_1}{\partial x} \frac{x}{y_1} + \frac{\partial y_2}{\partial x} \frac{x}{y_2} = S_x^{y_1} + S_x^{y_2} \tag{4}$$

Extending this we see that the sensitivity of a characteristic to the nth power is n times the sensitivity to the first power, that is, $S_x^{y^n} = nS_x^y$. Similarly, $S_{x^n}^y = S_x^y/n$.

Properties 10 through 11 *The sensitivity of a ratio of characteristics is the difference of the individual sensitivities.* Here we use properties 4 and 3 to obtain

$$S_x^{y_1/y_2} = S_x^{y_1} + S_x^{1/y_2} = S_x^{y_1} - S_x^{y_2} \tag{5}$$

Properties 12 through 14 *The sensitivity of a complex characteristic is also complex. The real part is the magnitude sensitivity, and the imaginary part is the phase sensitivity multiplied by the phase.*[1] To see this, let Y be the magnitude of y and ϕ be the phase. Thus $y = Ye^{j\phi}$. Using (1) and property 4 we obtain

$$S_x^{Ye^{j\phi}} = S_x^Y + S_x^{e^{j\phi}} = S_x^Y + \frac{\partial(\ln e^{j\phi})}{\partial x/x}$$

$$= S_x^Y + j\frac{\partial\phi}{\partial x/x} = S_x^Y + j\phi S_x^\phi \tag{6}$$

Table 3.1-1 contains a summary of the relations derived above as well as several other useful ones.

---

[1] This result assumes that x is real.

A usual problem encountered in any type of active or passive filter design is minimization of sensitivity. Thus, it is convenient to be able to determine the derivative of a given sensitivity function. For the case where y is a linear function of x we readily obtain

$$\frac{d}{dx} S_x^y = \frac{1}{x} S_x^y (1 - S_x^y) \tag{7}$$

Thus we see in this case that the derivative of a sensitivity function is readily evaluated without requiring differentiation. A similar (but more complicated expression) holds for the situation where the dependence of y on x is not linear (see Prob. 3-2).

3.2 FUNCTION SENSITIVITY

In this section we introduce the first of the specific sensitivity functions having the general relative form defined in (1) of Sec. 3.1. The quantity y in that relation is chosen to be the network function $N(s)$. The element x is usually chosen as some passive or active element in the circuit realization of the function. The sensitivity is called *function sensitivity*. It is defined as[1]

$$S_x^{N(s)} = \frac{\partial N(s)}{\partial x} \frac{x}{N(s)} \tag{1}$$

If the network function is written as a ratio of polynomials $A(s)$ and $B(s)$, thus

$$N(s) = \frac{A(s)}{B(s)} \tag{2}$$

then we may derive the following convenient form of (1):

$$S_x^{N(s)} = x \left[\frac{A'(s)}{A(s)} - \frac{B'(s)}{B(s)} \right] \tag{3}$$

where

$$A'(s) = \frac{\partial A(s)}{\partial x} \qquad B'(s) = \frac{\partial B(s)}{\partial x}$$

Under conditions of sinusoidal steady state, applying property 12 of Table 3.1-1 we find that

$$S_x^{N(j\omega)} = S_x^{|N(j\omega)|} + j \frac{\partial \arg N(j\omega)}{\partial x / x} \tag{4}$$

---

[1] This is also sometimes referred to as *classical* or *Bode sensitivity*. It was originally presented (as the reciprocal of the relation given here) in H. W. Bode, *Network Analysis and Feedback Amplifier Design*, D. Van Nostrand Company, Inc., Princeton, N.J., 1945, p. 52.

An alternate form of this expression which is frequently useful is obtained by defining an *attenuation function* $\alpha(\omega)$ and a *phase function* $\beta(\omega)$ as

$$\alpha(\omega) = \frac{1}{\ln |N(j\omega)|} \qquad \beta(\omega) = \arg N(j\omega) \tag{5}$$

In terms of these, (4) may be written

$$S_x^{N(j\omega)} = -\frac{\partial\alpha(\omega)}{\partial x/x} + j\frac{\partial\beta(\omega)}{\partial x/x} \tag{6}$$

Thus the real part of the sensitivity gives the change in the attenuation function, and the imaginary part gives the change in the phase function, both with respect to a normalized element change.

Example 3.2-1 *The Function Sensitivity of a Passive RLC Network* As an example of function sensitivity consider the series *RLC* network shown in Fig. 3.2-1. For this network

$$Y(s) = \frac{s(1/L)}{s^2 + s(R/L) + 1/LC} \tag{7}$$

Applying (3) we find

$$S_L^{Y(s)} = \frac{-s^2}{s^2 + s(R/L) + 1/LC}$$

$$S_C^{Y(s)} = \frac{1/LC}{s^2 + s(R/L) + 1/LC} \tag{8}$$

$$S_R^{Y(s)} = \frac{-sR/L}{s^2 + s(R/L) + 1/LC}$$

Further examining the sensitivity to R by inserting the nominal element values shown in Fig. 3.2-1 and letting $s = j\omega$ we obtain

$$S_R^{Y(j\omega)} = \frac{-\omega^2}{(3 - \omega^2)^2 + \omega^2} + j\frac{-\omega(3 - \omega^2)}{(3 - \omega^2)^2 + \omega^2} \tag{9}$$

Plots of the magnitude and phase of the network function for values of R of 1 and 1.1 Ω and of the sensitivity (for $R = 1\ \Omega$) are shown in Fig. 3.2-2. □

Ohms, henrys, farads

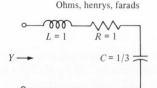

$L = 1$ $R = 1$

$Y \longrightarrow$ $C = 1/3$

Figure 3.2-1 The *RLC* network used in Example 3.2-1.

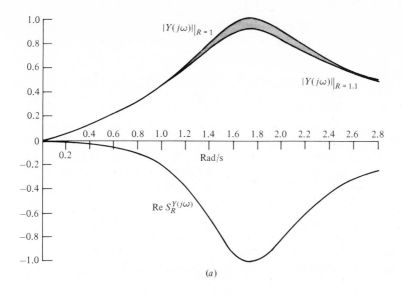

(a)

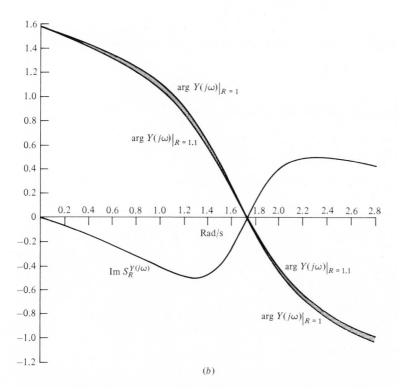

(b)

Figure 3.2-2 The function sensitivities for the network of Fig. 3.2-1. (a) The magnitude of a network function and the real part of the sensitivity for that function. (b) The argument of a network function and the imaginary part of the sensitivity for that function (arguments are specified in radians).

It is important to realize that the relations defined by sensitivity functions are exact as long as differential quantities are considered; that is, from (4), for the case where $x = R$, we may write

$$d \arg N(j\omega) = \operatorname{Im} S_R^{N(j\omega)} \frac{dR}{R} \qquad \frac{d\,|N(j\omega)|}{|N(j\omega)|} = \operatorname{Re} S_R^{N(j\omega)} \frac{dR}{R} \qquad (10)$$

When incremental quantities are used, however—i.e., when the quantities $d \arg N(j\omega)$, $d\,|N(j\omega)|$, and dR in (10) are replaced with $\Delta \arg N(j\omega)$, $\Delta\,|N(j\omega)|$, and ΔR—then the relations are only approximate. As an example of this, note that if we evaluate the sensitivity to R given in (9) at $\omega = 2$ rad/s, we obtain $S_R^{Y(j2)} = -0.8 + j0.4$. If we actually let R change value by 10 percent, i.e., go to a value of 1.1, we find that, considering actual changes in the magnitude and phase of $Y(j2)$, the sensitivity is $-0.747 + j0.369$. As the percentage change of R is made smaller, of course, the value approaches the one given above.

Another interesting observation concerning function sensitivity applies to the general second-order bandpass function. This has the form

$$N(s) = \frac{a_1 s}{s^2 + b_1 s + b_0} \qquad (11)$$

The sensitivity of this to the complex-frequency variable s is readily shown to be

$$S_s^{N(s)} = s\left(\frac{a_1}{a_1 s} - \frac{2s + b_1}{s^2 + b_1 s + b_0}\right) = \frac{-s^2 + b_0}{s^2 + b_1 s + b_0} \qquad (12)$$

Evaluating this sensitivity at resonance, i.e., at $s = j\sqrt{b_0}$, we find[1]

$$S_s^{N(s)}\bigg|_{s = j\sqrt{b_0}} = S_\omega^{N(j\omega)}\bigg|_{\omega = \sqrt{b_0}} = -j2\frac{\sqrt{b_0}}{b_1} = -j2Q \qquad (13)$$

where the second member is obtained from property 1 of Table 3.1-1, and where the network Q is defined as

$$Q = \frac{\sqrt{b_0}}{b_1} \qquad (14)$$

Thus we see that the sensitivity with respect to the complex or the real frequency variable is purely imaginary at resonance. Applying (4) to the above we conclude

$$\frac{\partial\,|N(j\omega)|}{\partial\omega}\bigg|_{\omega = \sqrt{b_0}} = 0 \qquad \frac{\partial \arg N(j\omega)}{\partial\omega/\omega}\bigg|_{\omega = \sqrt{b_0}} = -2Q \qquad (15)$$

These conclusions are illustrated in Fig. 3.2-2.

The development of function sensitivity given in this section is readily extended to determine the effect on the network function of variations of more than

---

[1] The term Q is discussed more fully in sec. 3.5.

one parameter. In this case, (1) of Sec. 3.2 may be written in the form

$$\frac{dN(s)}{N(s)} = d[\ln N(s)] = \sum_{i=1}^{n} S_{x_i}^{N(s)} \frac{dx_i}{x_i} \tag{16}$$

where n is the number of elements being considered. As an extension of this result, using (4) we note that

$$\frac{d\,|N(j\omega)|}{N(j\omega)} = \sum_{i=1}^{n} \mathrm{Re}\; S_{x_i}^{N(j\omega)} \frac{dx_i}{x_i} \tag{17}$$

$$d\,\arg N(j\omega) = \sum_{i=1}^{n} \mathrm{Im}\; S_{x_i}^{N(j\omega)} \frac{dx_i}{x_i} \tag{18}$$

Thus (16) compactly evaluates changes in both magnitude and phase. As an illustration, for Example 3.2-1 we find at resonance ($\omega = \sqrt{3}$), $S_R^{Y(j\sqrt{3})} = -1$, $S_L^{Y(j\sqrt{3})} = -j\sqrt{3}$, and $S_C^{Y(j\sqrt{3})} = -j\sqrt{3}$; thus

$$\frac{dY(s)}{Y(s)}\bigg|_{s=j\sqrt{3}} = \frac{d\,|Y(j\sqrt{3})|}{|Y(j\sqrt{3})|} + j\,d[\arg Y(j\sqrt{3})]$$

$$= (-1)\frac{dR}{R} + (-j\sqrt{3})\frac{dL}{L} + (-j\sqrt{3})\frac{dC}{C} \tag{19}$$

Some other observations concerning this circuit may be made from (19); namely, at resonance, the magnitude is unaffected by changes in the values of L or C, and the magnitude of the sensitivity to R is unity. In Sec. 3.6 we shall consider multiparameter sensitivity in more detail.

Some other useful applications of function sensitivity have been developed for the multiparameter case. The first of these specifies the worst-case deviation from the nominal magnitude characteristics of a given network realization when the network elements have a prescribed tolerance. If we let the individual nominal values of the n elements x_i be given as $x_i^{(\mathrm{nominal})}$, then we may define a *tolerance constant* ε by the requirement that

$$x_i^{(\mathrm{nominal})}(1 - \varepsilon) \le x_i \le x_i^{(\mathrm{nominal})}(1 + \varepsilon) \qquad i = 1, 2, \ldots, n \tag{20}$$

From (16) we see that the worst-case deviation will result when the elements take on extremal upper or lower values and that it will be proportional to ε. This, of course, assumes that only first-order effects are considered. Thus we may define a *worst-case magnitude sensitivity* $W^{|N(j\omega)|}$ by the relation

$$W^{|N(j\omega)|} = \sum_{i=1}^{n} \left| \mathrm{Re}\; S_{x_i}^{N(j\omega)} \right| \tag{21}$$

The maximum deviation from the nominal characteristic, i.e., the characteristic which results when the elements have their extreme values, is thus given as $\varepsilon W^{|N(j\omega)|}$. For any values of the network elements x_i satisfying (20), the magnitude characteristic will lie within definite bounds as illustrated in Fig. 3.2-3. A similar process may be used to define a worst-case phase sensitivity.

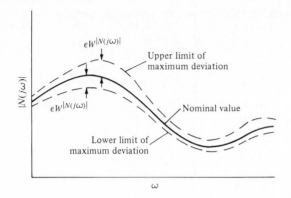

Figure 3.2-3 The worst-case magnitude sensitivity.

For some classes of networks, the worst-case magnitude sensitivity may be shown to have a lower bound. As an example of this, consider a network whose passive elements are limited to resistors, capacitors, and ideal transformers, and whose active elements are limited to gyrators, NICs, CCCSs, VCVSs, VCCSs, and CCVSs.[1] For such a class of networks, it may be shown that for any dimensionless transfer function $T(s)$

$$\sum_i S_{x_i}^{T(s)} = 2S_s^{T(s)} \tag{22}$$

where the x_i are taken to include only the passive elements of resistors and capacitors and the active elements of CCVSs and gyrators. This expression may be rewritten as

$$\sum_i S_{x_i}^{|T(j\omega)|} = 2S_\omega^{|T(j\omega)|} \tag{23}$$

Applying (21) to the above we obtain

$$W^{|T(j\omega)|} = \sum_i \left| S_{x_i}^{|T(j\omega)|} \right| \ge \left| \sum_i S_{x_i}^{|T(j\omega)|} \right| = \left| 2S_\omega^{|T(j\omega)|} \right| \tag{24}$$

[1] In terms of the transmission parameters of a two-port network, defined by the relations

$$\begin{bmatrix} V_1(s) \\ I_1(s) \end{bmatrix} = \begin{bmatrix} A & B \\ C & D \end{bmatrix} \begin{bmatrix} V_2(s) \\ -I_2(s) \end{bmatrix}$$

these elements are defined by the following properties:
Gyrator: $A = D = 0$, $C > 0$, $B > 0$
NIC (negative-impedance converter): $B = C = 0$, $(A/D) < 0$
CCCS (current-controlled current source): $A = B = C = 0$, $D \neq 0$
VCVS (voltage-controlled voltage source): $B = C = D = 0$, $A \neq 0$
    (See also sec. 6.1)
VCCS (voltage-controlled current source): $A = C = D = 0$, $B \neq 0$
CCVS (current-controlled voltage source): $A = B = D = 0$, $C \neq 0$

Comparing the first and last members of this expression we may define a lower bound $LW^{|T(j\omega)|}$ on the worst-case magnitude sensitivity as[1]

$$LW^{|T(j\omega)|} = \left| 2S_\omega^{|T(j\omega)|} \right| \tag{25}$$

Note from the right number of this equation that the lower bound is a function only of the network function $T(s)$, and thus it is independent of the particular synthesis technique used to realize the network function. A similar lower bound may be derived for worst-case phase sensitivity.

A second application of function sensitivity arises in applications where it is desired to use digital-computer-implemented optimization techniques to select element values so as to simultaneously minimize the sensitivities of a network to changes in all its elements. The optimization process operates on a scalar function indicating the level of minimization obtained. A frequent choice is that of the sum of the squares of the individual functions. Thus we may define a *quadratic magnitude sensitivity* $Q^{|N(j\omega)|}$ by the relation

$$Q^{|N(j\omega)|} = \sum_{i=1}^{n} \left(\mathrm{Re}\ S_{x_i}^{N(j\omega)} \right)^2 \tag{26}$$

It is also possible to use a scalar to simultaneously evaluate magnitude and phase sensitivity by defining a quadratic sensitivity as[2]

$$\sum_{i=1}^{n} S_{x_i}^{N(j\omega)} S_{x_i}^{*N(j\omega)} = \sum_{i=1}^{n} \left| S_{x_i}^{N(j\omega)} \right|^2 \tag{27}$$

We shall see examples of some of these types of sensitivity in the following chapters.

The use of function sensitivity provides a convenient means of illustrating an important property of the commonly used double-resistance-terminated lossless-ladder filters which are catalogued in App. A. For such filters, at any passband frequency of maximum gain, it may be shown that the source delivers maximum available power to the load. Thus at any such frequency a change, either plus or minus, in the value of any L or C component can only cause the gain to decrease. As a result, when only first-order effects are considered at these frequencies, the magnitude sensitivity, i.e., the real part of the function sensitivity, must be zero.[3] As an example of this, consider the fourth-order broadband bandpass filter shown

[1] M. L. Blostein, "Some Bounds on the Sensitivity in RLC Networks," *Proc. 1st Allerton Conf. Circuits and System Theory*, 1963, pp. 488–501.

[2] This was originally proposed by J. D. Schoeffler, "The Synthesis of Minimum Sensitivity Networks," *IEEE Trans. Circuit Theory*, vol. CT-11, no. 2, June 1964, pp. 271–276; and is sometimes referred to as *Schoeffler sensitivity*.

[3] H. J. Orchard, "Inductorless Filters," *Electronics Letters*, vol. 2, no. 6, June 1966, pp. 224–225.

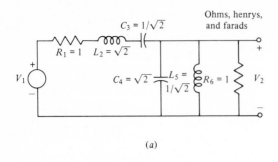

(a)

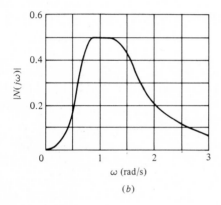

ω (rad/s)

(b)

Figure 3.2-4 A fourth-order broadband bandpass filter and its magnitude characteristic.

in Fig. 3.2-4a. We will let its voltage transfer function be specified as $N(s) = V_2(s)/V_1(s)$. As shown by its magnitude characteristic in Fig. 3.2-4b, it has a center frequency of 1 rad/s, a 3-dB-down bandwidth of 1 rad/s, and a Butterworth (maximally flat magnitude) characteristic. Thus its frequency of maximum gain is at the center frequency, 1 rad/s. Plots of $S^{|N(j\omega)|}$ for the various elements are given in Fig. 3.2-5. The zero-sensitivity property at the center frequency is readily seen in these plots. It should be noted, however, that as the passband of such a filter is narrowed, the magnitude sensitivities become larger. As an example of this, consider a filter similar to the one described above, but with element values $L_2 = C_4 = 10\sqrt{2}$ and $C_3 = L_5 = \frac{1}{10}\sqrt{2}$. The resulting magnitude characteristic is shown in Fig. 3.2-6. The center frequency is still 1 rad/s, but the filter is now narrow-band with a bandwidth of 0.1 rad/s. Plots of $S^{|N(j\omega)|}$ for the various elements are shown in Fig. 3.2-7. In comparing these plots with the ones given in Fig. 3.2-5, note that different ordinate scales have been used. From the plots we note that, despite the increased magnitude of the sensitivities at the edges of the passband, the sensitivity at the resonant frequency remains zero.

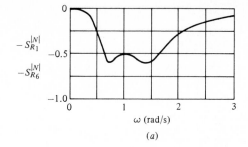

(a)

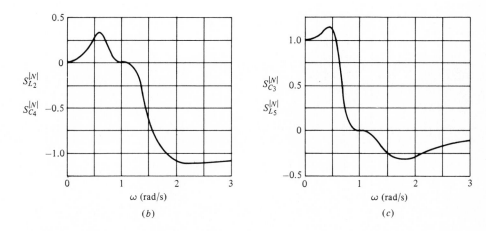

(b)

(c)

Figure 3.2-5 Plots of $S^{|N(j\omega)|}$ for the network of Fig. 3.2-4.

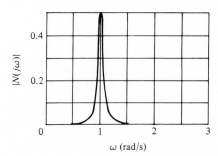

Figure 3.2-6 The magnitude characteristic of a narrow-band bandpass filter.

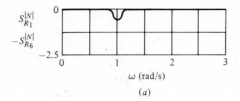

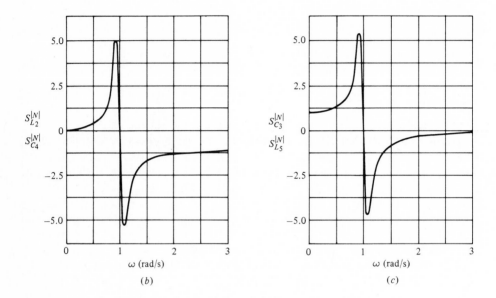

Figure 3.2-7 Plots of $S^{|N(j\omega)|}$ for the narrow-band bandpass filter.

3.3 COEFFICIENT SENSITIVITY

In general, a network function $N(s)$ for any active or passive lumped network is a ratio of polynomials having the form

$$N(s) = \frac{A(s)}{B(s)} = \frac{a_0 + a_1 s + a_2 s^2 + \cdots + a_m s^m}{b_0 + b_1 s + b_2 s^2 + \cdots + b_n s^n} \tag{1}$$

in which the coefficients a_i and b_i are real and are functions of the network elements x_i. Thus, for any arbitrary network element x we may define sensitivities which are relative in form, and which are called *coefficient sensitivities* as follows

$$S_x^{a_i} = \frac{\partial a_i}{\partial x} \frac{x}{a_i} \qquad S_x^{b_i} = \frac{\partial b_i}{\partial x} \frac{x}{b_i} \tag{2}$$

The manner in which the network function depends on any element x is a *bilinear dependence*.[1] Thus, $N(s)$ of (1) may also be written in the form

$$N(s) = \frac{A(s)}{B(s)} = \frac{E(s) + xF(s)}{C(s) + xD(s)} \tag{3}$$

where $C(s)$, $D(s)$, $E(s)$, and $F(s)$ are polynomials with real coefficients which are not functions of the network element x. This is true whether x is chosen to be the value of a passive resistor or capacitor, the gain of some amplifier or controlled source, etc.[2] For example, for the network shown in Fig. 3.2-1, using the indicated nominal values we may write

$$Y(s) = \frac{s}{(s^2 + 3) + Rs} = \frac{s/L}{s^2 + (1/L)(s + 3)} = \frac{s}{(s^2 + s) + 1/C} \tag{4}$$

Note that to insure a unique representation, in some of these cases we have used a reciprocal-valued element such as $1/C$ or $1/L$ for our variable x. From property 3 of Table 3.1-1, this simply multiplies the resulting sensitivity by a minus sign. Because of the bilinear-dependence property referred to above, there are only two ways in which a coefficient a_i (or b_i) may depend on a network element x. The first of these has the form $a_i = kx$, in which case, from property 2 of Table 3.1-1, $S_x^{a_i} = 1$. As an example of this, for the network shown in Fig. 3.2-1, from (7) of Sec. 3.2 we find the sensitivities given in Table 3.3-1. Note that these values are not unique. For example, if we multiply the numerator and denominator of the network function by LC, a different set of coefficient sensitivities are obtained. Thus, for uniqueness, we shall assume that the highest-degree coefficient of the denominator polynomial has been normalized to unity. The second possible dependence for a coefficient a_i (or b_i) is $a_i = k_0 + k_1 x$, in which case $S_x^{a_i} = k_1 x/(k_0 + k_1 x)$. In this latter situation we recognize two cases: (1) The parities of the terms k_0 and $k_1 x$ are the same, and thus the magnitude of the sensitivity is less than one; and (2) the terms have opposite parities, in which case the magnitude of the sensitivity is greater than one.

[1] A proof of this is given in L. P. Huelsman, *Theory and Design of Active RC Circuits*, McGraw-Hill Book Company, New York, 1968, Chap. 2.

[2] In the case of mutual inductance or the gyration constant of a gyrator, the quantity x may also actually be the square of the value of the element.

Table 3.3-1 Coefficient sensitivities for (7) of Sec. 3.2

| Element | Coefficient | | | |
	b_0	b_1	b_2	a_1
R	0	1	0	0
L	-1	-1	0	-1
C	-1	0	0	0

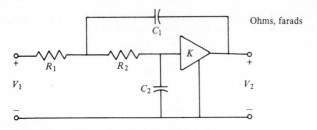

Figure 3.3-1 The active *RC* filter analyzed in Example 3.3-1.

Example 3.3-1 *Coefficient Sensitivities of an Active RC Filter*   Consider the network shown in Fig. 3.3-1 (in which the triangle represents an ideal VCVS of gain K). The voltage transfer function for this network (with only K retained as a variable and with $R_i = C_i = 1$) is

$$\frac{V_2(s)}{V_1(s)} = \frac{K}{s^2 + (3 - K)s + 1} \tag{5}$$

From (1) and (2) we find that

$$S_K^{b_1} = \frac{-K}{3 - K} \qquad S_K^{a_0} = 1 \tag{6}$$

All the other coefficient sensitivities for K are zero. In (6) above, obviously as $K \rightarrow 3$, $S_K^{b_1} \rightarrow \infty$. □

It should be noted that, in a case such as the one given in the preceding example, an infinite value for the sensitivity does not mean an infinite change in the value of the coefficient. It is rather the result of dividing by zero (the value of the coefficient when $K = 3$). A more meaningful measure of the change in such a case would be an *unnormalized coefficient sensitivity*[1]

$$US_x^{a_i} = \frac{\partial a_i}{\partial x/x}$$

from which we can compute the change of the coefficient as $da_i = US_x^{a_i} \, dx/x$. For the above example, $US_K^{b_1} = -K$. The coefficient sensitivity cases described above are summarized in Table 3.3-2.

The coefficient sensitivities defined in this section are readily related to the function sensitivity introduced in the preceding section by inserting (1) and (2) into (3) of Sec. 3.2. Thus we obtain

$$S_x^{N(s)} = \frac{\displaystyle\sum_{i=0}^{m} S_x^{a_i} a_i s^i}{A(s)} - \frac{\displaystyle\sum_{i=0}^{n} S_x^{b_i} b_i s^i}{B(s)} \tag{7}$$

---

[1] This sensitivity format is referred to as a *semirelative* sensitivity.

Table 3.3-2 Relations for coefficient sensitivities

Case	Form of a_i	Restrictions	$S_x^{a_i}$	$\lvert S_x^{a_i} \rvert$	$US_x^{a_i}$
1	kx		1	1	kx
2	$k_0 + k_1 x$	k_0 and $k_1 x$ of same sign	$k_1 x/(k_0 + k_1 x)$	< 1	$k_1 x$
3	$k_0 + k_1 x$	k_0 and $k_1 x$ of opposite sign	$k_1 x/(k_0 + k_1 x)$	> 1	$k_1 x$

As an example, using (7) of Sec. 3.2 we obtain

$$S_L^{Y(s)} = \frac{-1(1/L)s}{s/L} - \frac{(-1)(1/LC) + (-1)(R/L)s}{s^2 + s(R/L) + 1/LC} \tag{8}$$

which after reduction agrees with the result given in (8) of Sec. 3.2. Frequently, especially if digital-computational procedures are being implemented, it is easier to obtain the classical sensitivity by this means.

3.4 UNNORMALIZED ROOT SENSITIVITY

One of the most meaningful criteria that is used in determining the manner in which a network's properties vary as some element changes is a determination of how the zeros and poles of the network function, i.e., how the roots of the numerator and denominator polynomials, change as the value of the element changes. Thus we may define the *unnormalized* (semirelative) *root sensitivities*

$$US_x^{p_i} = \frac{\partial p_i}{\partial x/x} \qquad US_x^{z_i} = \frac{\partial z_i}{\partial x/x} \tag{1}$$

where p_i and z_i are the poles and zeros of the network function. Since both of these sensitivities refer to the roots of polynomials, we need to examine only the pole sensitivity in detail. The treatment of the zero sensitivity is identical.

Let $B(s)$ be the denominator polynomial of a network function $N(s)$. Because of the bilinear dependence discussed in Sec. 3.3, the dependence of this polynomial on any parameter x may be put in the form

$$B(s) = C(s) + xD(s) \tag{2}$$

where $C(s)$ and $D(s)$ are polynomials with real coefficients which are not functions of x. Evaluating (2) at any pole p_i of $N(s)$, we obtain

$$B(p_i) = C(p_i) + xD(p_i) = 0 \tag{3}$$

To determine the effect of a variation of x on p_i in this equation we may replace x by $x + \Delta x$, and p_i by $p_i + \Delta p_i$. Using a series expansion terminated after the first-order term, we see that $C(s + \Delta s) = C(s) + \Delta s C'(s)$, where $C'(s) = dC(s)/ds$.

Similarly, $D(s + \Delta s) = D(s) + \Delta s D'(s)$. Inserting these results in (3) we obtain

$$C(p_i) + \Delta p_i C'(p_i) + (x + \Delta x)[D(p_i) + \Delta p_i D'(p_i)] = 0 \qquad (4)$$

Retaining only first-order incremental terms, this may be put in the form

$$\frac{\Delta p_i}{\Delta x} = \frac{-D(p_i)}{B'(p_i)} \qquad (5)$$

In the limit as Δx approaches zero, we have

$$US_x^{p_i} = \frac{\partial p_i}{\partial x/x} = \frac{-xD(p_i)}{B'(p_i)} \qquad (6)$$

as the defining relation for our unnormalized pole sensitivity. An example follows.

Example 3.4-1 *Unnormalized Pole Sensitivities of an Active RC Filter* As an example of unnormalized pole sensitivity, consider the network function given in (5) of Sec. 3.3. The poles are

$$p_1 = \frac{K - 3}{2} + j\sqrt{1 - \frac{(3 - K)^2}{4}} = p_2^* \qquad (7)$$

From (1) or (6) we find

$$US_K^{p_1} = \frac{K}{2} + j\frac{K(3 - K)/4}{\sqrt{1 - (3 - K)^2/4}} = (US_K^{p_2})^* \qquad (8)$$

where, since the poles are complex conjugates, the sensitivities are also complex conjugates. As a further illustration, if the nominal value of K is 2, then $p_1 = p_2^* = -0.5 + j0.866$ and $US_K^{p_1} = US_K^{p_2*} = 1 + j0.577$. From these sensitivities we see that a 10 percent change in K (K going to a value of 2.2) produces $\Delta p_1 = 0.1 + j0.0577$; that is, the pole moves upward and to the right, the change in the real part being almost twice as great as the change in the imaginary part. The resulting value of p_1 as found from sensitivity calculations is $-0.4 + j0.924$. The exact value of p_1 is readily found to be $-0.4 + j0.9165$. A smaller perturbation of K would, of course, produce a closer correlation between the two results. □

The normalized root sensitivity introduced above effectively defines the root locus of a given pole position as a function of the value of the specific element. For second-order network functions, these loci will always be circles or straight lines. Some typical loci for various elements of the networks discussed in Examples 3.2-1 and 3.3-1 are given in Fig. 3.4-1 (only the upper half of the complex-frequency plane is shown).

The unnormalized root sensitivity is readily related to the coefficient sensitivity introduced in Sec. 3.3. To see this, in (2) we define

$$C(s) = c_0 + c_1 s + c_2 s^2 + \cdots \qquad D(s) = d_0 + d_1 s + d_2 s^2 + \cdots \qquad (9)$$

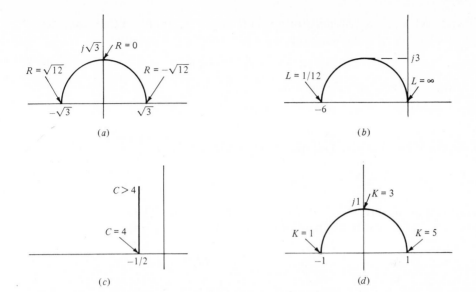

Figure 3.4-1 The loci of the poles of network functions produced by varying network parameters. (a) R in Fig. 3.2-1. (b) L in Fig. 3.2-1. (c) C in Fig. 3.2-1. (d) K in Fig. 3.3-1.

Thus the polynomial $B(s)$ of (2) may be written

$$B(s) = b_0 + b_1 s + b_2 s^2 + \cdots$$
$$= (c_0 + x\, d_0) + (c_1 + x\, d_1)s + (c_2 + x\, d_2)s^2 + \cdots \qquad (10)$$

An alternate form for $D(s)$ is thus readily seen to be

$$D(s) = \frac{\partial b_0}{\partial x} + \frac{\partial b_1}{\partial x} s + \frac{\partial b_2}{\partial x} s^2 + \cdots$$

$$= \frac{b_0}{x} S_x^{b_0} + \frac{b_1}{x} S_x^{b_1} s + \frac{b_2}{x} S_x^{b_2} s^2 + \cdots \qquad (11)$$

The polynomial $B'(s)$ is also readily found to be

$$B'(s) = b_1 + 2b_2 s + 3b_3 s^2 + \cdots \qquad (12)$$

Inserting (11) and (12) in (6), we obtain

$$US_x^{p_i} = \left. \frac{-\sum_{i=0}^{n} b_i s^i S_x^{b_i}}{\sum_{i=0}^{n-1} (i+1)b_{i+1} s^i} \right|_{s=p_i} \qquad (13)$$

The sensitivity definition given in (6) applies only to simple roots of $B(s)$, since if p_i is a multiple-order root, $B'(p_i) = 0$, and thus the sensitivity is infinite. This does *not*, of course, mean that an infinite change of the pole position results from a change in the element x. What it does mean can be seen by considering a network function $N(s)$ with a denominator polynomial of the form of (2), and replacing x by $x + \Delta x$. The new roots are then found by solving the equation

$$C(s) + (x + \Delta x)D(s) = 0 \tag{14}$$

To do this we first define a function

$$G(s) = \frac{xD(s)}{B(s)} \tag{15}$$

In terms of this function, (14) may be written in the form

$$1 + \frac{\Delta x}{x} G(s) = 0 \tag{16}$$

The poles of $G(s)$ are obviously the same as those of $N(s)$. Thus, assuming simple poles, the partial-fraction expansion of $G(s)$ will have the form

$$G(s) = \sum_{i=1}^{n} \frac{K_i^{(p)}}{s - p_i} + K_0^{(p)} \tag{17}$$

where the superscript (p) notation is used to indicate that the residues refer to poles and where the term $K_0^{(p)}$ only occurs if the order of $D(s)$ is the same as the order of $B(s)$. For values of s in the vicinity of the pole at p_i, the ith term in the expansion dominates. Thus from (17)

$$\lim_{s \to p_i} \left[1 + \frac{\Delta x}{x} G(s) \right] = 1 + \frac{\Delta x}{x} \frac{K_i^{(p)}}{s - p_i} = 1 + \frac{\Delta x}{x} \frac{K_i^{(p)}}{\Delta p_i} = 0 \tag{18}$$

where we have defined $s - p_i$ as Δp_i. From the second last member of the above, as Δx approaches zero, we find

$$US_x^{p_i} = \frac{\partial p_i}{\partial x/x} = -K_i^{(p)} \tag{19}$$

That is, the residues of $G(s)$ are the negative of the pole sensitivities. Now consider the case where a kth-order pole is located at p_1. The second last member of (18) now becomes

$$1 + \frac{\Delta x}{x} \left[\frac{K_{11}^{(p)}}{\Delta p_1} + \frac{K_{12}^{(p)}}{(\Delta p_1)^2} + \cdots + \frac{K_{1k}^{(p)}}{(\Delta p_1)^k} \right] = 0 \tag{20}$$

This may be written in the form

$$(\Delta p_1)^k + \frac{\Delta x}{x} [K_{11}^{(p)}(\Delta p_1)^{k-1} + K_{12}^{(p)}(\Delta p_1)^{k-2} + \cdots + K_{1k}^{(p)}] = 0 \tag{21}$$

The solution to this kth-order equation produces k values of Δp_1; that is, the kth-order root splits into k simple roots. For small values of Δp_1 we find that

$$\Delta p_1 = \left(\frac{-\Delta x}{x} K_{1k}^{(p)} \right)^{1/k} \tag{22}$$

Thus, the new simple roots are (for small changes of x) equiangularly spaced in a circle around p_1. An example follows.

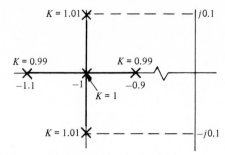

Figure 3.4-2 Changes in the second-order pole of Example 3.4-2.

Example 3.4-2 *An Active RC Filter with a Second-Order Pole*   As an example of a multiorder root, consider the network function given in (5) of Sec. 3.3. For a nominal value of $K = 1$, using the notation of (2), $B(s) = s^2 + 2s + 1$, $C(s) = s^2 + 3s + 1$, and $D(s) = -s$. Thus there is a second-order pole $p_1 = -1$. From (17)

$$G(s) = \frac{-s}{(s+1)^2} = \frac{-1}{s+1} + \frac{1}{(s+1)^2} \tag{23}$$

Thus $K_{12}^{(p)} = 1$ and $\Delta p_1 = \sqrt{-\Delta K}$. For $\Delta K = 0.01$ we obtain roots at $-1 \pm j0.1$, while for $\Delta K = -0.01$, the roots move to -0.9 and -1.1. These approximate pole locations are illustrated in Fig. 3.4-2.   □

A development similar to that given above holds for the zeros of a network function $N(s) = A(s)/B(s)$ with a numerator polynomial $A(s) = E(s) + xF(s)$. If we define (for the simple zero case)

$$H(s) = \frac{xF(s)}{A(s)} = \sum_{i=1}^{m} \frac{K_i^{(z)}}{s - z_i} + K_0^{(z)} \tag{24}$$

where the superscript (z) notation is used to indicate that the residues refer to zeros, then the zero sensitivities are obtained as

$$US_x^{z_i} = \frac{\partial z_i}{\partial x/x} = -K_i^{(z)} \tag{25}$$

Similarly, for a kth-order zero at z_1 we obtain (for small Δx)

$$\Delta z_1 = \left(-\frac{\Delta x}{x} K_{1k}^{(z)} \right)^{1/k}$$

where $K_{1k}^{(z)}$ is the coefficient of the highest-order term in the expansion for z_1 in $H(s)$.

The functions $G(s)$ and $H(s)$ defined above may be used to relate the unnormalized root sensitivities discussed in this section to the function sensitivity defined in Sec. 3.2. Using the general bilinear form for a network function

$$N(s) = \frac{A(s)}{B(s)} = \frac{E(s) + xF(s)}{C(s) + xD(s)} \tag{26}$$

and using (3) of Sec. 3.2, we obtain

$$S_x^{N(s)} = \frac{xF(s)}{A(s)} - \frac{xD(s)}{B(s)} = H(s) - G(s) \tag{27}$$

Substituting from the above, we obtain (for simple roots)

$$S_x^{N(s)} = \sum_{i=1}^{n} \frac{\mathrm{US}_x^{p_i}}{s - p_i} - \sum_{i=1}^{m} \frac{\mathrm{US}_x^{z_i}}{s - z_i} - K_0^{(p)} + K_0^{(z)} \tag{28}$$

Thus we see that the classical sensitivity may be expressed as a weighted sum of the unnormalized root sensitivities.

3.5 Q, ω_n, AND NORMALIZED ROOT SENSITIVITIES

In considering the performance of a network under conditions of sinusoidal steady state, the function sensitivity defined in (4) of Sec. 3.2 is frequently too cumbersome to be of much use, since it specifies the performance of the network over the entire frequency range. This is especially true in the bandpass case, in which it is more advisable to use criteria that emphasize the resonant, i.e., frequency-selective, nature of such functions. Such criteria are primarily the resonant frequency ω_n, at which the peak of the magnitude response occurs, and the relative sharpness or quality factor Q of that peak. This latter quantity is defined as

$$Q = \frac{\omega_n}{\mathrm{BW}} \tag{1}$$

where BW is the bandwidth defined as the difference between the frequencies at which the network function magnitude is 3 dB down from its peak magnitude at ω_n. In the frequently encountered second-order case these terms may be used to define a general bandpass function

$$N(s) = \frac{A(s)}{B(s)} = \frac{Hs}{s^2 + b_1 s + b_0} = \frac{Hs}{s^2 + s(\omega_n/Q) + \omega_n^2} \tag{2}$$

where HQ/ω_n is the gain at resonance $(s = j\omega_n)$. The quantities in the above expression are related to the pole position p_0, and to each other by the relations

$$p_0 = \sigma_0 + j\omega_0 = \frac{-b_1}{2} + j\sqrt{b_0 - \left(\frac{b_1}{2}\right)^2} = \frac{-\omega_n}{2Q} + j\frac{\omega_n}{2Q}\sqrt{4Q^2 - 1}$$

$$Q = \frac{|p_0|}{2|\sigma_0|} = \frac{\omega_n}{2|\sigma_0|} = \frac{\sqrt{b_0}}{b_1} \qquad \omega_n = \sqrt{b_0} \tag{3}$$

In terms of these quantities we may define Q and ω_n sensitivities as

$$S_x^Q = \frac{\partial Q/Q}{\partial x/x} \qquad S_x^{\omega_n} = \frac{\partial \omega_n/\omega_n}{\partial x/x} \tag{4}$$

Obviously these are relative in form. For the second-order case the above sensitivities are readily evaluated by relating them to the coefficient sensitivities using the expressions of (3) and those given in Table 3.1-1. Thus we obtain

$$S_x^Q = \tfrac{1}{2}S_x^{b_0} - S_x^{b_1} \qquad S_x^{\omega_n} = \tfrac{1}{2}S_x^{b_0} \tag{5}$$

Also, for the second-order bandpass case, the real part of the function sensitivity discussed in Sec. 3.2, which gives the sensitivity of the magnitude of the network function, can be related to the Q and ω_n sensitivities in two important cases.[1] The first of these occurs at resonance ($\omega = \omega_n$), for which it can be shown

$$S_x^{|N(j\omega_n)|} = \text{Re } S_x^{N(j\omega_n)} = S_x^Q - S_x^{\omega_n} \tag{6}$$

The second case is at the 3-dB-down frequencies $\omega_{3\,\text{dB}}$ which determine the bandwidth

$$S_x^{|N(j\omega_{3\,\text{dB}})|} = \text{Re } S_x^{N(j\omega_{3\,\text{dB}})} \approx -QS_x^{\omega_n} + \tfrac{1}{2}(S_x^Q - S_x^{\omega_n}) \tag{7}$$

Example 3.5-1 *Q and ω_n Sensitivities of Example Networks*   As an example of the determination of Q and ω_n sensitivities, using Table 3.3-1 for the coefficient sensitivities of the network of Fig. 3.2-1, and the relations of (5) we find

$$Q = \frac{1}{R}\sqrt{\frac{L}{C}} \qquad S_R^Q = -1 \qquad S_L^Q = \tfrac{1}{2} \qquad S_C^Q = -\tfrac{1}{2}$$

$$\omega_n = \frac{1}{\sqrt{LC}} \qquad S_R^{\omega_n} = 0 \qquad S_L^{\omega_n} = -\tfrac{1}{2} \qquad S_C^{\omega_n} = -\tfrac{1}{2} \tag{8}$$

Similarly, from Example 3.3-1 for the network shown in Fig. 3.3-1

$$Q = \frac{1}{3 - K} \qquad \omega_n = 1 \qquad S_K^Q = \frac{K}{3 - K} \qquad S_K^{\omega_n} = 0 \tag{9}$$

Note that in this latter example, as K increases ($K < 3$), Q and the Q sensitivity also increase, and that as $K \to 3$, they both go to infinity.   □

Another sensitivity which is frequently used to evaluate the characteristics of network functions, especially in the high-Q case, is a *normalized root sensitivity*. This differs from the relative sensitivity format in that the real and the imaginary parts are separately normalized. Thus, for a pole at $p_i = \sigma_i + j\omega_i$ we define the normalized root sensitivity as

$$NS_x^{p_i} = \frac{\partial \sigma_i / \sigma_i}{\partial x / x} + j\frac{\partial \omega_i / \omega_i}{\partial x / x} = S_x^{\sigma_i} + jS_x^{\omega_i} \tag{10}$$

---

[1] G. C. Temes and S. K. Mitra, *Modern Filter Theory and Design*, John Wiley & Sons, Inc., New York, 1973, p. 343.

where the σ_i and ω_i sensitivities are relative in form but the normalized root sensitivity is not. We readily obtain

$$NS_x^{p_i} = \text{Re } NS_x^{p_i} + j \text{ Im } NS_x^{p_i} = \frac{1}{\sigma_i} \text{ Re } US_x^{p_i} + j \frac{1}{\omega_i} \text{ Im } US_x^{p_i} \tag{11}$$

For the second-order case defined by (2) and (3), the normalized root sensitivity is readily related to the Q and ω_n sensitivities by the relations

$$\begin{aligned}
\text{Re } NS_x^{p_o} = S_x^{\sigma_0} = S_x^{\omega_n} - S_x^Q &\approx -S_x^Q \\
\text{Im } NS_x^{p_o} = S_x^{\omega_0} = S_x^{\omega_n} + S_x^Q/(4Q^2 - 1) &\approx S_x^{\omega_n}
\end{aligned} \tag{12}$$

where the approximate equalities hold for the high-Q case.

The techniques outlined above are readily extended to the third-order case. In this situation the denominator polynomial will have the form

$$D(s) = s^3 + d_2 s^2 + d_1 s + d_0 \tag{13}$$

This may be factored into a first-order term and a quadratic term. Thus, using the notation of (2) we may write

$$D(s) = (s + g)\left(s^2 + \frac{\omega_n}{Q} s + \omega_n^2\right) \tag{14}$$

Thus, for the third-order case we may define three sensitivity functions, namely, S_x^g, S_x^Q, and $S_x^{\omega_n}$. These are most conveniently calculated from the coefficient sensitivities $S_x^{d_i}$. The relations are readily found by first putting (14) in the form

$$D(s) = s^3 + s^2\left(g + \frac{\omega_n}{Q}\right) + s\left(\omega_n^2 + \frac{g\omega_n}{Q}\right) + g\omega_n^2 \tag{15}$$

Equating corresponding coefficients of (13) and (15), we obtain the relations

$$d_0 = g\omega_n^2$$

$$d_1 = \omega_n^2 + \frac{g\omega_n}{Q} \tag{16}$$

$$d_2 = g + \frac{\omega_n}{Q}$$

Taking the partial derivatives of these expressions we obtain

$$\begin{bmatrix} 2 & 0 & -1 \\ \left[2\omega_n^2 + \dfrac{g\omega_n}{Q}\right]\dfrac{1}{d_1} & -\dfrac{g\omega_n}{Qd_1} & \dfrac{g\omega_n}{Qd_1} \\ \dfrac{\omega_n}{Qd_2} & -\dfrac{\omega_n}{Qd_2} & \dfrac{g}{d_2} \end{bmatrix} \begin{bmatrix} S_x^{\omega_n} \\ S_x^Q \\ S_x^g \end{bmatrix} = \begin{bmatrix} S_x^{d_0} \\ S_x^{d_1} \\ S_x^{d_2} \end{bmatrix} \tag{17}$$

Solving the set of simultaneous equations we find that[1]

$$
\begin{bmatrix} S_x^{\omega_n} \\ S_x^{Q} \\ S_x^{g} \end{bmatrix} = \frac{1}{\Delta} \begin{bmatrix} (Q_g - \omega_n)d_0 & Q\omega_n^2 d_1 & -Q\omega_n^2 g d_2 \\ (2\omega_n Q^2 + gQ - \omega_n)d_0 & (\omega_n^2 Q - 2\omega_n gQ^2)d_1 & (Qg - 2\omega_n Q^2)\omega_n^2 d_2 \\ 2\omega_n^2 Q d_0/g & -2\omega_n^2 Q d_1 & 2\omega_n^2 Q g d_2 \end{bmatrix} \begin{bmatrix} S_x^{d_0} \\ S_x^{d_1} \\ S_x^{d_2} \end{bmatrix}
$$

$$(18)$$

where

$$
\Delta = 2\omega_n^2[Q(\omega_n^2 + g^2) - \omega_n g] \tag{19}
$$

As an example of the use of this relation, consider the normalized third-order Butterworth function. In this case, from Table 2.1-3, we have

$$
D(s) = s^3 + 2s^2 + 2s + 1 = (s + 1)(s^2 + s + 1) \tag{20}
$$

Thus, $d_0 = 1, d_1 = 2, d_2 = 2, g = 1, \omega_n = 1$, and $Q = 1$. Substituting these values in (18) we obtain

$$
\begin{bmatrix} S_x^{\omega_n} \\ S_x^{Q} \\ S_x^{g} \end{bmatrix} = \begin{bmatrix} 0 & 1 & -1 \\ 1 & -1 & -1 \\ 1 & -2 & 2 \end{bmatrix} \begin{bmatrix} S_x^{d_0} \\ S_x^{d_1} \\ S_x^{d_2} \end{bmatrix} \tag{21}
$$

For example, the Q sensitivity in this case is given as

$$
S_x^{Q} = S_x^{d_0} - S_x^{d_1} - S_x^{d_2} \tag{22}
$$

Example 3.5-2 *Sensitivities of a Third-Order Active RC Filter*   A third-order low-pass active *RC* network is shown in Fig. 3.5-1. The voltage transfer function for this circuit may be shown to be

$$
\frac{V_2(s)}{V_1(s)} = \frac{Kd_0}{s^3 + d_2 s^2 + d_1 s + d_0} \tag{23}
$$

where
$$
d_0 = G_1 G_2 G_3 S_1 S_2 S_3
$$

$$
d_1 = G_2 G_3 S_2 S_3 + G_2 G_3 S_1 S_2 + (G_2 G_3 S_1 S_3 + G_1 G_3 S_1 S_3)
$$
$$
\times (1 - K) + G_1 G_3 S_1 S_2 + G_1 G_2 S_1 S_2 \tag{24}
$$

$$
d_2 = G_1 S_1 + G_3 S_2 + G_2 S_2 + G_2 S_1 + G_3 S_3 (1 - K)
$$

and where, for convenience in taking the partial derivatives, we have used conductance $G_i = 1/R_i$ and susceptance $S_i = 1/C_i$. A solution for the Butterworth case of (20) is given by $K = 2, R_1 = 1.565, R_2 = 1.469, R_3 = 0.435$, and

---

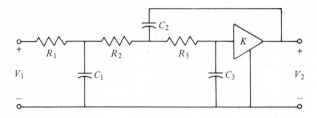

Figure 3.5-1 The third-order low-pass active *RC* network used in Example 3.5-2.

unity for all capacitors.[1] The coefficient sensitivities to the gain K are readily found to be

$$S_K^{d_0} = 0 \qquad S_K^{d_1} = -\frac{K(G_2 G_3 S_1 S_3 + G_1 G_3 S_1 S_3)}{d_1} = -3.0338$$

$$S_K^{d_2} = -\frac{K G_3 S_3}{d_2} = -2.2989$$

(25)

From (22) we see that

$$S_K^Q = 0 + 3.0338 + 2.2988 = 5.3327$$

(26)

□

Frequently in deriving a network function, it is more convenient to set the zero-degree coefficient to unity rather than the highest-degree one. In such a case the denominator polynomial $D(s)$ will have the form

$$D(s) = c_3 s^3 + c_2 s^2 + c_1 s + 1$$

(27)

The relations between the coefficient sensitivities of this denominator form and the one given in (13) are readily shown to be

$$S_x^{c_1} = S_x^{d_1} - S_x^{d_0} \qquad S_x^{d_0} = -S_x^{c_3}$$

$$S_x^{c_2} = S_x^{d_2} - S_x^{d_0} \qquad S_x^{d_1} = S_x^{c_1} - S_x^{c_3}$$

$$S_x^{c_3} = -S_x^{d_0} \qquad S_x^{d_2} = S_x^{c_2} - S_x^{c_3}$$

(28)

The sensitivities for Q, ω_n, and g given in terms of the d-coefficient sensitivities in (18) are readily modified using these expressions. For example, for the third-order Butterworth case the relations of (21) become

$$\begin{bmatrix} S_x^{\omega_n} \\ S_x^Q \\ S_x^g \end{bmatrix} = \begin{bmatrix} 1 & -1 & 0 \\ -1 & -1 & 1 \\ -2 & 2 & -1 \end{bmatrix} \begin{bmatrix} S_x^{c_1} \\ S_x^{c_2} \\ S_x^{c_3} \end{bmatrix}$$

(29)

[1] L. P. Huelsman, "An Equal-Valued-Capacitor Active RC Network Realization of a Third-Order Low-Pass Butterworth Characteristic," *Electronics Letters*, vol. 7, no. 10, May 20, 1971, pp. 271–272.

In this case, the Q sensitivity would be given as

$$S_x^Q = -S_x^{c_1} - S_x^{c_2} + S_x^{c_3} \tag{30}$$

One other application of the third-order sensitivity expressions derived above occurs for the case where a network realizes a second-order network function, but uses three independent reactive elements to do this. The resulting network function usually has a denominator polynomial of third order, but it also has a first-order factor in its numerator. In such a case a typical low-pass network function has the form

$$N(s) = \frac{A(s)}{D(s)} = \frac{H(s+g)}{s^3 + d_2 s^2 + d_1 s + d_0} \tag{31}$$

where the coefficients H, g, and d_i are functions of the element values. The design procedure for such a network will normally specify element values such that the numerator factor $(s+g)$ will also appear in the denominator; thus cancellation occurs, and the network function is effectively of second order.

Example 3.5-3 *A Second-Order Active RC Filter with Three Reactive Elements* A bandpass circuit with three independent capacitors is shown in Fig. 3.5-2.[1] If we choose $K = -(4Q - 1)$, the resulting voltage transfer function is

$$\frac{V_2(s)}{V_1(s)} = \frac{-(4Q-1)s(s+1)/2Q}{(s+1)[s^2 + s(1/Q) + 1]} = \frac{-(4Q-1)s/2Q}{s^2 + s(1/Q) + 1} \tag{32}$$

Comparing this expression with that given in (14) we see that $\omega_n = g = 1$, and thus we find that $d_0 = 1$, and $d_1 = d_2 = 1 + 1/Q$. The coefficient sensitivities with respect to the gain K are found to be $S_K^{d_0} = 0$, $S_K^{d_1} = S_K^{d_2} = -(4Q-1)/4Q(Q+1)$. Substituting these values in (18) we obtain $S_K^Q = 1 - 1/4Q$. Thus, the sensitivity is very low, although the gain of $-(4Q-1)$ required is high. □

---

[1] M. A. Soderstrand, M.S. thesis, University of California, Davis, 1969.

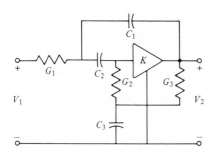

Figure 3.5-2 The second-order active *RC* filter discussed in Example 3.5-3.

3.6 MULTIPARAMETER STATISTICAL SENSITIVITY

In the preceding sections of this chapter various types of sensitivity were introduced. All these types had a common characterization, that a change in some overall network performance characteristic was related to a change in some particular network element. In this section we introduce a quite different type of sensitivity. It is a multiparameter sensitivity which provides a single (scalar) measure of the sensitivity properties of a given network. It takes into account not only the tolerance of any given element but also the manner in which a variation in one element is related to the variations in the other elements. It is called *statistical sensitivity*.[1,2]

Although the actual determination of a statistical sensitivity measure is usually fairly complex, there are several advantages that accrue from its use. For one thing, since it is a scalar quantity, it is directly usable as a measure of convergence when optimization techniques are used to minimize sensitivity. In addition, it has the capability of realistically taking into account the random nature of element value variations, as well as the manner in which the tolerances of different types of elements are correlated as a result of a particular filter fabrication technique. An example of this is the tracking of element value variations in resistors and capacitors that occurs due to their being fabricated on the same substrate in an integrated realization of a filter.

To begin our discussion of statistical sensitivity, let us define a general network transfer function as $T(j\omega, \mathbf{x})$, where $\mathbf{x}$ is a column vector of network elements x_i ($i = 1, 2, \ldots, k$). In addition, let $\Delta\mathbf{x}$ be a column vector of network element deviations Δx_i. These deviations are assumed to be random variables with zero mean and known statistics. The change in the network function caused by these deviations may now be defined as

$$\Delta T = T(j\omega, \mathbf{x} + \Delta\mathbf{x}) - T(j\omega, \mathbf{x}) \tag{1}$$

In terms of this quantity, for a given band of frequencies $\omega_1 \leq \omega \leq \omega_2$, we may now define a statistical sensitivity measure

$$M(\mathbf{x}) = E\left(\int_{\omega_1}^{\omega_2} \left|\frac{\Delta T}{T}\right|^2 d\omega\right) \tag{2}$$

---

[1] In the discussion that follows, an elementary knowledge of random variables and statistics is assumed. See, for example, G. R. Cooper and C. D. McGillem, *Probabilistic Methods of Signal and System Analysis*, Holt Rinehart and Winston, Inc., New York, 1971, Chaps. 2 and 3.

[2] A. L. Rosenblum and M. S. Ghausi, "Multiparameter Sensitivity in Active RC Networks," *IEEE Trans. Circuit Theory*, vol. CT-18, no. 6, November 1971, pp. 592–599.

where E is the expected value. To relate this more specifically to the network elements we first define the gradient of $T(j\omega, \mathbf{x})$ with respect to $\mathbf{x}$ as

$$\mathbf{V}_x T = \left[\frac{\partial T}{\partial x_1} \cdots \frac{\partial T}{\partial x_k}\right]^t \tag{3}$$

If we limit our discussion to the consideration only of first-order effects, then the quantity $\Delta T/T$ in (2) can be written as

$$\frac{\Delta T}{T} = \frac{[\mathbf{V}_x T]^t}{T} \Delta \mathbf{x} \tag{4}$$

In general, it is preferable to consider *normalized* element deviations rather than the absolute ones used in (4). To do this we now define a column vector $\Delta \hat{\mathbf{x}}$ of normalized element deviations and a matrix $\mathbf{D}$ of nominal element values. Thus

$$\Delta \mathbf{x} = \mathbf{D} \, \Delta \hat{\mathbf{x}} \tag{5}$$

where

$$\Delta \hat{\mathbf{x}} = \begin{bmatrix} \Delta x_1/x_1 \\ \vdots \\ \Delta x_k/x_k \end{bmatrix} \quad \text{and} \quad \mathbf{D} = \begin{bmatrix} x_1 & & \mathbf{0} \\ & x_2 & \\ & & \ddots \\ \mathbf{0} & & x_k \end{bmatrix} \tag{6}$$

In terms of $\mathbf{D}$ and $\mathbf{V}_x T$, we now define a column vector $\mathbf{d}$ as

$$\mathbf{d} = \mathbf{D} \frac{\mathbf{V}_x T}{T} \tag{7}$$

Note that the individual elements d_i of this vector are directly identifiable as the element function sensitivities introduced in Sec. 3.2, namely

$$d_i = \frac{x_i}{T} \frac{\partial T}{\partial x_i} = S_{x_i}^T \tag{8}$$

Inserting (5) in (4) and using (7) we find

$$\frac{\Delta T}{T} = \frac{[\mathbf{V}_x T]^t}{T} \mathbf{D} \, \Delta \hat{\mathbf{x}} = \mathbf{d}^t \, \Delta \hat{\mathbf{x}} \tag{9}$$

Using this result in (2) we obtain

$$M(\mathbf{x}) = E\left[\int_{\omega_1}^{\omega_2} (\mathbf{d}^t \, \Delta \hat{\mathbf{x}})^* \, \Delta \hat{\mathbf{x}}^t \mathbf{d} \, d\omega\right] \tag{10}$$

This result may be further simplified by noting that $\Delta \hat{\mathbf{x}}^* = \Delta \hat{\mathbf{x}}$, and defining

$$\mathbf{P} = E[\Delta \hat{\mathbf{x}} \, \Delta \hat{\mathbf{x}}^t] \tag{11}$$

Thus $\mathbf{P}$ is the $k \times k$ covariance matrix of the component tolerances. This is assumed to be known either from sample statistics or from given distributions. Combining (10) and (11) we obtain

$$M(\mathbf{x}) = \int_{\omega_1}^{\omega_2} (\mathbf{d}^t)^* \mathbf{Pd} \, d\omega \tag{12}$$

In this result note that if there are no cross-correlation terms, then the covariance matrix is diagonal, and we obtain

$$\mathbf{d}^{t*}\mathbf{Pd} = \sum_{i=1}^{k} |S_{x_i}^T|^2 \sigma_{x_i}^2 \tag{13}$$

where $\sigma_{x_i}^2$ is the variance of the element x_i; that is,

$$\sigma_{x_i}^2 = E\left[\left(\frac{\Delta x_i}{x_i}\right)^2\right] \tag{14}$$

If all the variances are taken as unity, then (13) gives the multiparameter sensitivity defined by Schoeffler [see (27) of Sec. 3.2].

In practice, the determination of the statistical sensitivity $M(\mathbf{x})$ defined in (12) can be expedited by letting the general transfer function $T(j\omega)$ be assumed to have the form

$$T(j\omega) = \left.\frac{N(s)}{D(s)}\right|_{s=j\omega}$$

$$= \left.\frac{b_n s^n + b_{n-1} s^{n-1} + \cdots + b_1 s + b_0}{s^n + a_{n-1} s^{n-1} + \cdots + a_1 s + a_0}\right|_{s=j\omega} \tag{15}$$

If we define the coefficient matrices $\mathbf{a}$ and $\mathbf{b}$ as

$$\mathbf{a} = [a_{n-1} \, a_{n-2} \, \cdots \, a_1 \, a_0]^t \qquad \mathbf{b} = [b_n \, b_{n-1} \, \cdots \, b_1 \, b_0]^t \tag{16}$$

then the quantities d_i defined in (8) are given as

$$\left.d_i\right|_{s=j\omega} = \frac{x_i}{T}\left[\frac{\partial T}{\partial b_n}\frac{\partial b_n}{\partial x_i} + \cdots + \frac{\partial T}{\partial b_0}\frac{\partial b_0}{\partial x_i}\right.$$

$$\left.+ \frac{\partial T}{\partial a_{n-1}}\frac{\partial a_{n-1}}{\partial x_i} + \cdots + \frac{\partial T}{\partial a_0}\frac{\partial a_0}{\partial x_i}\right]_{s=j\omega}$$

$$= \left.\frac{x_i}{T}\frac{\partial \mathbf{b}^t}{\partial x_i}\mathbf{V}_b T\right|_{s=j\omega} + \left.\frac{x_i}{T}\frac{\partial \mathbf{a}^t}{\partial x_i}\mathbf{V}_a T\right|_{s=j\omega} \tag{17}$$

where

$$\mathbf{V}_a T = [\partial T/\partial a_{n-1} \, \cdots \, \partial T/\partial a_0]^t \tag{18a}$$

$$\mathbf{V}_b T = [\partial T/\partial b_n \, \cdots \, \partial T/\partial b_0]^t \tag{18b}$$

For a given transfer function, the elements of these matrices are independent of the form of the realization or the values of the elements. If we now define a $k \times n$ matrix $\mathbf{C}_1$ and a $k \times (n + 1)$ matrix $\mathbf{C}_2$ as

$$
\mathbf{C}_1 = \begin{bmatrix} x_1 \dfrac{\partial \mathbf{a}^t}{\partial x_1} \\ \vdots \\ x_k \dfrac{\partial \mathbf{a}^t}{\partial x_k} \end{bmatrix} \quad \text{and} \quad \mathbf{C}_2 = \begin{bmatrix} x_1 \dfrac{\partial \mathbf{b}^t}{\partial x_1} \\ \vdots \\ x_k \dfrac{\partial \mathbf{b}^t}{\partial x_k} \end{bmatrix} \tag{19}
$$

then, using (18) and (19), the basic sensitivity measure of (12) may be written as

$$
M(\mathbf{x}) = \int_{\omega_1}^{\omega_2} \left(\frac{\nabla_a T}{T} \right)^{*t} \mathbf{C}_1^t \mathbf{P} \mathbf{C}_1 \left(\frac{\nabla_a T}{T} \right) d\omega
$$

$$
+ \int_{\omega_1}^{\omega_2} 2 \operatorname{Re} \left[\left(\frac{\nabla_a T}{T} \right)^{*t} \mathbf{C}_1^t \mathbf{P} \mathbf{C}_2 \left(\frac{\nabla_b T}{T} \right) \right] d\omega
$$

$$
+ \int_{\omega_1}^{\omega_2} \left(\frac{\nabla_b T}{T} \right)^{*t} \mathbf{C}_2^t \mathbf{P} \mathbf{C}_2 \left(\frac{\nabla_b T}{T} \right) d\omega \tag{20}
$$

This definition of statistical multiparameter sensitivity may be directly applied to a given network realization.

As an example of the use of the relation in (20), consider the network shown in Fig. 3.3-1. The general expression for the voltage transfer function of this network is

$$
T(s) = \frac{V_2(s)}{V_1(s)} = \frac{K G_1 S_1 G_2 S_2}{s^2 + s[G_1 S_1 + G_2 S_1 + G_2 S_2(1 - K)] + G_1 S_1 G_2 S_2} \tag{21}
$$

where $G_i = 1/R_i$ and $S_i = 1/C_i$ $(i = 1, 2)$. Assuming that the resonant frequency ω_n is normalized to 1 rad/s, the general form of the network function which is realized is

$$
T(s) = \frac{V_2(s)}{V_1(s)} = \frac{H a_0}{s^2 + a_1 s + a_0} = \frac{H}{s^2 + (1/Q)s + 1} \tag{22}
$$

Comparing (21) and (22) we see that

$$
G_1 S_1 + G_2 S_1 + G_2 S_2(1 - K) = \frac{1}{Q} \qquad G_1 S_1 G_2 S_2 = 1 \tag{23}
$$

To simplify our example, we shall consider only the first term in (20). Defining the parameter vector as $\mathbf{x} = [G_1 \, G_2 \, S_1 \, S_2 \, K]^t$, the $\mathbf{C}_1$ matrix of (19) becomes

$$
\mathbf{C}_1 = \begin{bmatrix} G_1 S_1 & G_1 S_1 G_2 S_2 \\ G_2 S_1 + G_2 S_2(1 - K) & G_1 S_1 G_2 S_2 \\ S_1(G_1 + G_2) & G_1 S_1 G_2 S_2 \\ G_2 S_2(1 - K) & G_1 S_1 G_2 S_2 \\ -K G_2 S_2 & 0 \end{bmatrix} \tag{24}
$$

If we now define $\alpha = G_1 S_1$, then using (23), the C_1 matrix of (24) may be written as

$$
C_1 =
\begin{bmatrix}
\alpha & 1 \\[2mm]
\dfrac{1}{Q} - \alpha & 1 \\[3mm]
\dfrac{1}{Q} + \dfrac{K-1}{\alpha} & 1 \\[3mm]
\dfrac{1-K}{\alpha} & 1 \\[3mm]
\dfrac{-K}{\alpha} & 0
\end{bmatrix}
\tag{25}
$$

We now assume that the design requires $Q = 10$ and that the frequency range of interest is defined by the relations

$$
\omega_1 = 1 - \frac{1}{Q} \qquad \omega_2 = 1 + \frac{1}{Q}
\tag{26}
$$

In addition, let us assume that the normalized variation in the resistance values $(\Delta R/R)$, the capacitance values $(\Delta C/C)$, and the gain values $(\Delta K/K)$ are taken to be uniformly distributed with a variance of 10^{-4}. For the case where these variations are uncorrelated, the covariance matrix is

$$
P =
\begin{bmatrix}
1 & 0 & 0 & 0 & 0 \\
0 & 1 & 0 & 0 & 0 \\
0 & 0 & 1 & 0 & 0 \\
0 & 0 & 0 & 1 & 0 \\
0 & 0 & 0 & 0 & 1
\end{bmatrix}
10^{-4}
\tag{27}
$$

Using the design values $\alpha = 0.7$ and $K = 1.5$, from (20) we find $M(\mathbf{x}) = 0.012$.

In the preceding example we showed how the statistical multiparameter sensitivity defined in (20) can be applied to a network in which the element tolerances are not correlated. We now consider a situation in which correlation is present. Such a situation occurs in most integrated-circuit realizations of active filters. One reason for this is that in such circuits the temperature coefficients of the resistors and the capacitors are inversely proportional to each other. This property reduces the effect of component tolerances on RC products. Since all the coefficients of a network function are algebraic combinations of such RC products, the coefficient sensitivities, and thus the function sensitivities, are reduced. When the resistor and capacitor tolerances are assumed to be random, then as a result of the inverse variation the correlation between these types of elements is negative, while the correlation between elements of the same type is positive. As an example of these effects for a realization of the filter shown in Fig. 3.3-1, let us assume a correlation coefficient of -0.7 for the resistor and capacitor tolerances. Thus we may write

$$E\left(\frac{\Delta R_i/R_i}{\sigma_{R_i}} \frac{\Delta C_j/C_j}{\sigma_{C_j}}\right) = -0.7 \tag{28}$$

where

$$\sigma_{x_i} = \sqrt{E(\Delta x_i/x_i)^2} \tag{29}$$

Similarly, let us assume a correlation coefficient of 0.7 for both the resistor and capacitor tolerances. Thus, for $i \neq j$

$$E\left(\frac{\Delta R_i/R_i}{\sigma_{R_i}} \frac{\Delta R_j/R_j}{\sigma_{R_j}}\right) = 0.7 \tag{30a}$$

$$E\left(\frac{\Delta C_i/C_i}{\sigma_{C_i}} \frac{\Delta C_j/C_j}{\sigma_{C_j}}\right) = 0.7 \tag{30b}$$

If we assume component variances of 10^{-4} as in the uncorrelated example given above, and also assume that there is no correlation between the passive element tolerances and the gain tolerance, the covariance matrix becomes

$$\mathbf{P} = \begin{bmatrix} 1 & 0.7 & -0.7 & -0.7 & 0 \\ 0.7 & 1 & -0.7 & -0.7 & 0 \\ -0.7 & -0.7 & 1 & 0.7 & 0 \\ -0.7 & -0.7 & 0.7 & 1 & 0 \\ 0 & 0 & 0 & 0 & 1 \end{bmatrix} 10^{-4} \tag{31}$$

For the same design parameters as used in the uncorrelated example, we find $M(\mathbf{x}) = 0.0072$, a reduction of sensitivity from the uncorrelated case. Thus we see that the statistical multiparameter sensitivity takes into account many aspects of circuit fabrication which are not treated by the more conventional sensitivities discussed in the preceding sections of this chapter.

3.7 SENSITIVITY COMPUTATIONS USING THE ADJOINT NETWORK

In practice, the determination of any of the sensitivities defined in Secs. 3.2 through 3.5 by "hand" methods may pose difficult computational problems for any but the simplest second-order network. Even for such networks, finding the network function with the elements expressed in literal form is usually tedious and error prone, and the difficulty of such a determination increases rapidly with the number of elements. Finding the partial derivatives provides additional tedium and increases the possibility of error still more. Thus, in general it is advantageous to use digital-computer methods to compute sensitivities. The most obvious method for doing this is to use one of the many available computer-aided design programs such as ASTAP, SYSCAP, SCEPTRE, or SPICE to make an analysis of the network with nominal element values, then repeat the analysis after having perturbed the value of one of the elements. In general, this is not a desirable procedure, since it requires a large number of analyses, i.e., one for each element.

In addition, the numerical accuracy of the result is poor since it depends on finding the difference of two nearly equal numbers. In this section we present a considerably more efficient method.

As a preliminary step in the development of an efficient method of sensitivity determination, we here introduce *Tellegen's theorem*. If we let $\mathbf{v}$ be the vector of branch voltages in a given network N, and $\mathbf{v}_n$ be the vector of (independent) node voltages, then from basic topological considerations we know that there exists an incidence matrix $\mathbf{A}$ which by Kirchhoff's voltage law has the property $\mathbf{v} = \mathbf{A}^t \mathbf{v}_n$.[1] In addition, if we let $\mathbf{i}$ be the vector of branch currents, then by Kirchhoff's current law $\mathbf{A}\mathbf{i} = \mathbf{0}$. From these relations we have

$$\mathbf{i}^t\mathbf{v} = \mathbf{v}^t\mathbf{i} = (\mathbf{A}^t\mathbf{v}_n)^t\mathbf{i} = \mathbf{v}_n^t\mathbf{A}\mathbf{i} = \mathbf{0} \tag{1}$$

This result says that the sum of the instantaneous powers delivered into all the branches of a given network is zero, i.e., that power is conserved. Now consider a second two-port network $\hat{N}$ with branch current vector $\hat{\mathbf{i}}$, which is required to have the same topology and the same branch numbering, but not necessarily the same branch elements as N. The incidence matrix $\hat{\mathbf{A}}$ for $\hat{N}$ is obviously the same as the one for N, that is, $\hat{\mathbf{A}} = \mathbf{A}$. Thus, $\hat{\mathbf{A}}\hat{\mathbf{i}} = \mathbf{A}\hat{\mathbf{i}} = \mathbf{0}$. We may now write

$$\mathbf{v}^t\hat{\mathbf{i}} = (\mathbf{A}^t\mathbf{v}_n)^t\hat{\mathbf{i}} = \mathbf{v}_n^t\mathbf{A}\hat{\mathbf{i}} = \mathbf{0} \tag{2}$$

Developments similar to the above produce the group of expressions

$$\mathbf{v}^t\hat{\mathbf{i}} = \mathbf{i}^t\hat{\mathbf{v}} = \hat{\mathbf{v}}^t\mathbf{i} = \hat{\mathbf{i}}^t\mathbf{v} = \mathbf{0} \tag{3}$$

Equation (3) says that the instantaneous sum of the products of the branch voltages (or currents) of a network N with the branch currents (or voltages) of a topologically equivalent network $\hat{N}$ is identically zero. This is Tellegen's theorem.

Example 3.7-1 *An Example of Tellegen's Theorem* Consider the network N shown in Fig. 3.7-1a. For a 2-V input step applied at $t = 0$, and for zero initial conditions, by routine circuit analysis we find that

$$
\begin{aligned}
i_1(t) &= \tfrac{2}{5}e^{-t} + \tfrac{3}{5}e^{-6t} - 1 & v_1(t) &= 2 \\
i_2(t) &= -\tfrac{2}{5}e^{-t} - \tfrac{3}{5}e^{-6t} + 1 & v_2(t) &= -\tfrac{4}{5}e^{-t} - \tfrac{6}{5}e^{-6t} + 2 \\
i_3(t) &= -\tfrac{4}{5}e^{-t} - \tfrac{1}{5}e^{-6t} + 1 & v_3(t) &= \tfrac{4}{5}e^{-t} + \tfrac{6}{5}e^{-6t} \\
i_4(t) &= \tfrac{2}{5}e^{-t} - \tfrac{2}{5}e^{-6t} & v_4(t) &= \tfrac{6}{5}e^{-t} - \tfrac{6}{5}e^{-6t} \\
i_5(t) &= \tfrac{2}{5}e^{-t} - \tfrac{2}{5}e^{-6t} & v_5(t) &= -\tfrac{2}{5}e^{-t} + \tfrac{12}{5}e^{-6t}
\end{aligned}
\tag{4}
$$

Now consider the circuit $\hat{N}$ shown in Fig. 3.7-1b. It has the same topology and branch numbering, but the branch elements are obviously quite different. For

---

[1] The matrix $\mathbf{A}$ used here is sometimes referred to as the *reduced* incidence matrix since the entries for the ground node are deleted.

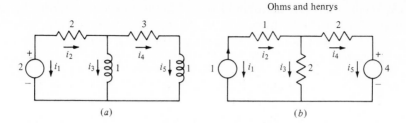

Ohms and henrys

Figure 3.7-1 Networks for illustrating Tellegen's theorem in Example 3.7-1.

it we find (for all time)

$$\hat{\mathbf{i}}^t = (-1, 1, \tfrac{3}{2}, -\tfrac{1}{2}, -\tfrac{1}{2}) \qquad \hat{\mathbf{v}}^t = (4, 1, 3, -1, 4) \tag{5}$$

All of the products given in (3) are readily shown to be zero for these two networks. ☐

We now introduce the concept of an *adjoint network*. We first note that the relations of Tellegen's theorem given in (3) also apply to transformed variables; i.e., we may write

$$\mathbf{V}^t\hat{\mathbf{I}} = \mathbf{I}^t\hat{\mathbf{V}} = \hat{\mathbf{V}}^t\mathbf{I} = \hat{\mathbf{I}}^t\mathbf{V} = \mathbf{0} \tag{6}$$

where by the use of capital letters we imply that the quantities are functions of the complex-frequency variable s. We now consider two networks N and $\hat{N}$ in which all independent sources have been removed to form external ports as illustrated in Fig. 3.7-2. We let $\mathbf{V}_p$ and $\mathbf{I}_p$ be the vectors of port variables of N and $\hat{\mathbf{V}}_p$ and $\hat{\mathbf{I}}_p$ be those of $\hat{N}$. Thus n-port open-circuit impedance matrices $\mathbf{Z}_{oc}$ and $\hat{\mathbf{Z}}_{oc}$ and n-port short-circuit admittance matrices $\mathbf{Y}_{sc}$ and $\hat{\mathbf{Y}}_{sc}$ are defined by the relations

$$\mathbf{V}_p = -\mathbf{Z}_{oc}\mathbf{I}_p \qquad \hat{\mathbf{V}}_p = -\hat{\mathbf{Z}}_{oc}\hat{\mathbf{I}}_p$$

$$\mathbf{I}_p = -\mathbf{Y}_{sc}\mathbf{V}_p \qquad \hat{\mathbf{I}}_p = -\hat{\mathbf{Y}}_{sc}\hat{\mathbf{V}}_p \tag{7}$$

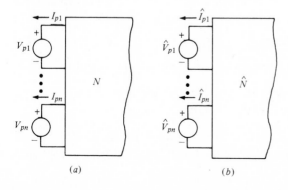

Figure 3.7-2 Two networks from which all independent sources have been removed.

where the minus signs are the result of the reference conventions chosen for the port variables (see Fig. 3.7-2). We now let the nonindependent source branches have variables $\mathbf{V}_b$ and $\mathbf{I}_b$ for N and $\hat{\mathbf{V}}_b$ and $\hat{\mathbf{I}}_b$ for $\hat{N}$. Thus we may define branch impedance matrices $\mathbf{Z}_b$ and $\hat{\mathbf{Z}}_b$ and branch admittance matrices $\mathbf{Y}_b$ and $\hat{\mathbf{Y}}_b$ by the relations

$$\mathbf{V}_b = \mathbf{Z}_b \mathbf{I}_b \qquad \hat{\mathbf{V}}_b = \hat{\mathbf{Z}}_b \hat{\mathbf{I}}_b$$
$$\mathbf{I}_b = \mathbf{Y}_b \mathbf{V}_b \qquad \hat{\mathbf{I}}_b = \hat{\mathbf{Y}}_b \hat{\mathbf{V}}_b \tag{8}$$

For some network configurations, either or both of these immittance matrices may not exist. In such a case, the addition of selected parasitic elements will remove this difficulty. Alternately, a hybrid matrix may be used to relate the branch variables. For N this may be put in the form

$$\begin{bmatrix} \mathbf{V}_{b1} \\ \mathbf{I}_{b2} \end{bmatrix} = \begin{bmatrix} \mathbf{H}_{11} & \mathbf{H}_{12} \\ \mathbf{H}_{21} & \mathbf{H}_{22} \end{bmatrix} \begin{bmatrix} \mathbf{I}_{b1} \\ \mathbf{V}_{b2} \end{bmatrix} \tag{9}$$

Similarly, for $\hat{N}$ we may write

$$\begin{bmatrix} \hat{\mathbf{V}}_{b1} \\ \hat{\mathbf{I}}_{b2} \end{bmatrix} = \begin{bmatrix} \hat{\mathbf{H}}_{11} & \hat{\mathbf{H}}_{12} \\ \hat{\mathbf{H}}_{21} & \hat{\mathbf{H}}_{22} \end{bmatrix} \begin{bmatrix} \hat{\mathbf{I}}_{b1} \\ \hat{\mathbf{V}}_{b2} \end{bmatrix} \tag{10}$$

We further assume that the controlling branches for controlled sources are either open circuits (for voltage control) or short circuits (for current control). Now, if N and $\hat{N}$ are linear time-invariant networks, then they are said to be adjoint (to each other) if the following hold:

1. The two networks have the same topology and ordering of branches; thus the two incidence matrices are equal, namely, $\mathbf{A} = \hat{\mathbf{A}}$. Thus Tellegen's theorem applies to the branch voltage and current variables of the two networks.
2. The branch impedance matrices and branch admittance matrices for the two networks are transposed. Thus

$$\mathbf{Z}_b^t = \hat{\mathbf{Z}}_b \qquad \mathbf{Y}_b^t = \hat{\mathbf{Y}}_b \tag{11}$$

As a result of this requirement, if no controlled sources are present in the two networks, they are identical. As a corollary of the above requirements, it may be shown that

$$\mathbf{Z}_{oc}^t = \hat{\mathbf{Z}}_{oc} \qquad \mathbf{Y}_{sc}^t = \hat{\mathbf{Y}}_{sc} \tag{12}$$

If a hybrid representation is used, the corresponding requirement is

$$\begin{bmatrix} \mathbf{H}_{11}^t & -\mathbf{H}_{21}^t \\ -\mathbf{H}_{12}^t & \mathbf{H}_{22}^t \end{bmatrix} = \begin{bmatrix} \hat{\mathbf{H}}_{11} & \hat{\mathbf{H}}_{12} \\ \hat{\mathbf{H}}_{21} & \hat{\mathbf{H}}_{22} \end{bmatrix} \tag{13}$$

3. Corresponding independent sources in both networks are of the same type, i.e., both current sources or both voltage sources, but they need not have the same value.

The second requirement given above provides the necessary information to construct the adjoint $\hat{N}$ of a given original network N. First, all the passive RLC elements are duplicated, since these represent diagonal elements in $\mathbf{Z}_b$ and $\mathbf{Y}_b$. Controlled sources are handled by observing the form of the $\mathbf{Z}_b$ or $\mathbf{Y}_b$ or other matrix. For example, for a VCCS of gain g with controlling voltage taken across branch i and output current at branch j in N, the result in $\mathbf{Y}_b$ is $y_{ij} = g$. Thus in $\hat{\mathbf{Y}}_b$ we must have $\hat{y}_{ji} = g$. This represents a VCCS with controlling voltage across branch j and output at branch i. Relations for other types of elements are readily obtained. A summary of the most important ones is given in Table 3.7-1. An example of the use of this table is shown in Fig. 3.7-3. Part a of this figure shows the original network, and part b shows the adjoint.

We now use Tellegen's theorem and the concept of an adjoint network to determine sensitivities. In the original network N, let the current variable vector $\mathbf{I}$

Table 3.7-1 Elements of the adjoint network

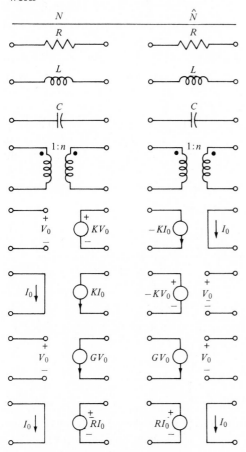

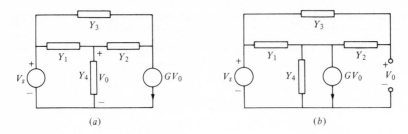

Figure 3.7-3 A network and its adjoint.

be partitioned into port currents $\mathbf{I}_p$ and nonport branch currents $\mathbf{I}_b$. Using a similar partitioning for the voltage variable vector $\mathbf{V}$ in N and the current and voltage vectors $\hat{\mathbf{I}}$ and $\hat{\mathbf{V}}$ in $\hat{N}$, we may write

$$
\begin{aligned}
\mathbf{I}^t &= (\mathbf{I}_p^t, \mathbf{I}_b^t) & \mathbf{V}^t &= (\mathbf{V}_p^t, \mathbf{V}_b^t) \\
\hat{\mathbf{I}}^t &= (\hat{\mathbf{I}}_p^t, \hat{\mathbf{I}}_b^t) & \hat{\mathbf{V}}^t &= (\hat{\mathbf{V}}_p^t, \hat{\mathbf{V}}_b^t)
\end{aligned}
\tag{14}
$$

Now, in the original network let the elements be perturbed. The resulting vector of currents may thus be written $\mathbf{I} + \Delta\mathbf{I}$. From Kirchhoff's current law we have $A(\mathbf{I} + \Delta\mathbf{I}) = 0$, and since $A\mathbf{I} = 0$, we also have

$$
A\,\Delta\mathbf{I} = \mathbf{0} \tag{15}
$$

Thus, $\Delta\mathbf{I}$ may be substituted for $\mathbf{I}$ in any of the relations in (6). By similar reasoning we find that we may substitute $\Delta\mathbf{V}$ for $\mathbf{V}$ in these relations. Making these substitutions and using (14), we obtain

$$
\begin{aligned}
\hat{\mathbf{V}}_p^t \, \Delta\mathbf{I}_p + \hat{\mathbf{V}}_b^t \, \Delta\mathbf{I}_b &= 0 \\
\hat{\mathbf{I}}_p^t \, \Delta\mathbf{V}_p + \hat{\mathbf{I}}_b^t \, \Delta\mathbf{V}_b &= 0
\end{aligned}
\tag{16}
$$

Subtracting the above two equations, we obtain

$$
\hat{\mathbf{V}}_p^t \, \Delta\mathbf{I}_p - \hat{\mathbf{I}}_p^t \, \Delta\mathbf{V}_p + \hat{\mathbf{V}}_b^t \, \Delta\mathbf{I}_b - \hat{\mathbf{I}}_b^t \, \Delta\mathbf{V}_b = 0 \tag{17}
$$

From (7) and (8), to a first-order approximation we have

$$
\begin{aligned}
\Delta\mathbf{V}_p &= -\Delta(\mathbf{Z}_{oc}\,\mathbf{I}_p) = -\Delta\mathbf{Z}_{oc}\,\mathbf{I}_p - \mathbf{Z}_{oc}\,\Delta\mathbf{I}_p \\
\Delta\mathbf{V}_b &= \Delta(\mathbf{Z}_b\,\mathbf{I}_b) = \Delta\mathbf{Z}_b\,\mathbf{I}_b + \mathbf{Z}_b\,\Delta\mathbf{I}_b
\end{aligned}
\tag{18}
$$

Substituting in (17) from the above, after some reduction we obtain

$$
\hat{\mathbf{I}}_p^t \, \Delta\mathbf{Z}_{oc}\,\mathbf{I}_p = \hat{\mathbf{I}}_b^t \, \Delta\mathbf{Z}_b\,\mathbf{I}_b \tag{19}
$$

Alternately, following a similar development we obtain

$$
\hat{\mathbf{V}}_p^t \, \Delta\mathbf{Y}_{sc}\,\mathbf{V}_p = \hat{\mathbf{V}}_b^t \, \Delta\mathbf{Y}_b\,\mathbf{V}_b \tag{20}
$$

The results of (19) and (20) provide explicit relations between the changes in the values of the branch elements and the changes in the values of the n-port parameters. For example in (19), if we excite network N only at port j and network $\hat{N}$ only at port i, then, using unity-valued excitation currents we may write

$$I_j = 1 \qquad I_k = 0 \qquad \hat{I}_i = 1 \qquad \hat{I}_k = 0 \tag{21}$$
$$ {}_{k \neq j} \phantom{\qquad \hat{I}_i = 1 \qquad} {}_{k \neq i}$$

Then (19) becomes

$$\Delta z_{ij} = \hat{\mathbf{I}}_b^t \, \Delta \mathbf{Z}_b \, \mathbf{I}_b \tag{22}$$

where $\mathbf{I}_b$ and $\hat{\mathbf{I}}_b$ are the branch currents in N and $\hat{N}$ respectively, which result from the indicated excitations. Thus, two analyses, one of $\hat{N}$ and one of N, provide information on all the sensitivities of the network function z_{ij}.

Example 3.7-2 *Sensitivity Determination by Use of an Adjoint Network* For the network shown in Fig. 3.7-4a it is desired to find the sensitivities of the input resistance R_{in} with respect to the network elements R_0, R_1, R_3, and R_5. To do this, we first redraw the network in the form shown in Fig. 3.7-4b. By routine circuit analysis for an input current $I_{p1} = 1$ we obtain the branch current vector $\mathbf{I}_b^t = (-1, -1, -\frac{6}{5}, \frac{1}{5}, \frac{1}{5})$. The adjoint network, constructed using Table 3.7-1, is shown in Fig. 3.7-4c. For an input current $\hat{I}_{p1} = 1$ we find the branch current vector $\hat{\mathbf{I}}_b^t = (-1, -1, -\frac{3}{5}, -\frac{2}{5}, -\frac{2}{5})$. The incremental branch impedance matrix has the form

$$\Delta \mathbf{Z}_b = \begin{bmatrix} \Delta R_1 & 0 & 0 & 0 & 0 \\ 0 & 0 & 0 & 0 & 0 \\ 0 & 0 & \Delta R_3 & 0 & 0 \\ 0 & \Delta R_0 & 0 & 0 & 0 \\ 0 & 0 & 0 & 0 & \Delta R_5 \end{bmatrix} \tag{23}$$

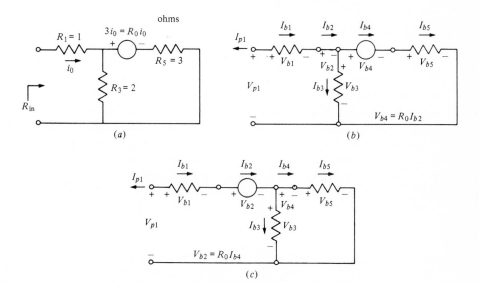

Figure 3.7-4 Use of the adjoint network to determine sensitivity in Example 3.7-2.

Thus, from (19) we obtain

$$\Delta R_{\text{in}} = \Delta z_{11} = \Delta R_1 + \frac{18}{25}\Delta R_3 - \frac{2}{25}\Delta R_5 + \frac{2}{5}\Delta R_0 \qquad (24)$$

From this result, in the limit as the incremental quantities approach zero we obtain

$$\frac{\partial R_{\text{in}}}{\partial R_1} = 1 \qquad \frac{\partial R_{\text{in}}}{\partial R_3} = \frac{18}{25} \qquad \frac{\partial R_{\text{in}}}{\partial R_5} = \frac{-2}{25} \qquad \frac{\partial R_{\text{in}}}{\partial R_0} = \frac{2}{5} \qquad (25)$$

The sensitivities are obviously easily found from the partial derivatives. These results may be verified by direct analysis of the original network. ☐

Several points deserve mention concerning the procedure given above in which the adjoint network was used to compute sensitivities. Computationally, the method is excellent, since it requires only two analyses of a given network to provide all the sensitivities for a given network immittance function, irrespective of the number of elements in the network. In addition, the results obtained were exact; i.e., the exact values of the partial derivatives are obtained. The procedure is readily extended to the sinusoidal steady-state case, in which the sensitivities at a given frequency are computed.[1,2] In such a case, the frequency-dependent elements in the branch immittance matrices will also include frequency; thus, in the incremental branch admittance matrices the frequency factor will also be present. For example, a branch with impedance $Z = j\omega L$ will produce an entry $\Delta Z = j\omega\,\Delta L$ in the incremental branch impedance matrix. Thus the factor $j\omega$ must be removed to obtain the partial derivative with respect to L.

3.8 SUMMARY

In this chapter we have introduced the general subject of sensitivity. In Sec. 3.1 a set of important relations which apply to any type of relative sensitivity function was derived. In Sec. 3.2 the concept of the sensitivity of an entire network function was discussed. Here it was shown that, for $s = j\omega$, such a function sensitivity has a real part which gives the sensitivity of the magnitude of a network function, and an imaginary part which is proportional to the phase sensitivity. Other types of sensitivity discussed in this chapter were the coefficient sensitivity (Sec. 3.3), the unnormalized root sensitivity (Sec. 3.4), the Q, ω_n, and normalized root sensitivity (Sec. 3.5), and multiparameter statistical sensitivity (Sec. 3.6). As the properties of these various sensitivities were developed, it was shown that interrelations existed

[1] S. W. Director and R. A. Rohrer, "The Generalized Adjoint Network and Network Sensitivities," *IEEE Trans. Circuit Theory*, vol. CT-16, no. 3, August 1969, pp. 318–323.

[2] Director and Rohrer, "Automated Network Design—the Frequency-Domain Case," *IEEE Trans. Circuit Theory*, vol. CT-16, no. 3, August 1969, pp. 330–337.

among them—for example, if the coefficient sensitivities were known, the function sensitivity could be directly derived from them, etc. In Sec. 3.7 it was shown how to calculate the sensitivities for an entire network by making two analyses, one of the original network and one of the adjoint network.

It is important to recognize how the tolerances of the individual elements of an active filter combine to give the tolerance of some overall network characteristic whose value is a function of the individual elements. To see this, let y be such a network characteristic, and let x_i $(i = 1, 2, \ldots, n)$ be the individual elements. We can express y as

$$y = f(x_1, x_2, \ldots, x_n) \tag{1}$$

The differential of y can thus be written

$$dy = \frac{\partial f}{\partial x_1} dx_1 + \frac{\partial f}{\partial x_2} dx_2 + \cdots + \frac{\partial f}{\partial x_n} dx_n \tag{2}$$

The fractional change in y can be found using the multiparameter sensitivity introduced in (16) of Sec. 3.2. Thus we may write

$$
\begin{aligned}
\frac{dy}{y} &= \frac{1}{y}\left(x_1 \frac{\partial f}{\partial x_1} \frac{dx_1}{x_1} + x_2 \frac{\partial f}{\partial x_2} \frac{dx_2}{x_2} + \cdots + x_n \frac{\partial f}{\partial x_n} \frac{dx_n}{x_n} \right) \\
&= S^y_{x_1} \frac{dx_1}{x_1} + S^y_{x_2} \frac{dx_2}{x_2} + \cdots + S^y_{x_n} \frac{dx_n}{x_n}
\end{aligned}
\tag{3}
$$

where the quantity dx_i/x_i $(i = 1, 2, \ldots, n)$ is the fractional change in the element x_i.

Example 3.8-1 *The Tolerance of a Resonant Frequency* A given network function has a denominator polynomial of the form $s^2 + (\omega_n/Q)s + \omega_n^2$. The circuit realization of this function is such that $\omega_n = 1/RC$, where R has a tolerance of $\pm 5\%$ and C has a tolerance of $\pm 10\%$. Following the above, we find that

$$\frac{d\omega_n}{\omega_n} = S^{\omega_n}_R \frac{dR}{R} + S^{\omega_n}_C \frac{dC}{C} = (-1)\frac{dR}{R} + (-1)\frac{dC}{C} \tag{4}$$

Thus, the worst-case fractional change in ω_n is ∓ 0.15. This corresponds to a percent change of ∓ 15 percent. $\square$

The nonideal parasitics of passive components which must be considered when using these components in active filter applications include tolerance limits, temperature dependence, frequency response, and noise. Of these, temperature behavior is typically the most important for both resistors and capacitors. In addition, for capacitors, the presence of shunt leakage as characterized by a dissipation factor may produce a nonideal frequency dependence, while for resistors, noise can be an important consideration.

The dependence of an individual network element upon any of its nonideal parasitics will have an influence on any overall network characteristic whose value is a function of this element. To see this, consider once more a network characteristic y which is a function of the elements $x_1, x_2, \ldots, x_n$ as given in (1). If the elements x_i are dependent on some nonideal parasitic designated as z, then y is also a function of z. Dividing (2) by dz yields

$$\frac{dy}{dz} = \frac{\partial f}{\partial x_1}\frac{dx_1}{dz} + \frac{\partial f}{\partial x_2}\frac{dx_2}{dz} + \cdots + \frac{\partial f}{\partial x_n}\frac{dx_n}{dz} \tag{5}$$

Typically the dependence of y upon z is treated as first-order effect and expressed as a fractional change in y per unit of z designated normally as *parts per million/unit of z*, or *ppm/unit of z*. Multiplying both members of (5) by $1/y$ we obtain the mathematical formulation of such a dependence as

$$\frac{dy/y}{dz} = \frac{1}{y}\left(\frac{\partial f}{\partial x_1}\frac{dx_1}{dz} + \frac{\partial f}{\partial x_2}\frac{dx_2}{dz} + \cdots + \frac{\partial f}{\partial x_n}\frac{dx_n}{dz}\right) \tag{6}$$

This general form can be used to examine the fractional change of a function y with respect to a parameter z.

Example 3.8-2 *Temperature Dependence of a Finite-Gain Amplifier*   The gain of a noninverting finite-gain amplifier is given as $K = 1 + R_2/R_1$. If R_2 and R_1 are both functions of temperature, it is desired to find the fractional change of K in ppm/°C. It is assumed that $K = 10$ and $R_1 = 1$ kΩ. Thus $R_2 = 9$ kΩ is required to obtain $K = 10$. In addition, the temperature parasitic behavior of R_1 and R_2 is given as $dR_1/dT = 2000$ ppm/°C and $dR_2/dT = 1000$ ppm/°C. From (6) we obtain

$$\frac{dK/K}{dT} = \frac{1}{K}\left(-\frac{R_2}{R_1}\frac{dR_1/R_1}{dT} + \frac{R_2}{R_1}\frac{dR_2/R_2}{dT}\right) = \frac{R_2}{R_1 + R_2}\left(\frac{dR_2/R_2}{dT} - \frac{dR_1/R_1}{dT}\right)$$

$$= 0.9(1000 - 2000) = -900 \text{ ppm/°C} \qquad\qquad \square$$

Tables 3.8-1 and 3.8-2 give some properties of various kinds of resistors and capacitors. Of these, the metal-film type of resistor is a common choice for active filter realizations. For capacitors, mica, polystyrene, polypropylene, and NPO ceramic types are usually used.

The results and conclusions of the studies of the various sensitivities made in this chapter will be of great use to us in the chapters that follow. In general, sensitivity measures provide the single most important technique for evaluating various types of active and passive filter realizations. Thus we shall find many references to sensitivity in the remaining chapters of this book.

Table 3.8-1 Typical characteristics of popular active filter resistors*

Resistor type	Thick film	Thin film		Discrete			
	Cermet	Tantalum nitride (oxynitride)	Special	Wirewound	Carbon composition	Metal film	Carbon film
Temp. coefficient ppm/°C	±100	−100 ± 20 (−200 ± 20)	−10	±5	±1500	0 to +50	0 to +50
Temp. range over which TC is linear (°C)	Nonlinear	0–100	−55 to +125	+25 to +85	Nonlinear	+25 to +85	Nonlinear
Practical resistance range (Ω)	3Ω–1MΩ	10Ω–1MΩ	100Ω–0.3M	100Ω–100kΩ	10Ω–2MΩ	10Ω–1MΩ	10Ω–2MΩ
Practical precision (%)	0.5	0.1	0.02	0.01	5.0	0.5	1.0
Aging (%/20 yr)	1.0	0.1	.01	0.01	5.0	0.1	2–5.0
Maximum (Ω/square)	10^7	300					
Remarks	Special inks & processes can improve stability & precision		TCR = ±1 ppm/°C and aging of 10 ppm/yr are available	Special winding methods can reduce inductance		0.1% aging and TCR = ±1 ppm/°C are available	

* From K. L. Su, "Active Filters," *Circuits and Systems*, vol. 10, no. 5, October 1976, pp. 2–8.

Table 3.8-2 Typical characteristics of capacitors suitable for active filter applications

Characteristic	Capacitor types							
	Mica	Poly-styrene	Mylar	Poly-propylene	NPO ceramic	High-K ceramic	Thin-film tantalum	Poly-carbonate
Q (1 kHz)	1000	5000	200	1000	1000	40	400	500
Temperature coefficient (ppm/°C)	$+35 \pm 35$	-120 ± 30	$+600$	-200 to -500	0 ± 30	$-50,000$ to $+20,000$	$+200 \pm 25$	$+40$
Temperature range over which TC is linear, °C	-55 to $+125$	-40 to $+85$	Nonlinear	Nonlinear	-55 to $+150$	Nonlinear	-40 to $+65$	-35 to $+125$
Aging in % for 20 yr	$+0.1$	$+0.2$	$+2$	± 0.3	± 0.05	$+15$	-0.5	$+0.2$
Minimum tolerance (%)	± 1	± 0.5	± 20	± 0.5	± 0.25	± 10	± 5	± 0.25
Capacitance range (pF–µF)	1–0.1	500 to 10	5000 to 1	500 to 10	10 to 0.05	100 to 20	1000 to 200	1000 to 10

PROBLEMS

3-1 *(Sec. 3.1)* *(a)* Derive property 7 in Table 3.1-1.
   (b) Repeat for property 14.
   (c) Repeat for property 15.

3-2 *(Sec. 3.1)* Show that the derivative of a sensitivity function S_x^y, for the case where y is a function of x which is not necessarily linear, is given as[1]

$$\frac{d}{dx}S_x^y = \frac{1}{x}S_x^y(1 + S_x^{dy/dx} - S_x^y)$$

3-3 *(Sec. 3.2)* *(a)* Draw plots comparable with those given in Fig. 3.2-2 for the case where the inductance L of Example 3.2-1 is the parameter which is varied.
   (b) Repeat for the capacitance C.

3-4 *(Sec. 3.2)* *(a)* Find the sensitivity $S_x^{N(s)}$ for the network shown in Fig. P3-4, where $N(s)$ is the open-circuit voltage transfer function $V_2(s)/V_1(s)$ and x is the inductance L.
   (b) Repeat for the capacitance C.
   (c) Repeat for the resistance R.
   (d) Repeat for the conductance G.

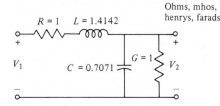

Figure P3-4

3-5 *(Sec. 3.2)* Show that the relations given in (15) of Sec. 3.2 for the network used in Example 3.2-1 agree with the plots given in Fig. 3.2-2.

3-6 *(Sec. 3.2)* Find the multiparameter variation $dN(j1)/N(j1)$ for the network function defined in Prob. 3-4 with respect to the network parameters L, C, G, and R.

3-7 *(Sec. 3.3)* *(a)* Find the coefficient sensitivities for the network function $N(s)$ defined in Prob. 3-4 with respect to the inductance L.
   (b) Repeat for the capacitance C.
   (c) Repeat for the resistance R.
   (d) Repeat for the conductance G.

3-8 *(Sec. 3.3)* *(a)* Use the results of Prob. 3-7 to obtain the function sensitivity for $N(s)$ with respect to the inductance L obtained in Prob. 3-4.
   (b) Repeat for the capacitance C.
   (c) Repeat for the resistance R.
   (d) Repeat for the conductance G.

3-9 *(Sec. 3.3)* Use the coefficient sensitivities determined in (6) of Sec. 3.3 to find the function sensitivity with respect to K for the circuit shown in Fig. 3.3-1. Verify the result by directly determining the sensitivity.

3-10 *(Sec. 3.4)* Verify the relations given in (8) of Sec. 3.4.

3-11 *(Sec. 3.4)* Find the unnormalized pole sensitivities for the network of Example 3.2-1 for the parameters R, L, and C.

---

[1] J. Gorski-Popiel, "Classical Sensitivity—A Collection of Formulas," *IEEE Trans. Circuit Theory*, vol. CT-10, no. 2, June 1962, pp. 300–302.

3-12 *(Sec. 3.4)* *(a)* Prove that the loci of the poles of the network function of Example 3.2-1 with respect to variations in the resistance R is a circle.

(b) Repeat for the inductance L.

(c) Show that for variations in the capacitance C, the loci are straight lines.

3-13 *(Sec. 3.4)* *(a)* Find the unnormalized pole sensitivity for the network function $N(s)$ defined in Prob. 3-4 with respect to the inductance L.

(b) Repeat for the capacitance C.

(c) Repeat for the resistance R.

(d) Repeat for the conductance G.

3-14 *(Sec. 3.4)* Use the coefficient sensitivities given in Table 3.3-1 to find the unnormalized pole sensitivities for the network of Example 3.2-1, for the parameters R, L, and C. Compare the results with those obtained in Prob. 3-11.

3-15 *(Sec. 3.4)* Use the coefficient sensitivities determined in Prob. 3-7 to find the unnormalized pole sensitivities for the network shown in Fig. P3-4 for the parameters L, C, R, and G. Compare the results with those obtained from Prob. 3-13.

3-16 *(Sec. 3.4)* Use the coefficient sensitivities determined in (6) of Sec. 3.3 to find the unnormalized pole sensitivity with respect to K for the circuit shown in Fig. 3.3-1. Compare the result with that given in (8) of Sec. 3.4.

3-17 *(Sec. 3.4)* *(a)* For the network shown in Fig. 3.2-1, assume the element values $L = 1$, $C = 1$, and $R = 2$. Use the techniques for handling multiple-order poles presented in Sec. 3.4 to determine the change in pole positions for a positive 1 percent change in R.

(b) Repeat for a negative 1 percent change in R.

3-18 *(Sec. 3.4)* Repeat Prob. 3-17 using C as the parameter which is changed.

3-19 *(Sec. 3.4)* Repeat Prob. 3-17 using L as the parameter which is changed.

3-20 *(Sec. 3.4)* Starting with the unnormalized pole sensitivities found in Prob. 3-11, derive the function sensitivities given in Example 3.2-1.

3-21 *(Sec. 3.5)* Find the Q and ω_n sensitivities of the network described in Example 3.2-1, with respect to the elements R, L, and C.

3-22 *(Sec. 3.5)* *(a)* Find the Q and ω_n sensitivities for the network defined in Prob. 3-4 with respect to the inductance L.

(b) Repeat for the capacitance C.

(c) Repeat for the resistance R.

(d) Repeat for the conductance G.

3-23 *(Sec. 3.5)* The third-order low-pass network configuration shown in Fig. 3.5-1, with a network function as defined in Example 3.5-2, is used to realize a third-order Chebyshev characteristic with 0.5-dB ripple. The required element values are $K = 2$, $R_1 = 0.2681 \ \Omega$, $R_2 = 2.778 \ \Omega$, $R_3 = 1.876 \ \Omega$, and unity for all the capacitors.[1] Find the Q and ω_n sensitivities (with respect to K).

3-24 *(Sec. 3.6)* *(a)* Find the statistical multiparameter sensitivity over a frequency range 0 to 1 rad/s for the voltage transfer function of the network shown in Fig. P3-24. Treat only the denominator and assume that the normalized variations in the element values are uncorrelated and have a uniform distribution with a variance of 10^{-3}.

(b) Repeat for the case where the correlation coefficient between the elements is -0.7.

[1] R. S. Aikens, "Canonic Active RC Networks," M.S. thesis, University of Arizona, Tucson, 1972.

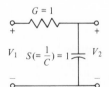

Figure P3-24

3-25 (*Sec. 3.6*) (*a*) Find the statistical multiparameter sensitivity over a frequency range 0 to 1 rad/s for the voltage transfer function of the network shown in Fig. P3-25. Treat only the denominator and assume that the normalized variations in the element values are uncorrelated and have a uniform distribution with a variance of 10^{-3}.

(*b*) Repeat for the case where the correlation coefficients between the elements are all -0.7.

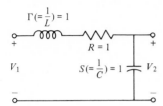

Figure P3-25

3-26 (*Sec. 3.7*) Verify the time-domain analysis for the branch currents and voltages given in Example 3.7-1.

3-27 (*Sec. 3.7*) Use the adjoint network method to find the sensitivity of $Y(j\omega)$ to the parameters R, L, and C for the network defined in Example 3.2-1.

3-28 (*Sec. 3.7*) Find the sensitivity of the transfer admittance for the network shown in Fig. 3.7-4 to the various circuit elements. The transfer admittance is defined as $Y_{21} = -I_{p2}/V_{p1}$, where port 2 is defined at the right of the network by deleting the vertical line which represents a short circuit.

3-29 (*Sec. 3.7*) Verify the sensitivities found in Example 3.7-2 by finding an expression for R_{in} and taking appropriate partial derivatives.

FOUR

RC-AMPLIFIER FILTERS—I

In this and the following chapters we will consider the general subject of active filters. More specifically, we will present methods for realizing all types of network functions through the use of filter circuits comprising both active and passive elements, the latter being restricted exclusively to resistors and capacitors. Such filters are referred to collectively as *active RC* or *inductorless* filters. There are many reasons why active *RC* filters are attractive, and indeed may be preferable to their purely passive *RLC* counterparts. For example, active *RC* filters usually weigh less and require less space than do passive ones. This is an important consideration in satellite and other airborne applications. As another example, active filters can be fabricated in microminiature form using integrated circuit techniques. Thus they can be mass-produced inexpensively. On the other hand, since it is not possible to "integrate" an inductor, passive circuits can only be produced using discrete components. This is usually far more expensive. For these and many other reasons, many traditional passive filtering applications, notably in the telecommunications industry, have now been modified so as to use active filters exclusively. As a result, annual production of active filters is now in the millions, and many companies offer these items as an off-the-shelf standard item. In this chapter we shall introduce one of the most well-known types of active *RC* filters, the *RC-amplifier* filter.

4.1 THE *RC*-AMPLIFIER FILTER

There are two general methods of using active *RC* filters to realize network functions. The first of these is the *cascade* method. The method is so named because the network function to be realized is first factored into a product of second-order terms.[1] Each term is then individually realized by an active *RC* circuit, and a cascade or series connection of the circuits is then used to realize the overall network function. The individual active *RC* circuits, of course, must be designed in such a manner that they do not interact with each other when the cascade connection is made; i.e., they must be isolated from each other. The second general method of using active *RC* circuits to realize network functions is the *direct* method, in which a single circuit is used to realize the entire network function. A discussion of this method will be given in Chap. 6.

The cascade method of using active *RC* circuits to realize network functions, as described above, offers many advantages to the network designer. First of all, since any given active *RC* circuit is only required to realize a second-order network function, its configuration is usually relatively simple and the number of elements required is usually small. As a result, the synthesis procedure needed for determining the values of the elements is straightforward, and the implementation of additional constraints, such as the use of standard element values or the minimization of sensitivity, is usually easy to achieve. As another advantage, each active *RC* circuit may be individually tuned to make certain that it exactly realizes its specific second-order factor. This, of course, is far easier than trying to tune a circuit in which all the elements interact, as is the case when the direct method of realization is used.

Now let us consider the active element of an active *RC* filter. Although theoretically any type of controlled source may be used as the active element, in practice the VCVS (*voltage-controlled voltage source*) has proven to be the preferred one. Ideally, the VCVS is a two-port device characterized by the following properties: (1) infinite input impedance, (2) zero output impedance, and (3) an output voltage which is linearly proportional to the input voltage, the constant of proportionality being referred to as the *gain*. A model and a circuit symbol for a VCVS are given in Fig. 4.1-1. The VCVS is also referred to as a *voltage amplifier*, or more simply just as an *amplifier*. The gain constant may be positive, in which case the VCVS is said to be *noninverting*; or negative, in which case it is said to be *inverting*. Among the reasons for the popularity of the VCVS as the active element of active *RC* filters is the ease with which it can be realized using an operational amplifier. As an example, the noninverting VCVS can be realized using a

---

[1] If an odd-order function is to be realized, either a passive first-order circuit or a third-order active *RC* circuit will be needed in the cascade. This will be discussed further in Sec. 4.5.

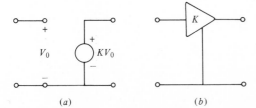

(a)   (b)

Figure 4.1-1 A voltage-controlled voltage source.

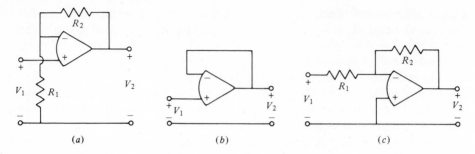

Figure 4.1-2 (*a*) A noninverting VCVS realization. (*b*) A voltage follower. (*c*) An inverting VCVS realization.

differential-input operational amplifier in the circuit shown in Fig. 4.1-2*a*. The gain of the resulting VCVS is given by the relation

$$\frac{V_2}{V_1} = \frac{R_1 + R_2}{R_1} = K \tag{1}$$

Obviously the gain will always be greater than unity. A unity-gain noninverting VCVS can be realized as shown in Fig. 4.1-2*b*. This circuit is also referred to as a *voltage follower*. The inverting VCVS can be realized as shown in Fig. 4.1-2*c*. For this circuit the gain is

$$\frac{V_2}{V_1} = \frac{-R_2}{R_1} = K \tag{2}$$

We now define a general active *RC* filter configuration consisting of a three-port passive *RC* network combined with a VCVS, as shown in Fig. 4.1-3. This circuit will be referred to as an *RC-amplifier* filter. The network function of interest for this circuit is the voltage transfer function $V_2(s)/V_1(s)$. Note that due to the zero output impedance of the VCVS, the network function realized by this filter will be completely independent of any other network connected at the terminals where $V_2(s)$ is defined. Thus, the *RC*-amplifier filter provides the necessary isolation required in applying the cascade synthesis method.

The *RC*-amplifier filter configuration shown in Fig. 4.1-3 may be analyzed by first considering the three-port passive network to be defined by a set of short-

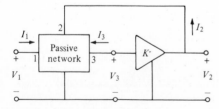

Figure 4.1-3 An *RC*-amplifier filter configuration.

circuit admittance or *y* parameters. Thus, for the passive network portion of Fig. 4.1-3 we may write

$$\begin{bmatrix} I_1(s) \\ I_2(s) \\ I_3(s) \end{bmatrix} = \begin{bmatrix} y_{11}(s) & y_{12}(s) & y_{13}(s) \\ y_{21}(s) & y_{22}(s) & y_{23}(s) \\ y_{31}(s) & y_{32}(s) & y_{33}(s) \end{bmatrix} \begin{bmatrix} V_1(s) \\ V_2(s) \\ V_3(s) \end{bmatrix} \tag{3}$$

where, due to the passive nature of the network, $y_{ij}(s) = y_{ji}(s)$; that is, the *y* matrix is symmetric. If we now consider the effect of the VCVS, we also have the equations

$$I_3(s) = 0 \qquad V_2(s) = K V_3(s) \tag{4}$$

Combining (3) and (4) we obtain the overall voltage transfer function for the *RC*-amplifier filter as

$$\frac{V_2(s)}{V_1(s)} = \frac{-K y_{31}(s)}{y_{33}(s) + K y_{32}(s)} \tag{5}$$

In most of the *RC*-amplifier filters to be discussed in this chapter, we will further restrict the passive network to have the form shown within the dashed lines in Fig. 4.1-4. To find the *y* parameters for this network we first write the nodal equations

$$\begin{bmatrix} I_1 \\ I_2 \\ 0 \\ 0 \end{bmatrix} = \begin{bmatrix} Y_1 & 0 & 0 & -Y_1 \\ 0 & Y_2 + Y_6 & -Y_6 & -Y_2 \\ 0 & -Y_6 & Y_3 + Y_4 + Y_6 & -Y_3 \\ -Y_1 & -Y_2 & -Y_3 & Y_1 + Y_2 + Y_3 + Y_5 \end{bmatrix} \begin{bmatrix} V_1 \\ V_2 \\ V_3 \\ V_4 \end{bmatrix} \tag{6}$$

where, for convenience, we have deleted the (*s*) notation. The left members of the third and fourth equations have been set to zero, since no external currents are applied to node 3 or 4. If the fourth equation is solved for V_4 we obtain

$$V_4 = \frac{Y_1 V_1 + Y_2 V_2 + Y_3 V_3}{D(s)} \tag{7}$$

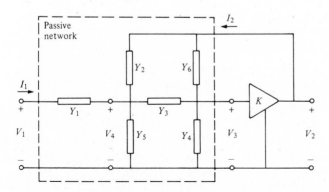

Figure 4.1-4 An *RC*-amplifier filter with a specific form of passive network.

where

$$D(s) = Y_1 + Y_2 + Y_3 + Y_5 \tag{8}$$

Substituting (7) into (6) we find

$$
\begin{bmatrix} I_1 \\ I_2 \\ 0 \end{bmatrix} = \begin{bmatrix} Y_1 - \dfrac{Y_1^2}{D(s)} & \dfrac{-Y_1 Y_2}{D(s)} & \dfrac{-Y_1 Y_3}{D(s)} \\[2ex] \dfrac{-Y_1 Y_2}{D(s)} & Y_2 + Y_6 - \dfrac{Y_2^2}{D(s)} & -Y_6 - \dfrac{Y_2 Y_3}{D(s)} \\[2ex] \dfrac{-Y_1 Y_3}{D(s)} & -Y_6 - \dfrac{Y_2 Y_3}{D(s)} & Y_3 + Y_4 + Y_6 - \dfrac{Y_3^2}{D(s)} \end{bmatrix} \begin{bmatrix} V_1 \\ V_2 \\ V_3 \end{bmatrix} \tag{9}
$$

By comparison with (3), we see that these relations define the y parameters of the passive network shown in Fig. 4.1-4. If we now restrict each of the admittances Y_i to being either a single resistor or a single capacitor, then the polynomials which are the numerators of these parameters can be at most of second order. Thus, for the y parameters used in (5) we may write

$$-y_{31}(s) = \frac{Y_1 Y_3}{D(s)} = \frac{N_{31}(s)}{D(s)} = \frac{a_2 s^2 + a_1 s + a_0}{D(s)} \tag{10a}$$

$$-y_{32}(s) = \frac{Y_6(Y_1 + Y_2 + Y_3 + Y_5) + Y_2 Y_3}{D(s)} = \frac{N_{32}(s)}{D(s)} = \frac{b_2 s^2 + b_1 s + b_0}{D(s)} \tag{10b}$$

$$y_{33}(s) = \frac{(Y_3 + Y_4 + Y_6)(Y_1 + Y_2 + Y_5) + Y_3(Y_4 + Y_6)}{D(s)}$$

$$= \frac{N_{33}(s)}{D(s)} = \frac{c_2 s^2 + c_1 s + c_0}{D(s)} \tag{10c}$$

The relations which determine the coefficients of the polynomials $N_{31}(s)$, $N_{32}(s)$, and $N_{33}(s)$ of (10) involve only sums and products of the admittances of passive, i.e., positive-valued, resistors and capacitors. Thus, the coefficients of these polynomials can only be positive (or zero). Since this is true, we may now rewrite the general voltage transfer function given in (5) in terms of these positive-coefficient polynomials as

$$\frac{V_2(s)}{V_1(s)} = \frac{KN_{31}(s)}{N_{33}(s) - KN_{32}(s)} \tag{11}$$

From this expression, we see that if the VCVS gain is positive, i.e., if the VCVS is noninverting, the decomposition of the denominator polynomial of the voltage transfer function will be of a type referred to as a *difference decomposition*. In this case, the overall voltage transfer function will also be noninverting. Similarly, a *sum decomposition* will result if the VCVS gain is negative, in which case the voltage transfer function will be inverting. In practice, the use of a noninverting VCVS has been found to give better results. Some of the reasons for this will be

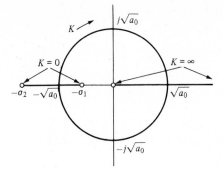

Figure 4.1-5 A difference decomposition.

found in the following sections of this chapter. Thus, the difference decomposition is the more important one. Of the many possible forms for the difference decomposition,[1] the one usually encountered in RC-amplifier filters is the one in which a second-order denominator polynomial $P(s)$ is decomposed as

$$P(s) = s^2 + a_1 s + a_0 = (s + \sigma_1)(s + \sigma_2) - K\alpha s \qquad (12)$$

where a_1 and a_0 determine the resulting pole locations, and where σ_1, σ_2, and α are functions of the various passive network elements. Such a decomposition produces a circular root locus as a function of K. Its general form is shown in Fig. 4.1-5. From the figure we see that when K is increased beyond the value $(\sigma_1 + \sigma_2)/\alpha$, the poles of the network function move into the right-half plane and the network becomes unstable. For a frequency-normalized denominator polynomial having the form $P(s) = s^2 + (1/Q)s + 1$, the optimum difference decomposition is $(s + 1)^2 - K\alpha s$. This produces a lower bound for S_K^Q of $2Q - 1$.[2] Since this value is relatively large, RC-amplifier realizations are in general suitable only for realizing network functions with low values of Q.

4.2 LOW-PASS SINGLE-AMPLIFIER POSITIVE-GAIN FILTERS

In the preceding section we presented a basic RC-amplifier filter configuration suitable for use in the realization of second-order voltage transfer functions. The circuit is given in Fig. 4.1-4. In this section we will show how the basic structure shown in the figure can be used to realize low-pass functions. The general procedure that we shall use will be extended to include other types of filter functions in the following sections.

---

[1] A detailed treatment of sum and difference decompositions may be found in L. P. Huelsman, *Theory and Design of Active RC Circuits*, McGraw-Hill Book Company, New York, 1968, secs. 2.6 and 2.7.

[2] I. M. Horowitz, "Optimization of Negative-Impedance Conversion Methods of Active-RC Synthesis," *IRE Trans. Circuit Theory*, vol. CT-6, no. 3, September 1960, pp. 352–354.

The general form of the second-order low-pass voltage transfer function can be written as

$$\frac{V_2(s)}{V_1(s)} = \frac{H_0 \omega_n^2}{s^2 + (\omega_n/Q)s + \omega_n^2} = \frac{K N_{31}(s)}{N_{33}(s) - K N_{32}(s)} \tag{1}$$

where H_0 is the gain at direct current, ω_n is the undamped natural frequency, and Q is the quality factor originally defined in Sec. 3.5. The right member of (1) is taken from (11) of Sec. 4.1. If the poles of the network function of (1) are located at $p_0 = \sigma_0 \pm j\omega_0$, then the relations between the quantities ω_n and Q and the pole locations are given as

$$p_0 = \sigma_0 \pm j\omega_0 = -\frac{\omega_n}{2Q} \pm j\frac{\omega_n}{2Q}\sqrt{4Q^2 - 1} \tag{2}$$

These relations are further illustrated in Fig. 4.2-1.

Let us now compare (1) with the relations given in (10) of Sec. 4.1. Looking first at the numerators, from (10a) of Sec. 4.1, we see that the coefficients a_1 and a_2 of $N_{31}(s)$ must be zero. Thus, the quantities Y_1 and Y_3 must be constants; i.e., they must represent the admittances of resistors. As such they may be written as

$$Y_1 = G_1 \qquad Y_3 = G_3 \tag{3}$$

Now consider the denominators of (1). The denominator of the right member may be regarded as specifying the decomposition of the denominator polynomial $s^2 + (\omega_n/Q)s + \omega_n^2$ of the center member. Since only second-order functions are being considered, and since $N_{33}(s)$ and $N_{32}(s)$ are respectively the numerators of an *RC* driving-point admittance function and an *RC* ladder network transfer admittance, their zeros are restricted to the negative-real axis. Thus, the *RC*-

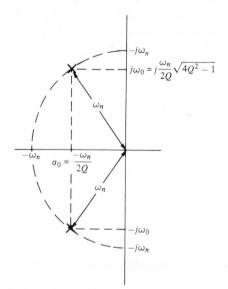

Figure 4.2-1 Relations between parameters defining complex poles.

amplifier circuit of Fig. 4.1-4 produces complex-conjugate poles using the difference decomposition of (12) of Sec. 4.1. As a numerical example, in (1), let ω_n be normalized to unity, $N_{33}(s) = s^2 + 3s + 1 = (s + 0.382)(s + 2.618)$, and $N_{32}(s) = s$. Equating the denominators of (1) we see that

$$s^2 + (1/Q)s + 1 = (s + 0.382)(s + 2.618) - Ks \tag{4}$$

The locus of the roots of this expression [the poles of the network function defined in (1)] are plotted as a function of the VCVS gain K in Fig. 4.2-2. In the figure we see that as the gain K is increased, the poles of (1) start from the s-plane locations $s = -0.382, -2.618$ ($K = 0$). They then move toward each other, becoming coincident at $s = -1$ ($K = 1$). Next, they break apart and become complex-conjugate, finally crossing the $j\omega$ axis at $s = \pm j1$ ($K = 3$). Further increases in the gain produce an unstable network.

The remaining elements in the low-pass RC-amplifier filter are now determined by noting that in (10) of Sec. 4.1 we require $c_i \neq 0$ ($i = 0, 1, 2$) in order that $N_{33}(s)$ be of second order with negative-real zeros. In addition, for $N_{32}(s)$ to have a simple zero at the origin, we require $b_0 = b_2 = 0$ and $b_1 \neq 0$. Using these results we obtain

$$N_{32} = Y_6(G_1 + Y_2 + G_3 + Y_5) + Y_2 G_3 = b_1 s \tag{5a}$$

$$N_{33}(s) = (G_3 + Y_4 + Y_6)(G_1 + Y_2 + Y_5) + G_3(Y_4 + Y_6)$$
$$= c_2 s^2 + c_1 s + c_0 \tag{5b}$$

For $b_0 = b_2 = 0$, we see from (5a) that

$$Y_2 = sC_2 \qquad Y_6 = 0 \qquad \text{thus} \qquad N_{32}(s) = sC_2 G_3 \tag{6}$$

This obviously realizes the required zero at the origin of $N_{32}(s)$. Using this result, (5b) now becomes

$$N_{33}(s) = (G_3 + Y_4)(G_1 + sC_2 + Y_5) + G_3 Y_4 \; = \; c_2 s^2 + c_1 s + c_0 \tag{7}$$

For $c_2 \neq 0$, we require $Y_4 = sC_4$. Substituting this in (7) we obtain

$$N_{33}(s) = (G_3 + sC_4)(G_1 + sC_2 + Y_5) + sG_3 C_4 \tag{8}$$

$$G_3 Y_4 = c_1 \, 1$$

$$Y_4 = \frac{c_1}{G_3} \, 1$$

$$= c_4 \, 1$$

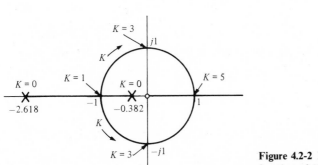

Figure 4.2-2 Locus of the roots of (4) as a function of K.

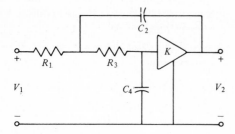

Figure 4.2-3 A low-pass Sallen and Key filter.

We now note that setting $Y_5 = 0$, and thus obtaining a desirable simplification of the network, still allows us to satisfy the requirement $c_i \neq 0$. Doing this we obtain

$$N_{33}(s) = (G_3 + sC_4)(G_1 + sC_2) + sG_3C_4$$

$$= s^2C_2C_4 + s(G_3C_2 + G_1C_4 + G_3C_4) + G_1G_3 \qquad (9)$$

The resulting realization for the low-pass RC-amplifier filter is shown in Fig. 4.2-3. The transfer function for this is readily shown to be

$$\frac{V_2(s)}{V_1(s)} = \frac{G_1G_3K}{s^2C_2C_4 + s(G_3C_2 + G_1C_4 + G_3C_4 - KG_3C_2) + G_1G_3} \qquad (10)$$

A more conventional form of (10) is achieved by dividing the numerator and denominator by C_2C_4 and using $R_i = 1/G_i$. The result is

$$\frac{V_2(s)}{V_1(s)} = \frac{K/R_1R_3C_2C_4}{s^2 + s(1/R_3C_4 + 1/R_1C_2 + 1/R_3C_2 - K/R_3C_4) + 1/R_1R_3C_2C_4} \qquad (11)$$

Comparing (1) and (11) gives the following set of equations:

$$\omega_n = \frac{1}{\sqrt{R_1R_3C_2C_4}} \qquad (12a)$$

$$\frac{1}{Q} = \sqrt{\frac{R_3C_4}{R_1C_2}} + \sqrt{\frac{R_1C_4}{R_3C_2}} + (1 - K)\sqrt{\frac{R_1C_2}{R_3C_4}} \qquad (12b)$$

$$H_0 = K \qquad (12c)$$

Since there are five unknowns, namely, R_1, C_2, R_3, C_4, and K, and only three equations, a unique solution may be obtained by specifying

$$R_1 = R_3 = R \qquad (13a)$$

$$C_2 = C_4 = C \qquad (13b)$$

Substituting (13) into (12) yields

$$\omega_n = \frac{1}{RC} \qquad \frac{1}{Q} = 3 - K \qquad H_0 = K \qquad (14)$$

A set of design equations is now easily derived as (design 1)

$$RC = \frac{1}{\omega_n} \qquad K = 3 - \frac{1}{Q} \tag{15}$$

In this case H_0 is no longer a free parameter but is constrained to the value K. The realization shown in Fig. 4.2-3 is often called the *low-pass Sallen and Key active RC filter*.[1]

> **Example 4.2-1** *An Equal-Resistance Equal-Capacitance Low-Pass Filter (design 1)* A low-pass filter is desired in which $\omega_n = 6283$ rad/s (1 kHz) and $Q = 0.7071$ (a Butterworth characteristic). From (15) we obtain
>
> $$RC = \frac{1}{6283} = 1.59 \times 10^{-4} \qquad K = 1.586 \tag{16}$$
>
> If we select $C = 0.1 \ \mu F$, then $R = 1.59$ kΩ. The resulting realization has $H_0 = 1.586$. ☐

A different design procedure (design 2) from the one using (13) occurs if we set $K = 1$. From (12) we obtain

$$\frac{1}{Q} = \sqrt{\frac{R_3 C_4}{R_1 C_2}} + \sqrt{\frac{R_1 C_4}{R_3 C_2}} \tag{17}$$

We now define two parameters n and m as

$$n = \frac{R_3}{R_1} \qquad m = \frac{C_4}{C_2} \tag{18}$$

It should be noted that n and m are the ratios of the resistor and capacitor values respectively. Furthermore, let us define

$$R_1 = R \qquad C_2 = C \tag{19}$$

Equation (11) can now be written as

$$\frac{V_2(s)}{V_1(s)} = \frac{1/mnR^2C^2}{s^2 + (1/RC)[(n + 1)/n]s + 1/mnR^2C^2} \tag{20}$$

Equating (1) and (20) results in the following design formulas

$$\omega_n = \frac{1}{\sqrt{mn} \ RC} \qquad \frac{1}{Q} = (n + 1)\sqrt{\frac{m}{n}} \tag{21}$$

From the expression for $1/Q$, it can be shown that for any given value of m the maximum value of Q will occur when $n = 1$. This case is not optimum, however,

---

[1] R. P. Sallen and E. L. Key, "A Practical Method of Designing RC Active Filters," *IRE Trans. Circuit Theory*, vol. CT-2, March 1955, pp. 74–85. Their paper is one of the most referenced in the field of active filters.

for if $n = 1$, Q is equal to $\frac{1}{2}\sqrt{m}$. For most values of Q, this will produce excessively high capacitor ratios. A more practical approach is to select a value of m compatible with standard capacitor values such that

$$m \leq \frac{1}{4Q^2} \tag{22}$$

n can then be calculated from

$$n = \left(\frac{1}{2mQ^2} - 1\right) \pm \frac{1}{2mQ^2}\sqrt{1 - 4mQ^2} \tag{23}$$

This equation provides two values of n for any given Q and m. The values are readily shown to be reciprocal; thus, the use of either one produces the same element spread.

Example 4.2-2 *A Unity-Gain Low-Pass Filter* (*design 2*)   It is desired to use a unity-gain VCVS to realize the voltage transfer function

$$\frac{V_2(s)}{V_1(s)} = \frac{0.988}{s^2 + 0.179s + 0.988} \tag{24}$$

where the complex-frequency variable has been normalized (see Sec. 1.4) by a radian frequency of 10^{-4} rad/s. From (1) we obtain $\omega_{n(\text{normalized})} = 0.994$ rad/s and $Q = 5.553$. From (22) we must select $m \leq 0.0081$. Choosing $m = 0.001$, from (23) we find $n = 0.0329$, 30.397. If we select $n = 30.397$, then from (21) with $\omega_n = 10^4\omega_{nn}$ we find $RC = 5.7703 \times 10^{-4}$. If we select $C_2 = C = 0.1\ \mu\text{F}$, then $C_4 = 100\ \text{pF}$, $R_1 = R = 5.77\ \text{k}\Omega$, and $R_3 = 175.4\ \text{k}\Omega$.   □

Another practical choice for the low-pass single-amplifier positive-gain active filter is the one in which both capacitors are equal-valued and the gain of the VCVS is set to 2.0. The equal-valued capacitors are convenient because they can be impedance normalized to a "stock" value. The gain of 2.0 is attractive because it can be precisely obtained by using equal-valued feedback resistors around an operational amplifier. From (1) and (11) for $C_2 = C_4 = C$ and $K = 2$ we readily derive (design 3)

$$R_1 = \frac{Q}{\omega_n C}$$

$$R_3 = \frac{1}{R_1 \omega_n^2 C^2} \quad \text{or} \quad R_3 = \frac{R_1}{Q^2} \tag{25}$$

Example 4.2-3 *An   Equal-Capacitance   Gain-of-2   Low-Pass   Filter* (*design 3*)   To realize the normalized (Butterworth) voltage transfer function

$$\frac{V_2(s)}{V_1(s)} = \frac{2}{s^2 + \sqrt{2}\,s + 1} \tag{26}$$

where a normalization of 10^4 rad/s has been made, from (25) using $C = 1$, $\omega_n = 1$, and $Q = 1/\sqrt{2}$, we find $R_1 = 1/\sqrt{2}$ and $R_3 = \sqrt{2}$. Frequency denormalizing, and applying an additional impedance denormalization of 10^3, we obtain the design values $R_1 = 0.707$ kΩ, $R_3 = 1.414$ kΩ, $C_2 = C_4 = 0.1$ μF, and $K = 2$. $\square$

Since as pointed out in connection with (12) there are only three network function specifications, namely, ω_n, Q, and H_0, while there are five network parameters, there are obviously many other design procedures than those given above which may be specified for the Sallen and Key low-pass filter.[1] The three described in the preceding paragraphs are typical of the majority of these. Some of the other procedures may be found in the Problems.

Now let us consider the sensitivity aspects of the Sallen and Key filter. For added generality we will use the circuit shown in Fig. 4.2-4, in which the gain-determining resistors used in connection with an operational amplifier are shown explicitly as R_A and R_B and where the VCVS gain $K = 1 + R_B/R_A$. Using the sensitivity definitions of Sec. 3.5 and the relations of (12) and (1), we obtain

$$S_{R_1}^Q = -\frac{1}{2} + Q\sqrt{\frac{R_3 C_4}{R_1 C_2}} = -S_{R_3}^Q \tag{27a}$$

$$S_{C_2}^Q = -\frac{1}{2} + Q\left(\sqrt{\frac{R_1 C_4}{R_3 C_2}} + \sqrt{\frac{R_3 C_4}{R_1 C_2}}\right) = -S_{C_4}^Q \tag{27b}$$

$$S_K^Q = QK\sqrt{\frac{R_1 C_2}{R_3 C_4}} \tag{27c}$$

$$S_{R_A}^Q = -Q(K-1)\sqrt{\frac{R_1 C_2}{R_3 C_4}} = -S_{R_B}^Q \tag{27d}$$

$$S_{R_1, R_3, C_2, C_4}^{\omega_n} = -\tfrac{1}{2} \tag{27e}$$

$$S_{K, R_A, R_B}^{\omega_n} = 0 \tag{27f}$$

[1] J. V. Wait, L. P. Huelsman, and G. A. Korn, *Introduction to Operational Amplifier Theory and Applications*, McGraw-Hill Book Company, New York, 1975, chap. 4.

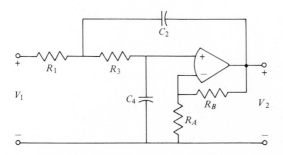

Figure 4.2-4 A low-pass Sallen and Key filter using an operational amplifier.

Table 4.2-1 Sensitivities for three designs of Sallen and Key low-pass *RC*-amplifier filters

Sensitivity	Design 1	Design 2	Design 3
$S_{R_1}^Q = -S_{R_3}^Q$	$-\frac{1}{2} + Q$	$-\frac{1}{2} + Q\sqrt{mn}$	$\frac{1}{2}$
$S_{C_2}^Q = -S_{C_4}^Q$	$-\frac{1}{2} + 2Q$	$-\frac{1}{2} + Q\left(\sqrt{\dfrac{m}{n}} + \sqrt{mn}\right)$	$\frac{1}{2} + Q^2$
S_K^Q	$3Q - 1$	$\dfrac{Q}{\sqrt{mn}}$	$2Q^2$
$S_{R_A}^Q = -S_{R_B}^Q$	$1 - 2Q$	0	$-Q^2$
$S_{R_1, R_3, C_2, C_4}^{\omega_n}$	$-\frac{1}{2}$	$-\frac{1}{2}$	$-\frac{1}{2}$
$S_{K, R_A, R_B}^{\omega_n}$	0	0	0

The values of these sensitivities for designs 1, 2, and 3 are given in Table 4.2-1. This table emphasizes the fact that the ω_n sensitivities are identical for all three designs, but that the Q sensitivities vary widely.

It is frequently possible to reduce the sensitivities cataloged in Table 4.2-1, although a price is usually exacted for such a reduction. Frequently this price is an increase in the spread of element values. For example, choosing ratios of 1 : 10 for the resistors R_1 and R_3 and 10 : 1 for the capacitors C_2 and C_4, rather than the 1 : 1 ratios specified in design 1, considerably lowers many of the Q sensitivities. A similar result is obtained by choosing R_1 and R_3 equal in design 2. Finally the value of the gain K may be chosen in such a way as to obtain the best compromise between the active and passive sensitivities. Examples of many of these situations may be found in the Problems.

4.3 BANDPASS AND HIGH-PASS SINGLE-AMPLIFIER POSITIVE-GAIN FILTERS

In the preceding section we used the general second-order *RC*-amplifier configuration described at the beginning of this chapter to develop a low-pass filter circuit. In this section we will use similar procedures to develop bandpass and high-pass circuits. We first consider the bandpass case. The general second-order bandpass voltage transfer function has the form

$$\frac{V_2(s)}{V_1(s)} = \frac{H_0(\omega_n/Q)s}{s^2 + (\omega_n/Q)s + \omega_n^2} = \frac{KN_{31}(s)}{N_{33}(s) - KN_{32}(s)} \tag{1}$$

In this equation H_0 is the maximum magnitude of the network function in the passband. It is also referred to as the *gain at the resonant frequency*. The quantities ω_n and Q have the same significance as they did in the low-pass case, namely, ω_n is the *undamped natural frequency* and Q is the *quality factor*. These quantities are related to the poles of the network function as described in (2) of Sec. 4.2 and in Fig. 4.2.-1. The right member of (1) is taken from (11) of Sec. 4.1.

Comparing the numerator of (1) with (10a) of Sec. 4.1 we see that the coefficients a_0 and a_2 of $N_{31}(s)$ must be zero. Thus, either Y_1 or Y_3 must be a capacitor and the other a resistor. Let us first choose

$$Y_1 = G_1 \qquad Y_3 = sC_3 \tag{2}$$

Thus $N_{31}(s) = sG_1 C_3$. Following the same procedure used in connection with (5) of Sec. 4.2, we obtain

$$N_{32}(s) = Y_6(G_1 + Y_2 + sC_3 + Y_5) + sC_3 Y_2 \tag{3a}$$

$$N_{33}(s) = (sC_3 + Y_4 + Y_6)(G_1 + Y_2 + Y_5) + sC_3(Y_4 + Y_6) \tag{3b}$$

The denominator decomposition must be a different one having the same general form as used for the low-pass filter. To obtain this we choose

$$Y_2 = G_2 \qquad Y_6 = 0 \qquad \text{thus} \qquad N_{32}(s) = sC_3 G_2 \tag{4}$$

For these choices we obtain

$$N_{33}(s) = (sC_3 + Y_4)(G_1 + G_2 + Y_5) + sC_3 Y_4 \tag{5}$$

Finally, to make $N_{33}(s)$ a second-order polynomial, as required by the decomposition, we choose $Y_4 = G_4$ and $Y_5 = sC_5$. Thus we obtain

$$N_{33}(s) = s^2 C_3 C_5 + s(G_1 C_3 + G_2 C_3 + G_4 C_3 + G_4 C_5) + G_4(G_1 + G_2) \tag{6}$$

The overall voltage transfer function now becomes

$$\frac{V_2(s)}{V_1(s)} = \frac{sKG_1 C_3}{s^2 C_3 C_5 + s(G_1 C_3 + G_2 C_3 + G_4 C_3 + G_4 C_5 - KG_2 C_3) + G_4(G_1 + G_2)} \tag{7}$$

This may also be put in the form

$$\frac{V_2(s)}{V_1(s)} =$$

$$\frac{sK/R_1 C_5}{s^2 + s(1/R_1 C_5 + 1/R_2 C_5 + 1/R_4 C_5 + 1/R_4 C_3 - K/R_2 C_5) + (1/R_4 C_3 C_5)(1/R_1 + 1/R_2)} \tag{8}$$

The final circuit configuration is shown in Fig. 4.3-1. It is usually referred to as a *Sallen and Key bandpass filter circuit.* Solving the above equation for the parameters of the network function given in (1) we obtain

$$\omega_n = \left(\frac{1 + R_1/R_2}{R_1 R_4 C_3 C_5}\right)^{1/2} \tag{9a}$$

$$\frac{1}{Q} = \frac{[1 + (R_1/R_2)(1 - K)]\sqrt{R_4 C_3/R_1 C_5} + \sqrt{R_1 C_3/R_4 C_5} + \sqrt{R_1 C_5/R_4 C_3}}{\sqrt{1 + R_1/R_2}} \tag{9b}$$

$$H_0 = \frac{K/R_1 C_5}{1/R_1 C_5 + 1/R_2 C_5 + 1/R_4 C_5 + 1/R_4 C_3 - K/R_2 C_5} \tag{9c}$$

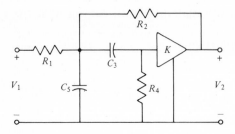

Figure 4.3-1 A bandpass Sallen and Key filter.

One design procedure for obtaining a unique solution for the values of the network elements is to choose an equal-valued-resistor equal-valued-capacitor one. Thus (design 1)

$$R_1 = R_2 = R_4 = R \qquad C_3 = C_5 = C \tag{10}$$

In this case, the relations of (9) become

$$\omega_n = \frac{\sqrt{2}}{RC} \qquad Q = \frac{\sqrt{2}}{4 - K} \qquad H_0 = \frac{K}{4 - K} \tag{11}$$

Solving for RC and K we obtain

$$RC = \frac{\sqrt{2}}{\omega_n} \qquad K = 4 - \frac{\sqrt{2}}{Q} \tag{12}$$

Note that for K to be positive, Q must be greater than $\sqrt{2}/4$.

Example 4.3-1 *An Equal-Resistance Equal-Capacitance Bandpass Filter (design 1)* It is desired to realize a bandpass filter in which $Q = 10$ and $\omega_n = 10^4$ rad/s. From (12) we find that $RC = \sqrt{2} \times 10^{-4}$s and $K = 3.8586$. Choosing C as 0.1 μF, we obtain $R = 1.414$ kΩ. The resulting value of H_0 is 27.289. □

There are many possible choices for the element values of the filter shown in Fig. 4.3-1, other than those given by (12). Another practical design results if the VCVS gain is constrained to a value of 2, which, as has been pointed out, is a value readily obtained with great stability by using two equal-valued precision resistors as feedback around an operational amplifier. In this case we may select (design 2)

$$R_1 = C_3 = C_5 = 1 \qquad K = 2 \tag{13}$$

For these choices, equating (1) and (8) we obtain

$$\omega_n^2 = \left(1 + \frac{1}{R_2}\right)\frac{1}{R_4} \tag{14a}$$

$$\frac{\omega_n}{Q} = 1 + \frac{2}{R_4} - \frac{1}{R_2} \tag{14b}$$

This set of simultaneous nonlinear equations is readily solved to satisfy specific values of ω_n and Q.

Example 4.3-2 *An Equal-Capacitance Gain-of-2 Bandpass Filter* (*design 2*)   It is desired to design an impedance and frequency normalized bandpass filter in which $\omega_n = 1$ and $Q = 2$. The capacitors are to be of unity value, and the VCVS is to have a gain of 2. Using (14), and for convenience letting $G_i = 1/R_i$, we obtain

$$1 = (1 + G_2)G_4 \tag{15a}$$

$$0.5 = 1 + 2G_4 - G_2 \tag{15b}$$

Solving the first equation for G_4, substituting the result in the second equation, and rearranging the terms we obtain $G_2^2 + 0.5G_2 - 2.5 = 0$. Solving this equation for G_2 and substituting the result in (15a) we find $R_2 = 0.7403\ \Omega$ and $R_4 = 2.3508\ \Omega$.   □

A different bandpass filter circuit from the one described above is obtained by letting $Y_1 = sC_1$ and $Y_3 = G_3$ rather than using the choices of (2). Following through the remaining steps of the realization procedure yields a circuit configuration with three capacitors. As such it is of less practical interest than the one described above. Further details concerning this circuit may be found in the problems.

The second type of filter to be considered in this section is the high-pass one. The general second-order high-pass voltage transfer function has the form

$$\frac{V_2(s)}{V_1(s)} = \frac{H_0 s^2}{s^2 + (\omega_n/Q)s + \omega_n^2} = \frac{KN_{31}(s)}{N_{33}(s) - KN_{32}(s)} \tag{16}$$

In this equation, H_0 is the gain at infinite frequency, and ω_n, Q, and the right member are as defined for (1). The filter circuit which realizes this network function could be developed in a manner similar to that used for the low-pass and bandpass functions. Instead, we here select a simpler approach. We first note that this type of network function is readily related to the low-pass one by using the low-pass to high-pass transformation introduced in Sec. 2.4; namely, we let $s = 1/p$, where s is the original low-pass complex frequency variable and p is the new high-pass complex frequency variable. We may now directly use low-pass filter realizations to derive high-pass filter realizations by applying this frequency transformation to the elements. In addition, we then apply an impedance normalization of $1/p$. This latter, of course, leaves the voltage transfer function invariant since it is dimensionless. For any (low-pass) resistor of value R, the procedure is as follows:

$$\boxed{Z_{\text{LP}}(s) = R} \longrightarrow \boxed{Z_{\text{HP}}(p) = R} \longrightarrow \boxed{Z_{z\ \text{transformed}}(p) = R/p} \tag{17}$$

Low-pass to   Impedance
high-pass   normalization of $1/p$
transformation

Table 4.3-1 A transformation of network elements under the low-pass to high-pass transformation with an additional impedance transformation of $1/s$.

Low-pass	High-pass
R (resistor)	$C = 1/R$ (capacitor)
C (capacitor)	$R = 1/C$ (resistor)
V_0 , KV_0 (VCVS)	V_0 , KV_0 (VCVS)

The result of the two transformations is obviously a capacitor of value $1/R$ in the high-pass realization. Similarly, for a (low-pass) capacitor of value C we obtain

$$\boxed{Y_{\text{LP}}(s) = sC} \longrightarrow \boxed{Y_{\text{HP}}(p) = C/p} \longrightarrow \boxed{Y_{z\,\text{transformed}}(p) = C} \qquad (18)$$

$$\underset{\substack{\text{Low-pass to}\\ \text{high-pass}\\ \text{transformation}}}{} \qquad \underset{\substack{\text{Impedance}\\ \text{normalization of } 1/p}}{}$$

Thus, the result of the two transformations in this case is a resistor of value $1/C\ \Omega$. Since the gain of the VCVS is dimensionless and not a function of frequency, the VCVS remains invariant under the two transformations. These results are summarized in Table 4.3-1. Applying this table to the network of Fig. 4.2-3 gives the positive-gain high-pass realization shown in Fig. 4.3-2. Thus the choices for the passive elements of Fig. 4.1-4 are seen to be $Y_1 = sC_1$, $Y_2 = G_2$, $Y_3 = sC_3$, $Y_4 = G_4$, and $Y_5 = Y_6 = 0$. Substitution of these values into $N_{31}(s)$, $N_{32}(s)$, and $N_{33}(s)$ as defined in (10) of Sec. 4.1 permits the transfer function of (16) to be found as

$$\frac{V_2(s)}{V_1(s)} = \frac{s^2 C_1 C_3 K}{s^2 C_1 C_3 + s(G_2 C_3 + G_4 C_1 + G_4 C_3 - K G_2 C_3) + G_2 G_4} \qquad (19)$$

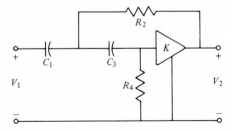

Figure 4.3-2 A high-pass Sallen and Key filter.

Dividing the numerator and denominator of this by $C_1 C_3$, we obtain

$$\frac{V_2(s)}{V_1(s)} = \frac{s^2 K}{s^2 + s(1/R_2 C_1 + 1/R_4 C_3 + 1/R_4 C_1 - K/R_2 C_1) + 1/R_2 R_4 C_1 C_3} \quad (20)$$

Comparing this with (16) results in

$$\omega_n = \frac{1}{\sqrt{R_2 R_4 C_1 C_3}} \quad (21a)$$

$$\frac{1}{Q} = \sqrt{\frac{R_4 C_3}{R_2 C_1}} + \sqrt{\frac{R_2 C_1}{R_4 C_3}} + \sqrt{\frac{R_2 C_3}{R_4 C_1}} - K\sqrt{\frac{R_4 C_3}{R_2 C_1}} \quad (21b)$$

$$H_0 = K \quad (21c)$$

Following the design 1 procedure for the low-pass network in Sec. 4.2 by choosing $R_2 = R_4 = R$ and $C_1 = C_3 = C$ gives (design 1)

$$\omega_n = \frac{1}{RC} \quad (22a)$$

$$\frac{1}{Q} = 3 - K \quad (22b)$$

$$H_0 = K \quad (22c)$$

which are identical with (14) of Sec. 4.2. Thus a high-pass realization with $Q = 0.7071$ and $\omega_n = 6280$ rad/s can be obtained by using the R, C, and K values of Example 4.2-1 as $1/C$, $1/R$, and K respectively in the circuit of Fig. 4.3-2.

As with the positive gain low-pass realization of Sec. 4.2, a second high-pass filter design may be obtained by setting $K = 1$. Assuming that $m = C_3/C_1$ and $n = R_4/R_2$, and letting $C_1 = C$ and $R_2 = R$, (20) becomes

$$\frac{V_2}{V_1} = \frac{s^2}{s^2 + \frac{s}{RC}\frac{1}{n}\frac{m+1}{m} + \frac{1}{mn(RC)^2}} \quad (23)$$

Comparing this with (16) we get the design formulas (design 2)

$$\omega_n = \frac{1}{\sqrt{mn}\,RC} \qquad \frac{1}{Q} = \frac{m+1}{\sqrt{mn}} \quad (24)$$

From these it can be shown that for any given value of n the minimum value of $1/Q$ will occur when $m = 1$. Since it is usually desirable to have a minimum $1/Q$ for any given n, we will let $m = 1$. In this case (24) simplifies to

$$\omega_n = \frac{1}{\sqrt{n}\,RC} \qquad Q = \frac{\sqrt{n}}{2} \quad (25)$$

Example 4.3-3 *A Unity-Gain High-Pass Filter* (*design 2*)   It is desired to use a unity-gain VCVS to realize the voltage transfer function

$$\frac{V_2(s)}{V_1(s)} = \frac{s^2}{s^2 + 0.179s + 0.988} \tag{26}$$

where the complex frequency variable has been normalized by a value of 10^{-4} rad/s. Q and $\omega_{n\,(\text{normalized})}$ have the same values found in Example 4.2-2, namely, 5.553 and 0.994 rad/s. From (25) we find $n = 4Q^2 = 123.34$, and $RC = 1/\omega_n\sqrt{n} = 0.09058$. Denormalizing, $RC = 0.09058 \times 10^{-4}$. If we select $C = 0.01\ \mu F$, then $R = R_2 = 905.8\ \Omega$ and $R_4 = 111.72\ k\Omega$.   □

Another practical set of values for the high-pass single-amplifier positive-gain *RC* amplifier filter shown in Fig. 4.3-2 is the one in which both capacitors are equal-valued and the gain of the VCVS is set to 2.0. Thus, for the normalized case we have $C_1 = C_3 = C$ and $K = 2$. From (20) for this case we find that (design 3)

$$\omega_n = \frac{1}{C\sqrt{R_2 R_4}} \qquad \frac{\omega_n}{Q} = \frac{1}{C}\left(\frac{2}{R_4} - \frac{1}{R_2}\right) \tag{27}$$

These equations can be solved for values of R_2 and R_4 in a manner similar to that used in Example 4.3-2.

Now let us consider some of the sensitivity aspects of the bandpass and high-pass filter circuit realizations given in this section. A comparison of the denominator of the low-pass voltage transfer function given in (11) of Sec. 4.2 with the denominators of the bandpass and high-pass functions given in (8) and (20) of this section readily shows that the type of polynomial decomposition used to obtain the complex-conjugate poles is the same in all three cases. Thus, the root loci for all three filters have the general form shown in Fig. 4.2-2. As such, sensitivity analyses of the bandpass and high-pass designs will yield results similar to those given in Table 4.2-1 for the low-pass types. For example, the ω_n sensitivities to K are zero for all three types of filters. Further details of such analyses are covered in the Problems.

In this section and the preceding one, we have presented a discussion of the positive-gain Sallen and Key structures for low-pass, bandpass, and high-pass network functions. The advantages of these structures in general are that they are characterized by easy-to-use design relations; the designer has good control over the element values and their spread; and it is possible to use low values of VCVS gain, which is convenient since such values are readily stabilized. There are also some disadvantages, of which the major one is the high sensitivities that result when these circuits are used to realize high-Q functions. This is readily apparent not only from sensitivity analyses such as those given in Table 4.2-1, but also from the shape of their root loci, as shown in Fig. 4.2-2. Obviously, in this figure, when the poles are "high-Q," i.e., located close to the $j\omega$ axis, a very small increase in the gain will move them into the right half of the complex-frequency plane and thus make the circuit unstable. One solution to this sensitivity problem that has been proposed is the use of negative-gain filter structures, i.e., ones using inverting VCVSs. Such filters have root loci that do not cross into the right-half plane. Not only are such filters always stable, but their sensitivities are usually lower than those of the positive-gain ones. Unfortunately, the negative-gain filters also have several disadvantages: Their element value spreads are frequently large; their design methods are often more complex; and they usually require large values of VCVS gain, which are more difficult to stabilize than low ones. In general, these negative points outweigh the positive ones. As a result, negative-gain filters are not widely used. Some examples of them may be found in the Problems.

4.4 FILTERS REALIZING COMPLEX-CONJUGATE ZEROS

In this section we consider active filter realizations for completely general second-order transfer functions. Such functions include the low-pass, bandpass, and highpass ones as special cases. They are usually referred to as *biquadratic filter functions*. The general form for the second-order biquadratic voltage transfer function is

$$\frac{V_2(s)}{V_1(s)} = H\frac{s^2 + b_1 s + b_0}{s^2 + a_1 s + a_0} = H\frac{s^2 + (\omega_z/Q_z)s + \omega_z^2}{s^2 + (\omega_p/Q_p)s + \omega_p^2} \tag{1}$$

where H is a constant, ω_z and ω_p are the zero and pole undamped frequencies, and Q_z and Q_p are the Q's of the complex zeros and poles.[1] It will be assumed that the zeros may be real or complex and that they may be anywhere in the complex-frequency plane, including the right-half portion of it. Equation (1) includes the elliptic function defined in Sec. 2.3. In this case it has the form

$$\frac{V_2(s)}{V_1(s)} = H\frac{s^2 + b_0}{s^2 + a_1 s + a_0} = H\frac{s^2 + \omega_z^2}{s^2 + (\omega_p/Q_p)s + \omega_p^2} \tag{2}$$

where the zeros are located on the $j\omega$ axis.

The first type of biquadratic filter we shall consider is a single-amplifier finite-gain realization which realizes (2).[2,3] The circuit is shown in Fig. 4.4-1. It has the general form shown in Fig. 4.1-3, with a network function given as (5) of Sec. 4.1. From that function we see that the zeros of $y_{31}(s)$ determine the zeros of the transfer function. Since $y_{31}(s)$ is the transfer admittance of an RC network, its zeros may be on the $j\omega$ axis, as is needed here. The network required to realize this function is a modification of the twin-T circuit.[4] The modification consists of

[1] Formally, Q is defined only for complex poles; however, it is convenient here to extend its usage to zeros as well.

[2] W. J. Kerwin and L. P. Huelsman, "The Design of High-Performance Active RC Bandpass Filters," *Proc. IEEE Intern. Conv. Rec.*, part 10, March 1966, pp. 74–80.

[3] Kerwin, "An Active RC Elliptic Function Filter," *IEEE Region 6 Conf. Rec.*, vol. 2, April 1966, pp. 640–641.

[4] L. G. Cowles, "The Parallel-T Resistance-Capacitance Network," *Proc. IRE*, vol. 40, December 1952, pp. 1712–1717.

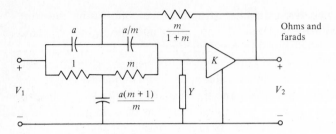

Figure 4.4-1 A finite-gain realization for biquadratic network functions.

adding a single resistor or capacitor as the element $Y(s)$ in Fig. 4.4-1. If $Y = 1/R$, the circuit realizes (2) with the restriction $a_0 > b_0$, that is, that the zeros be closer than the poles to the origin of the complex-frequency plane. The transfer function for this restriction is given as (case 1)

$$\frac{V_2(s)}{V_1(s)} = \frac{K(s^2 + 1/a^2)}{s^2 + [(m+1)/a][1/R + (2-K)/m]s + [1 + (m+1)/R]/a^2} \tag{3}$$

where a, m, and K are defined in Fig. 4.4-1, and where m may be chosen arbitrarily to control the spread of element values. Equating (2) and (3) gives the design relations

$$a = \sqrt{\frac{1}{b_0}} \tag{4a}$$

$$R = \frac{m+1}{a_0/b_0 - 1} \tag{4b}$$

$$K = 2 + \frac{m}{m+1}\left(\frac{a_0}{b_0} - 1 - \frac{a_1}{\sqrt{b_0}}\right) \tag{4c}$$

$$H = K \tag{4d}$$

If $Y = aCs$, then the circuit of Fig. 4.4-1 realizes (2) with the restriction $b_0 > a_0$, that is, that the poles be closer to the origin than the zeros are. The transfer function for this restriction is (case 2)

$$\frac{V_2(s)}{V_1(s)} = \frac{\{K/[(m+1)C + 1]\}(s^2 + 1/a^2)}{s^2 + s\{(m+1)[C + (2-K)/m]/a[(m+1)C + 1]\} + 1/a^2[(m+1)C + 1]} \tag{5}$$

Equating (2) and (5) gives the following design equations

$$a = \frac{1}{\sqrt{b_0}} \tag{6a}$$

$$C = \frac{b_0/a_0 - 1}{m+1} \tag{6b}$$

$$K = 2 + \frac{m}{m+1}\left(\frac{b_0}{a_0} - 1 - \frac{a_1\sqrt{b_0}}{a_0}\right) \tag{6c}$$

$$H = \frac{a_0}{b_0}K \tag{6d}$$

The factor m can be chosen arbitrarily as before.

Example 4.4-1 *A Low-Pass Elliptic Filter*   It is desired to realize the elliptic voltage transfer function

$$\frac{V_2(s)}{V_1(s)} = \frac{H(s^2 + 7.464102)}{s^2 + 0.998942s + 1.170077} \tag{7}$$

In Table 2.3-1 this function is shown to have a 1-dB ripple in the passband, and at least 17 dB of attenuation for all frequencies greater than 2 rad/s. Obviously, case 2 applies; and from (6), using $m = 0.2$, we find $a = 0.36603$, $C = 4.4826$, $K = 2.5078$, and $H = 0.39312$. The circuit is shown in Fig. 4.4-2. Since, from Example 2.4-2 a value of H of 0.139713 corresponds with a maximum passband gain of unity, the maximum passband gain of this realization is $0.39312/0.139713 = 2.8138$. □

The second type of biquadratic filter to be considered in this section is the two-amplifier finite-gain type. One of the reasons for considering such a structure is the complexity of the design equations for the preceding (single-amplifier filter) when the transmission zeros are located off the $j\omega$ axis. The two-amplifier finite-gain structure discussed here is shown in Fig. 4.4-3.[1] This realization uses an inverting amplifier with unity gain and a noninverting amplifier with a gain of

[1] L. S. Bobrow and S. L. Hakimi, "A Note on Active-RC Realization of Voltage Transfer Functions," *IEEE Trans. Circuit Theory*, vol. CT-11, no. 4, December 1964, pp. 493–494. See also R. J. A. Paul, "Active Network Synthesis Using One-Port RC Networks," *Proc. IEEE*, vol. 113, no. 1, January 1966, pp. 83–86; and S. K. Mitra, U.S. Patent 3,401,352, September 1967.

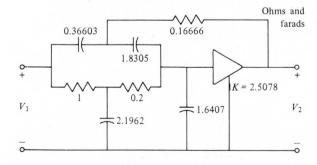

Figure 4.4-2 The low-pass elliptic filter realization of Example 4.4-1.

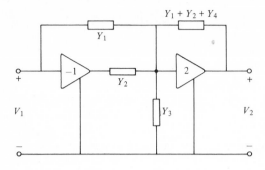

$$Y_1 + Y_2 + Y_4$$

Figure 4.4-3 A two-amplifier finite-gain realization for biquadratic network functions.

two. Both of these are readily implemented using operational amplifiers. Its transfer function is

$$\frac{V_2(s)}{V_1(s)} = \frac{2(Y_1 - Y_2)}{Y_3 - Y_4} \tag{8}$$

The values of the admittances can be found by dividing both the numerator and denominator of (1) by a factor $s + c(c > 0)$, which makes possible partial fraction expansions which can be realized by passive *RC* networks. For example, for the numerator we get

$$\frac{H(s^2 + b_1 s + b_0)}{s + c} = Hs + \frac{Hb_0}{c} + \frac{k_b s}{s + c} = 2(Y_1 - Y_2) \tag{9}$$

The quantity k_b is the residue of the pole at $s = -c$ and is given as

$$k_b = \frac{H(c^2 - cb_1 + b_0)}{-c} \tag{10}$$

It may be positive or negative depending upon the relative values of $c, b_1,$ and b_0. If k_b is positive, then $Y_2 = 0$, the inverter is not present, and we have

$$Y_1(s) = \frac{Hs}{2} + \frac{Hb_0}{2c} + \frac{1}{2/k_b + 2c/k_b s} \tag{11}$$

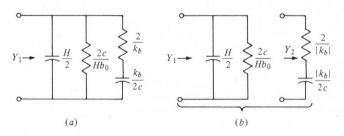

Figure 4.4-4 Realizations for $Y_1(s)$ and $Y_2(s)$ in Fig. 4.4-3.

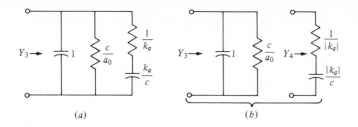

Figure 4.4-5 Realizations for $Y_3(s)$ and $Y_4(s)$ in Fig. 4.4-3.

The realization of Y_1 in this case is shown in Fig. 4.4-4a. If k_b is negative, then Y_1 and Y_2 become

$$Y_1(s) = \frac{Hs}{2} + \frac{Hb_0}{2c} \qquad Y_2(s) = \frac{1}{2/|k_b| + 2c/|k_b|s} \qquad (12)$$

Their realizations are shown in Fig. 4.4-4b. A partial-fraction expansion of the denominator of (1) divided by $s + c$ is expressed as

$$\frac{s^2 + a_1 s + a_0}{s + c} = s + \frac{a_0}{c} + \frac{k_a s}{s + c} = Y_3 - Y_4 \qquad (13)$$

The quantity k_a is the residue of the pole at $s = -c$ and is given as

$$k_a = \frac{c^2 - a_1 c + a_0}{-c} \qquad (14)$$

If k_a is positive, then $Y_4 = 0$, and Y_3 is expressed as

$$Y_3(s) = s + \frac{a_0}{c} + \frac{1}{1/k_a + c/k_a s} \qquad (15)$$

A realization is shown in Fig. 4.4-5a. If k_a is negative, then

$$Y_3(s) = s + \frac{a_0}{c} \qquad Y_4(s) = \frac{1}{1/|k_a| + c/|k_a|s} \qquad (16)$$

and they are realized as shown in Fig. 4.4-5b. Usually, since c is free to be chosen, it can be selected so as to make either k_a or k_b equal to zero.

Example 4.4-2 *An All-Pass Function* It is desired to use the configuration of Fig. 4.4-3 to realize the following all-pass transfer function:

$$\frac{V_2(s)}{V_1(s)} = \frac{s^2 - 2s + 1}{s^2 + 2s + 1} \qquad (17)$$

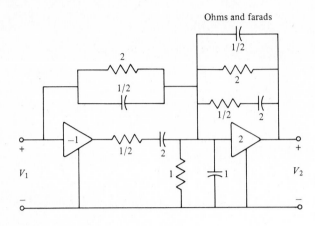

Figure 4.4-6 Realization of the all-pass filter of Example 4.4-2.

The choice $c = +1$ will simplify the network. For this choice the numerator partial-fraction expansion is

$$\frac{s^2 - 2s + 1}{s + 1} = s + 1 - \frac{4s}{s + 1} \tag{18}$$

Hence $Y_1(s) = (s + 1)/2$ and $Y_2(s) = 2s/(s + 1)$. The denominator partial-fraction expansion is

$$\frac{s^2 + 2s + 1}{s + 1} = s + 1 \tag{19}$$

Thus, $Y_3(s) = s + 1$ and $Y_4(s) = 0$. The final realization is shown in Fig. 4.4-6. □

It should be noted that the circuit given in Fig. 4.4-3 can also be used to realize higher-order network functions. In this case the numerator and denominator must be divided by an expression $(s + c_1)(s + c_2) \cdots (s + c_k)$, where k is less than the degree of the numerator or the denominator (whichever is higher).

The third type of biquadratic filter realization to be considered in this section is the single-amplifier infinite-gain one. It is shown in Fig. 4.4-7.[1] It uses a single differential-input operational amplifier. The transfer function of this filter is easily found to be

$$\frac{V_2(s)}{V_1(s)} = \frac{Y_1(Y_2 + Y_b + Y_3) - Y_2(Y_1 + Y_a + Y_4)}{Y_3(Y_1 + Y_a + Y_4) - Y_4(Y_2 + Y_b + Y_3)} \tag{20}$$

[1] J. S. Brugler, "RC Synthesis with Differential Input Operational Amplifiers," Stanford Electronics Labs Rep. 6560-4, June 1966.

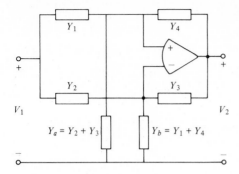

$Y_a = Y_2 + Y_3$

$Y_b = Y_1 + Y_4$

Figure 4.4-7 A single-amplifier infinite-gain realization for biquadratic network functions.

If $Y_1 + Y_a + Y_4 = Y_2 + Y_b + Y_3$ or $Y_a = Y_2 + Y_3$ and $Y_b = Y_1 + Y_4$, then (20) becomes

$$\frac{V_2(s)}{V_1(s)} = \frac{Y_1 - Y_2}{Y_3 - Y_4} \tag{21}$$

Any common factors in $Y_1 + Y_4$ and $Y_2 + Y_3$ may be dropped to simplify the synthesis. Except for the overall multiplying factor of 2, (8) and (21) are exactly alike. The mechanics of the design procedure are thus almost identical. The details are left to the reader as an exercise.

The fourth type of biquadratic filter realization to be discussed in this section uses two operational amplifiers. It is shown in Fig. 4.4-8.[1] Analysis of this circuit results in

$$\frac{V_2(s)}{V_1(s)} = \frac{Y_1 - Y_2}{Y_3 - Y_4} \tag{22}$$

which is the same as (21). Thus the synthesis procedure outlined for the filter shown in Fig. 4.4-7 also applies to this configuration.

[1] W. P. Lovering, "Analog Computer Simulation of Transfer Functions," *Proc. IEEE*, vol. 53, no. 3, March 1965, pp. 306–307.

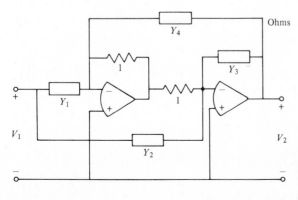

Figure 4.4-8 A two-amplifier infinite-gain realization for biquadratic network functions.

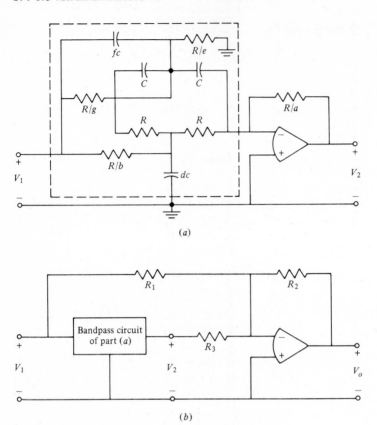

(a)

(b)

Figure 4.4-9 A single-amplifier infinite-gain realization for biquadratic network functions.

One other single-operational-amplifier biquadratic filter realization will also be considered here. As shown in Fig. 4.4-9a, it uses twin-T networks as its passive components.[1] Its voltage transfer function is

$$\frac{V_2(s)}{V_1(s)} = -\frac{fs^2 + (g/T)s + (b/T^2)}{s^2 + (ad/T)s + [1 + a(2 + b)]/T^2} \tag{23}$$

where

$$b + 2 = g + e$$

$$f + 2 = d \tag{24}$$

$$T = RC$$

[1] T. Hamilton and A. Sedra, "A Single Amplifier Biquad Active Filter," *Proc. Intern. Symp. Circuits and Systems*, April 1972, pp. 355–359.

From (1) and (23) we get

$$\omega_p = \frac{\sqrt{1 + a(2 + b)}}{T} \qquad Q_p = \frac{\sqrt{1 + a(2 + b)}}{ad}$$

$$\omega_z = \frac{\sqrt{b}}{\sqrt{f}\,T} \qquad Q_z = \frac{\sqrt{bf}}{g} \tag{25}$$

$$H_0 = f$$

If H_0, ω_p, ω_z, Q_p, and Q_z are specified, then the above equations can be solved for the parameters a, b, e, f, g, R, and C. Table 4.4-1 shows some of the standard cases of second-order filters using this method of realization. The circuit shown in Fig. 4.4-9a must be modified as shown in Fig. 4.4-9b in order to realize zeros in the right-half plane. If $R_1 = R_3$, then the transfer function for Fig. 4.4-9b becomes

$$\frac{V_o(s)}{V_1(s)} = -\frac{R_2}{R_1}\left[\frac{s^2 - (2a/T)s + (1 + 2a)/T^2}{s^2 + (2a/T)s + (1 + 2a)/T^2}\right] \tag{26}$$

In this case an all-pass realization is obtained. The design procedure is given in the last row of Table 4.4-1. Some other biquadratic filter realizations may be found in Sec. 5.2.

Table 4.4-1 Design procedure for the filter of Fig. 4.4-9

Type	Choice of parameters	Transfer function
Low-pass	$f = g = 0$ $d = 2$ $b + 2 = e$	$\dfrac{V_2}{V_1} = -\dfrac{b/T^2}{s^2 + \dfrac{2a}{T}s + \dfrac{1 + (2 + b)a}{T^2}}$
High-pass	$g = b = 0$ $e = 2$ $2 + f = d$	$\dfrac{V_2}{V_1} = -\dfrac{fs^2}{s^2 + \dfrac{ad}{T}s + \dfrac{1 + 2a}{T^2}}$
Bandpass	$f = b = 0$ $d = 2$ $g + e = 2$	$\dfrac{V_2}{V_1} = -\dfrac{(g/T)s}{s^2 + \dfrac{2a}{T}s + \dfrac{1 + 2a}{T^2}}$
Notch	$g = 0$ $e = b + 2$ $g + e = 2$	$\dfrac{V_2}{V_1} = -\dfrac{fs^2 + b/T^2}{s^2 + \dfrac{ad}{T}s + \dfrac{1 + (2 + b)a}{T^2}}$
All-pass	$f = b = 0$ $d = 2$ $g + e = 2$ $g = 4a$	$\dfrac{V_o}{V_1} = \dfrac{-R_2}{R_1}\dfrac{s^2 - \dfrac{2a}{T}s + \dfrac{1 + 2a}{T^2}}{s^2 + \dfrac{2a}{T}s + \dfrac{1 + 2a}{T^2}}$

4.5 HIGHER-ORDER FILTERS

Most of the synthesis techniques presented in the previous sections of this chapter were restricted to second-order filter realizations. As pointed out in Sec. 4.1, however, such realizations may be cascaded to achieve higher-order ones, since the use of a VCVS as the output element provides the necessary isolation. Such a cascade will have the form shown in Fig. 4.5-1. The overall voltage transfer function $T(s)$ will be given by

$$T(s) = T_1(s) \, T_2(s) \cdots T_{n/2}(s) \tag{1}$$

where n is the order of the overall filter and has been assumed to be even, and where $T_i(s)$ is the voltage transfer function of an individual section which has the form

$$T_i(s) = \frac{a_{2i}s^2 + a_{1i}s + a_{0i}}{s^2 + (\omega_{ni}/Q_i)s + \omega_{ni}^2} \tag{2}$$

There are several important variations that should be considered in applying the cascade method. These are treated briefly in this section.

A first consideration in the application of the cascade method of realizing higher-order functions occurs if the function to be realized is of odd order. In such a case, there are two possible techniques that may be applied. The first of these is to add a passive first-order circuit as one of the elements of the cascade. This will realize the negative-real pole associated with the odd-order function. Depending on whether or not a zero at the origin is also to be realized, the passive network will have the form shown in Fig. 4.5-2a, for which the voltage transfer function (with *no* zero at the origin) is

$$\frac{V_2(s)}{V_1(s)} = \frac{1/RC}{s + 1/RC} \tag{3}$$

or the form shown in Fig. 4.5-2b, which has the voltage transfer function (*with a zero at the origin*)

$$\frac{V_2(s)}{V_1(s)} = \frac{s}{s + 1/RC} \tag{4}$$

Either circuit can conveniently be added as the last element of the cascade, in which case the output VCVS of the preceding stage provides the necessary isolation. Alternately, the passive circuit can be located elsewhere in the cascade, in which case suitable isolating amplifiers must be provided.

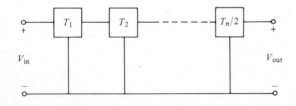

Figure 4.5-1 A cascade of second-order filter realizations.

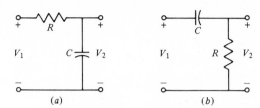

(a)   (b)   **Figure 4.5-2** First-order filter sections.

A second technique which may be used when the function to be realized is of odd order (excluding the case where the order is 1) is the use of a single third-order active *RC* circuit of the Sallen and Key type to realize both a negative-real pole and a pair of complex-conjugate ones. Such a circuit is shown in Fig. 4.5-3.[1] To make the circuit as practical as possible all the capacitors have been chosen equal to a normalized value of unity, as shown. For this constraint, the voltage transfer function of the circuit is

$$\frac{V_2(s)}{V_1(s)} = \frac{K}{R_1 R_2 R_3 s^3 + [2R_1 R_3 + R_1 R_2 + R_2 R_3 (2 - K)]s^2 + [R_1 + R_3 + (R_2 + R_3)(2 - K)]s + 1} \tag{5}$$

To determine the values R_i of the resistors and the gain value K of the noninverting VCVS, we begin by noting that for a third-order low-pass voltage transfer function having the form

$$\frac{V_2(s)}{V_1(s)} = \frac{H}{a_3 s^3 + a_2 s^2 + a_1 s + 1} \tag{6}$$

the following set of simultaneous (nonlinear) equations results:

$$a_3 = R_1 R_2 R_3 \tag{7a}$$

$$a_2 = 2R_1 R_3 + R_2 R_3 + R_1 R_2 (2 - K) \tag{7b}$$

$$a_1 = R_1 + R_3 + (R_1 + R_2)(2 - K) \tag{7c}$$

The solution of these equations is readily accomplished by numerical methods. As an example of the interactions that occur among the variables, the results of a

[1] L. P. Huelsman, "An Equal-Valued-Capacitor Active RC Network Realization of a Third-Order Low-Pass Butterworth Characteristic," *Electronics Letters*, vol. 7, no. 10, May 20, 1971, pp. 271–272.

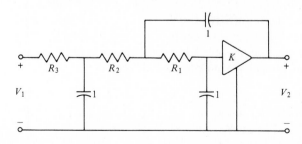

Figure 4.5-3 A third-order low-pass Sallen and Key type filter.

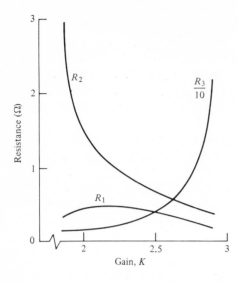

Figure 4.5-4 Design chart for using the filter of Fig. 4.5-3 to obtain a Butterworth characteristic.

series of such solutions made for various values of K for the third-order Butterworth case $(a_3 = 1, a_2 = 2, a_1 = 2)$ is given in Fig. 4.5-4. From the figure we see that solutions exist for a range of K from approximately 1.86 to 2.99, although the required resistance values have a somewhat extreme ratio near the limits of this range. Of special interest is the solution for $K = 2$, a value of gain which has been shown to be readily and precisely produced by an operational amplifier with two equal-valued feedback resistors. For this value of gain, the resulting resistor values for the third-order Butterworth case are $R_1 = 1.565$, $R_2 = 1.469$, and $R_3 = 0.435$ Ω. It is readily shown that for $K = 2$, a solution giving positive resistor values exists for any choice of the coefficients a_i of (6) for which the denominator polynomial of the network function is Hurwitz.[1] Table 4.5-1 gives design information for cascaded realizations of various orders of normalized Butterworth filters. For a given order n, each entry gives the resistor values for a second-order circuit having the form shown in Fig. 4.2-3 (if only R_1 and R_3 are specified) or a third-order circuit having the form shown in Fig. 4.5-3 (if R_1, R_2, and R_3 are specified). The table assumes the use of unity-valued capacitors, and a gain of 2.0 for the VCVSs. In Tables 4.5-2 and 4.5-3 similar information is given for the realization of Chebyshev filters.

At this point, the use of even higher-order RC amplifier stages might be considered. In general, however, the sensitivity of such stages becomes quite large as compared with those of the lower-order stages of the cascade. As an example of this, consider a fourth-order single-amplifier filter having the general form shown in Fig. 4.5-5a and the cascade of the two second-order single amplifiers having the

---

[1] S. Tirtoprodjo, "Constraint Removal for Huelsman's Equal-Valued-Capacitor Active RC Circuit," *Electronic Letters.* vol. 7, no. 16, 1971, pp. 448–449.

Table 4.5-1 Butterworth realizations using cascaded second- and third-order RC-amplifier filters

n	R_1	R_2	R_3
2	0.70711		1.41421
3	1.56520	1.46940	0.43480
4	1.30656		0.76537
	0.54120		1.84776
5	1.61803		0.61803
	2.10994	0.93280	0.50809
6	1.93185		0.51764
	0.70711		1.41421
	0.51764		1.93185
7	2.24698		0.44504
	0.80194		1.24698
	2.28449	0.84595	0.51745
8	2.56292		0.39018
	0.89998		1.11114
	0.60134		1.66294
	0.50980		1.96157
9	2.87939		0.34730
	1.00000		1.00000
	0.65270		1.53209
	2.35892	0.81451	0.52046
10	3.19623		0.31287
	1.10134		0.90798
	0.70711		1.41421
	0.56116		1.78201
	0.50623		1.97538

form shown in Fig. 4.5-5b. If both configurations are used to realize a Butterworth characteristic and all the amplifier gains are changed by 5 percent from their nominal values, the results are as shown in Fig. 4.5-6. Obviously, the cascade of second-order sections provides superior performance. As the order of the filter realized by a cascade increases, the overall sensitivities can become quite large. As an example of this, in Fig. 4.5-7 the magnitude sensitivity to the gain K of all the amplifiers in a given realization is plotted for fifth-, seventh-, and ninth-order Butterworth realizations. These sensitivity curves have been minimized by letting the third-order section in the cascade realize the complex-conjugate pole pair with the lowest Q. To show why this is important, in Fig. 4.5-8 a similar plot is given for

Table 4.5-2 Chebyshev $\frac{1}{2}$-dB ripple realizations using cascaded second- and third-order *RC*-amplifier filters

n	R_1	R_2	R_3
2	0.70145		0.94026
3	1.87657	2.77768	0.26806
4	2.85139		0.32976
	1.18108		2.37557
5	4.46576		0.21619
	3.43569	3.04181	0.55393
6	6.43914		0.15181
	2.35689		0.71912
	1.72536		3.69170
7	8.77144		0.11220
	3.13049		0.47193
	4.91761	3.88785	0.80423
8	11.46261		0.08621
	4.02513		0.33512
	2.68951		1.03671
	2.28006		4.98097
9	14.51264		0.06828
	5.04019		0.25135
	3.28975		0.67171
	6.37762	4.82524	1.04759
10	17.92153		0.05539
	6.17534		0.19612
	3.96481		0.47427
	3.14649		1.33583
	2.83849		6.25989

ninth-order filters in which the third-order filter realizes other complex-conjugate pole pairs.[1] Obviously, the high-Q pair should not be selected for realization by the third-order section in such a cascade.

As another consideration in the application of the cascade method of realizing higher-order functions, we note that the individual stages $T_i(s)$ need not necessarily all be of the same type. For example, if n is even and $T(s)$ is a higher-order

---

[1] M. Hanlon, "The Effects of Pole Pairing in Circuit Sensitivity," class project report, University of Arizona, Tucson, 1978.

Table 4.5-3 Chebyshev 1-dB ripple realizations using cascaded second- and third-order RC-amplifier filters

n	R_1	R_2	R_3
2	0.91097		0.99567
3	2.27516	3.64424	0.24549
4	3.58330		0.28289
	1.48425		2.41140
5	5.58919		0.18103
	4.03696	3.92009	0.50845
6	8.04104		0.12553
	2.94322		0.60920
	2.15459		3.72173
7	10.93877		0.09209
	3.90400		0.39199
	5.73554	4.98496	0.73885
8	14.28235		0.07043
	5.01530		0.27557
	3.35111		0.87546
	2.84094		5.00983
9	18.07178		0.05560
	6.27626		0.20549
	4.09654		0.55661
	7.41686	6.17387	0.96277
10	22.30704		0.04501
	7.68648		0.15974
	4.93502		0.38928
	3.91646		1.12661
	3.53309		6.28949

bandpass function, then one choice is to make half of the individual stages be high-pass and the other half be low-pass. This arbitrary selection of the zeros associated with the individual sections provides an additional advantage. If the individual transfer function has both complex zeros and complex poles, then it has been shown that the function sensitivity of the ith stage $S_x^{N_i(s)}$ can be reduced by selecting the poles and zeros to be realized by that stage as far apart as possible.[1]

[1] G. S. Moschytz, "Second-Order, Pole-Zero Pair Selection for nth-Order Minimum Sensitivity Networks," *IEEE Trans. Circuit Theory*, vol. CT-17, November 1970, pp. 527–534.

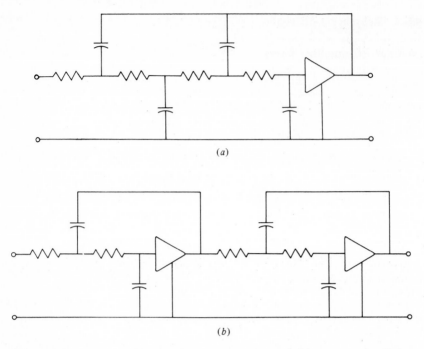

(a)

(b)

Figure 4.5-5 Realization of a fourth-order low-pass function by (a) a fourth-order filter and (b) two second-order filters.

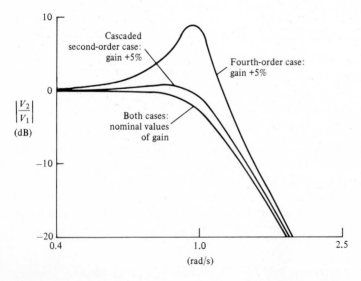

Figure 4.5-6 Effects of 5 percent gain changes on the filters of Fig. 4.5-5.

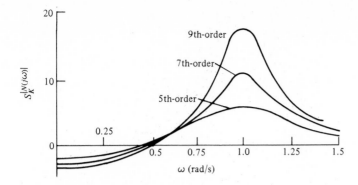

Figure 4.5-7 Sensitivity to gain of cascade Butterworth realizations.

Another important consideration—the decision as to which poles and zeros each cascaded section is to realize—is illustrated as follows. Suppose that a fourth-order bandpass function consisting of two complex-pole pairs and two zeros at the origin (and two at infinity) as shown in Fig. 4.5-9a is to be realized. Such a bandpass function can be decomposed into the cascade of a low-pass second-order stage and a high-pass second-order stage. Now let us see what difference the choice of poles for these stages makes. We first use the high-pass stage to realize the poles at $-\sigma_0 \pm j\omega_1$ and the low-pass to realize the poles at $-\sigma_0 \pm j\omega_2$. A

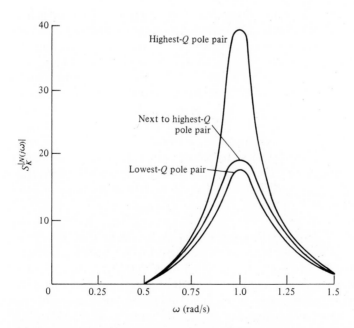

Figure 4.5-8 Sensitivity to gain for various pole pairings.

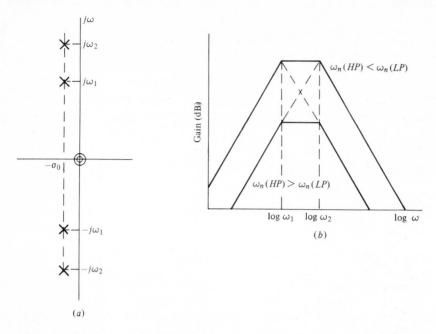

Figure 4.5-9 Cascading a low-pass and a high-pass filter.

Bode plot of the result is shown in Fig. 4.5-9*b* as the curve labeled $\omega_n(\text{HP}) < \omega_n(\text{LP})$. If we reverse the choice, then the curve labeled $\omega_n(\text{HP}) > \omega_n(\text{LP})$ results. While the poles and zeros of the overall transfer functions are exactly the same in both cases, the gain in the passband is higher for $\omega_n(\text{HP}) < \omega_n(\text{LP})$ than it is for $\omega_n(\text{HP}) > \omega_n(\text{LP})$. The reason is easy to see by examining the individual stages. In the $\omega_n(\text{HP}) > \omega_n(\text{LP})$ case, both the high-pass and the low-pass stages provide attenuation in the passband, while for the $\omega_n(\text{HP}) < \omega_n(\text{LP})$ case, neither stage is contributing loss in the passband. Thus, to achieve large gain it is preferable to realize the poles closest to the origin with high-pass stages and the poles farthest from the origin with low-pass stages. Furthermore, to achieve the largest dynamic range, the stages with the lowest Q should be placed first. This prevents overdriving of the amplifiers. Otherwise, even though the overall gain may be unity, the individual high-Q stages may have sufficient gain that the signal would be first amplified and then attenuated in the following low-Q stages.

Example 4.5-1 *A High-Order Bandpass Filter* A bandpass filter having a maximally flat magnitude response is to be designed using cascaded second-order bandpass stages. The center frequency is to be 3000 Hz with a -3-dB bandwidth of 600 Hz and a -30-dB bandwidth of not more than 1500 Hz. The bandpass gain is to be unity. As a first step, we must find the number of second-order bandpass stages required. Since the ratio of the 30-dB bandwidth to the 3-dB bandwidth is $1500/600 = 2.5$, we may first determine the

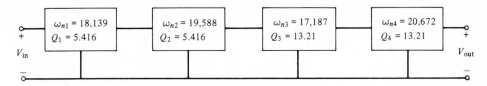

Figure 4.5-10 The order of the cascaded stages in Example 4.5-1.

order of a prototype normalized low-pass filter with a -3-dB bandwidth of 1 rad/s and a minimum attenuation of 30-dB at 2.5 rad/s. Using the nomograph of Fig. 2.1-4 we find that the order of the low-pass prototype filter required is 4, which means that four second-order bandpass stages will be required to meet the filter specification. Table 2.1-3 gives the normalized low-pass poles as

$$s_{n1,2} = 0.38268 \pm j0.92388 \tag{8a}$$

$$s_{n3,4} = 0.92388 \pm j0.38268 \tag{8b}$$

Using the low-pass to bandpass transformation of Sec. 2.4, and frequency denormalizing, we obtain the bandpass poles

$$p_1, p_1^* = -1.67458 \times 10^3 \pm j18.0616 \times 10^3 \tag{9a}$$

$$p_2, p_2^* = -1.80836 \times 10^3 \pm j19.5042 \times 10^3 \tag{9b}$$

$$p_3, p_3^* = -0.65492 \times 10^3 \pm j17.1747 \times 10^3 \tag{9c}$$

$$p_4, p_4^* = -0.78775 \times 10^3 \pm j20.6576 \times 10^3 \tag{9d}$$

From the above roots we can calculate ω_{ni} and Q_i for each individual second-order stage. Thus

$$\omega_{n1} = 18,139 \text{ rad/s} \qquad Q_1 = 5.416 \tag{10a}$$

$$\omega_{n2} = 19,588 \text{ rad/s} \qquad Q_2 = 5.416 \tag{10b}$$

$$\omega_{n3} = 17,187 \text{ rad/s} \qquad Q_3 = 13.121 \tag{10c}$$

$$\omega_{n4} = 20,672 \text{ rad/s} \qquad Q_4 = 13.121 \tag{10d}$$

The order of the cascaded stages is shown in Fig. 4.5-10. The gain at resonance of each stage is assumed to be unity. $\qquad \square$

4.6 FINITE-GAIN AMPLIFIERS

In the preceding sections of this chapter we presented an introduction to the properties of *RC*-amplifier filters. In this section we consider the amplifier itself. Specifically we consider how the nonideal properties of the operational amplifier affect the overall characteristics of the finite-gain amplifier.

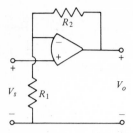

Figure 4.6-1 A noninverting finite-gain amplifier.

The general circuit configuration for a noninverting finite-gain amplifier is shown in Fig. 4.6-1. As a first analysis, we will assume that the operational amplifier has infinite input impedance and zero output impedance. Letting $A(s)$ be the overall gain we find that

$$A(s) = \frac{V_o(s)}{V_s(s)} = \frac{A_d(s)}{1 + A_d(s)[R_1/(R_1 + R_2)]} \tag{1}$$

Using a dominant-pole model for the operational amplifier's differential-mode gain [see (17) of App. B],

$$A_d(s) = \frac{GB}{s + \omega_a} = \frac{A_0 \omega_a}{s + \omega_a} \tag{2}$$

where A_0 is the dc gain, ω_a is the bandwidth, and GB is the gain-bandwidth product or unity-gain bandwidth. Substituting (2) into (1) gives

$$A(s) = \frac{V_o(s)}{V_s(s)} = \frac{GB}{s + \omega_a[1 + A_0 R_1/(R_1 + R_2)]} \approx \frac{GB}{s + GBR_1/(R_1 + R_2)} \tag{3}$$

where the approximation holds if $A_0 R_1/(R_1 + R_2) > 1$. The magnitude and phase of (3) can be written as

$$|A(j\omega)| = \frac{GB}{\sqrt{\omega^2 + [GBR_1/(R_1 + R_2)]^2}} \tag{4a}$$

$$\arg[A(j\omega)] = -\tan^{-1}\left[\frac{\omega}{GB}\left(1 + \frac{R_2}{R_1}\right)\right] \tag{4b}$$

When $\omega < GBR_1/(R_1 + R_2)$, (4) may be approximated as

$$|A(j\omega)| \approx 1 + \frac{R_2}{R_1} \tag{5a}$$

$$\arg[A(j\omega)] \approx -\frac{\omega}{GB}\left(1 + \frac{R_2}{R_1}\right) \tag{5b}$$

Equation (5b) is of considerable importance in illustrating the influence of the operational amplifier's frequency response upon the characteristics of the active filter. For example, suppose $1 + R_2/R_1 = 3$ and $\omega = GB/30$. From (5b) we see

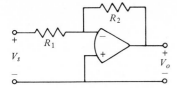

Figure 4.6-2 An inverting finite-gain amplifier.

that the finite-gain amplifier produces a phase lag of 5.73°. This phase lag may greatly affect the performance of the overall realization.

The general circuit configuration for an inverting finite-gain amplifier is shown in Fig. 4.6-2. The voltage transfer function can be written as

$$A(s) = \frac{V_o(s)}{V_s(s)} = \frac{-A_d(s)[R_2/(R_1 + R_2)]}{1 + A_d(s)[R_1/(R_1 + R_2)]} \tag{6}$$

Substituting (2) into (6) results in

$$A(s) = \frac{V_o(s)}{V_s(s)} = \frac{-[R_2/(R_1 + R_2)]\text{GB}}{s + \omega_a[1 + A_0 R_1/(R_1 + R_2)]} \approx \frac{-[R_2/(R_1 + R_2)]\text{GB}}{s + \text{GB}[R_1/(R_1 + R_2)]} \tag{7}$$

where the approximate equality holds if $A_0 R_1/(R_1 + R_2) > 1$. The magnitude and phase response of (7) can be expressed as

$$|A(j\omega)| = \frac{\text{GB}R_2/(R_1 + R_2)}{\sqrt{\omega^2 + [\text{GB}R_1/(R_1 + R_2)]^2}} \tag{8a}$$

$$\arg[A(j\omega)] = \pi - \tan^{-1}\left[\frac{\omega}{\text{GB}}\left(1 + \frac{R_2}{R_1}\right)\right] \tag{8b}$$

When $\omega < \text{GB}R_1/(R_1 + R_2)$, (8) may be approximated as

$$|A(j\omega)| \approx \frac{R_2}{R_1} \tag{9a}$$

$$\arg[A(j\omega)] \approx \pi - \frac{\omega}{\text{GB}}\left(1 + \frac{R_2}{R_1}\right) \tag{9b}$$

Comparing (3) and (7), we note that the poles of the inverting and noninverting finite-gain amplifier are similar. However, there is one difference which becomes especially apparent for low values of amplifier gain. For example, consider a unity-gain amplifier. In the noninverting configuration of Fig. 4.6-1, this requires $R_1 = \infty$ and $R_2 = 0$. Thus, the pole of the transfer function given in (3) is located at $s = -\text{GB}$. For the inverting configuration shown in Fig. 4.6-2, however, for unity gain $R_1 = R_2$ is required. From (7), the pole of the voltage transfer function of this circuit is located at $s = -\text{GB}/2$. Thus, the frequency limitation of the inverting finite-gain amplifier is half that of the noninverting one. Obviously, as the ratio R_2/R_1 becomes large, the pole locations of the two types of amplifiers approach the same value, namely, $s = -\text{GB}R_1/R_2$.

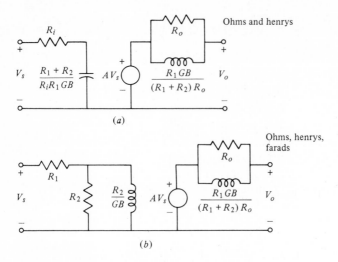

Figure 4.6-3 Effect of operational amplifier input and output resistances on finite-gain amplifiers. (*a*) Noninverting. (*b*) Inverting.

Now let us consider the case where the input and output impedances of the operational amplifier are not ideal. In this situation the frequency dependence of $A_d(s)$ causes the input and output impedances of the finite-gain amplifiers to also be frequency dependent. In this case [see (11) and (12) of App. B], the equivalent amplifier circuits are shown in Fig. 4.6-3. In the noninverting case shown in Fig. 4.6-3*a*, $A(s)$ is given by (3). In the inverting case shown in Fig. 4.6-3*b*, $A(s)$ is given by (7). These circuits assume that R_1 or R_2 is greater than R_o, and that R_1 or R_2 is less than R_i. Other details may be found in the Problems. Exact equivalent circuits for these amplifiers have been derived in the literature.[1]

In most active filter applications, the input impedances of the inverting and noninverting finite-gain amplifiers have little effect on the overall filter performance. An exception is found in the case of negative-gain realizations, in which the input impedance is approximately equal to R_1 and, as a result, can load the *RC* feedback network. Another problem with the negative-gain realization is that the ratio R_2/R_1 in Fig. 4.6-2 must be large for high-Q realizations (a Q of 10 requires $R_2/R_1 \cong 900$). However, (8*b*) shows that to keep the phase lag caused by an inverting amplifier less than 6° for a Q of 10, the maximum filter design frequency is approximately GB/8500. For a GB of 1 MHz, this would limit the filter's frequency to approximately 100 Hz. These and other effects account for the lack of popularity of negative-gain filters. Similarly, the output impedances of the noninverting and inverting finite-gain amplifiers have little influence on overall active filter performance. The reason is that these impedances are generally

[1] K. Soundararajan and K. Ramakrishna, "Characteristics of Nonideal Operational Amplifiers," *IEEE Trans. Circuits and Systems*, vol. CAS-21, no. 1, January 1974, pp. 69–75.

very small because the inner feedback loop consisting of the resistors R_1 and R_2 acts to reduce any inherent output impedance of the operational amplifier.

One of the operational amplifier parameters which may have a noticeable effect on the overall performance of active filters using finite-gain amplifiers is the slew rate. Slew-rate limitations can affect the small-signal behavior of finite-gain amplifiers when the signal level is large and/or the frequency is high. To see this, consider the model for the influence of slew on the open-loop response of an operational amplifier developed in App. B. It has been shown[1] that when feedback is applied around the circuit in Fig. B-12, in which $f(V_i)$ is modeled as shown in Fig. B-13, large-signal amplitudes can cause a distortion in both the amplitude and phase response of the amplifier. The analysis of this distortion requires the solution of a set of nonlinear equations and is not given here. An important result of this analysis, however, is that the small-signal phase shift of the inverting amplifier can be expressed as

$$\arg\left[A(j\omega)\right] = \pi - \tan^{-1}\left[\frac{(\omega/\mathrm{GB})(1 + R_2/R_1)}{N(A)}\right] \tag{10}$$

where $N(A)$ is a describing function[2] and A is the peak amplitude of the sinusoidal voltage applied at V_i in Fig. B-12. When there is no slew-generated distortion, then $N(A) = 1$. However, as slew influences the response, $N(A)$ becomes less than unity and causes an *increase in the small-signal phase lag*. This can lead to signal-stability problems in some types of active filters. The influence of the large-signal distortion described above is shown graphically in Fig. 4.6-4 for a unity-gain inverting amplifier. The parameter M_G/δ used in the figure relates the peak amplitude M_G of the input signal to the threshold level δ defined in Fig. B-13. Both the theoretically predicted and typical actual observed values are given. If the gain of the amplifier is changed from -1 to -10, the characteristics shown in Fig. 4.6-5 result. The large-signal characteristics of the noninverting amplifier are not illustrated here because of the fact that its slewing characteristics are more complex than those of the inverting amplifier.[3]

One may reduce large-signal distortion either by selecting operational amplifiers with a higher slew rate or by preventing the operational amplifiers from slewing. FET input operational amplifiers not only have high slew rates but the value of δ is larger than for BJT operational amplifiers. An operational amplifier may be prevented from slewing by preventing V_i from ever exceeding δ. However, this is a costly solution since slew occurs when the input amplitude is above δ *and when the slope of the output waveform exceeds SR*. Unfortunately when the input signal amplitude is above δ and the output slope is less than SR, slew does not occur and the input is being needlessly limited. A better solution is to precede the

[1] P. E. Allen, "Slew Induced Distortion in Operational Amplifiers," *IEEE J. Solid-State Circuits*, vol. SC-12, no. 1, February 1977, pp. 39–44.

[2] J. G. Truxal, *Automatic Feedback Control System Synthesis*, McGraw-Hill Book Company, New York, 1955.

[3] J. Solomon, "The Monolithic Op Amp: A Tutorial Study," *IEEE J. Solid-State Circuits*, vol. SC-9, December 1974, pp. 314–332.

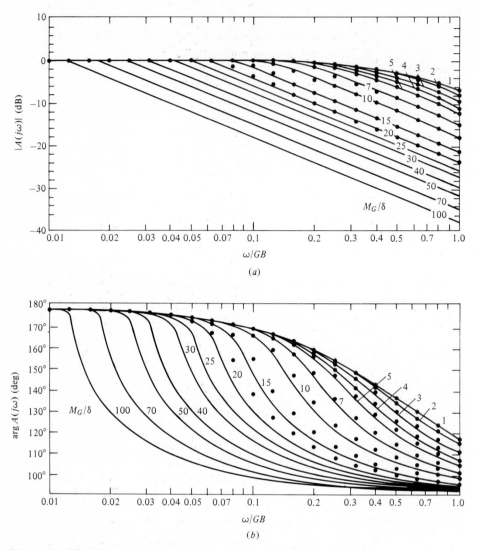

Figure 4.6-4 Effect of large-signal distortion in a unity-gain inverting amplifier.

operational amplifier with a slope-limiting circuit so that slopes greater than SR at the output cannot occur.[1] Unfortunately such a slope-limit circuit introduces its own distortion in the phase response. In many cases, the active filter designer will find the slew rate to be the final limit in performance of a given realization. It should be noted that the slew rate will always influence a realization, even though in a given application it may not actually cause instability.

[1] P. E. Allen, "Large Signal Influences on Single Amplifier Active Filters," *Proc. 20th Midwest Symp. Circuits and Systems*, August 1977, pp. 289–294.

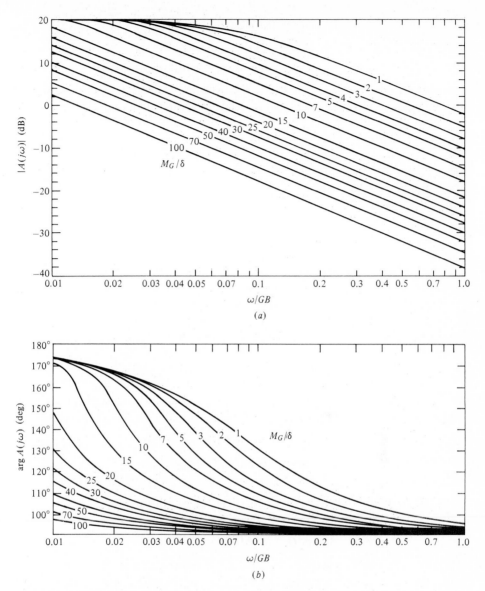

Figure 4.6-5 Effect of large-signal distortion in an inverting amplifier with a gain of 10.

4.7 FREQUENCY-DEPENDENT SENSITIVITY

This section introduces a new concept in sensitivity analysis, namely, sensitivity to a frequency-dependent network parameter. In active filter realizations, the gain of the active network elements exhibits such a frequency dependence. More specifically, the operational amplifiers used to synthesize these networks have a differential-mode gain characterized by the dominant-pole model

$$A_d(s) = \frac{GB}{s + \omega_a} = \frac{A_0 \omega_a}{s + \omega_a} \tag{1}$$

where A_0 is the dc gain, ω_a is the bandwidth, and GB is the gain-bandwidth product or unity-gain bandwidth. At frequencies greater than ω_a, (1) simplifies to

$$A_d(s) \simeq \frac{GB}{s} \tag{2}$$

The sensitivity of the network poles to GB will be studied in this section. The type of sensitivity that best lends itself to such a study is the *relative complex root sensitivity*, defined as

$$S_{GB}^{p_i} = \frac{\partial p_i/p_i}{\partial GB/GB} = -\frac{1}{p_i GB} \frac{\partial p_i}{\partial (1/GB)} \tag{3}$$

where p_i is the location of one of the poles of the transfer function for the network. This will be referred to as a *frequency-dependent sensitivity*.

Now let us consider a general method for calculating $S_{GB}^{p_i}$. For the case where $A_d(s)$ can be represented by (2), then the denominator $D(s)$ of an RC-amplifier realization can be written as the following power series:

$$D(s) = P_0(s) + P_1(s)\frac{s}{GB} + P_2(s)\left(\frac{s}{GB}\right)^2 + \cdots \tag{4}$$

where the coefficients $P_i(s)$ of the series are polynomials in the complex frequency variable s. If only the first two terms of (4) are retained, then $D(s)$ is approximated as

$$D(s) \simeq P_0(s) + \frac{s}{GB} P_1(s) \tag{5}$$

This may be written as

$$D(s) \simeq k\left(s^2 + \frac{\omega_n}{Q}s + \omega_n^2\right) + \frac{s}{GB}(q_2 s^2 + q_1 s + q_0) \tag{6}$$

where k, q_2, q_1, and q_0 are functions of ω_n, Q, and the parameters of the realization. Obviously, as GB approaches infinity, $D(s)$ approaches the standard second-order transfer-function denominator. Using the techniques of Sec. 3.4, from (5) we obtain,

$$\frac{\partial p_i}{\partial (1/GB)} = \frac{-sP_1(s)}{dD(s)/ds}\bigg|_{s=p_i} \approx \frac{-sP_1(s)}{dP_0(s)/ds}\bigg|_{s=p_i} \tag{7}$$

A more useful form of (7) is

$$\frac{\partial p_i}{\partial GB} = -\left(\frac{1}{GB}\right)^2 \frac{\partial p_i}{\partial (1/GB)} = \left(\frac{1}{GB}\right)^2 \frac{sP_1(s)}{dD(s)/ds}\bigg|_{s=p_i} \approx \left(\frac{1}{GB}\right)^2 \frac{sP_1(s)}{dP_0(s)/ds}\bigg|_{s=p_i} \tag{8}$$

Applying (8) to (6) we obtain

$$\text{Re } \frac{\partial p_i}{\partial GB} = \frac{-\omega_n^2}{2kGB^2}\left(\frac{q_2}{Q^2} - \frac{q_1}{Q\omega_n} - q_2 + \frac{q_0}{\omega_n^2}\right) \tag{9a}$$

$$\text{Im } \frac{\partial p_i}{\partial GB} = \frac{-\omega_n^2}{2k\sqrt{(4Q^2 - 1)}GB^2}\left(\frac{q_2}{Q^2} - \frac{q_1}{Q\omega_n} - q_2 + \frac{q_0}{\omega_n^2} - 2q_2 + \frac{2Qq_1}{\omega_n}\right) \tag{9b}$$

These equations may be simplified by defining

$$x = q_2 - q_1 \frac{Q}{\omega_n} \qquad y = q_2 - \frac{q_0}{\omega_n^2} \tag{10}$$

Then for (9) we obtain

$$\text{Re } \frac{\partial p_i}{\partial GB} = \frac{-\omega_n^2}{2kGB^2}\left(\frac{x}{Q^2} - y\right) \tag{11a}$$

$$\text{Im } \frac{\partial p_i}{\partial GB} = \frac{-\omega_n^2}{2kGB^2\sqrt{4Q^2 - 1}}\left(\frac{x}{Q^2} - 2x - y\right) \tag{11b}$$

Substituting these in (3) gives

$$S_{GB}^{p_i} = \frac{\omega_n^2}{2kp_iGB}\left|\left(\frac{x}{Q^2} - y\right) + j\left(\frac{x/Q^2 - 2x - y}{\sqrt{4Q^2 - 1}}\right)\right| \tag{12}$$

Since $p_i = (\omega_2/2Q)(-1 + j\sqrt{4Q^2 - 1})$, it may be shown that

$$S_{GB}^{p_i} = \frac{\omega_n}{2QkGB}\left(-x + j\frac{2Q^2y - x}{\sqrt{4Q^2 - 1}}\right) \tag{13}$$

The real and imaginary components of (13) may thus be expressed as

$$\text{Re } S_{GB}^{p_i} = \frac{-\omega_n x}{2kQGB} \tag{14a}$$

$$\text{Im } S_{GB}^{p_i} = \frac{\omega_n(2Q^2y - x)}{2kQGB\sqrt{4Q^2 - 1}} \tag{14b}$$

Solving for $S_{GB}^{\omega_n}$ and S_{GB}^{Q} from (12) in Sec. 3.5 (the fact that normalized root sensitivity is used in this relationship makes no difference) gives

$$S_{GB}^{Q} = \frac{4Q^2 - 1}{4Q^2}(\text{Im } S_{GB}^{p_i} - \text{Re } S_{GB}^{p_i}) \tag{15a}$$

$$S_{GB}^{\omega_n} = \frac{\text{Re } S_{GB}^{p_i}}{4Q^2} + \frac{4Q^2 - 1}{4Q^2}(\text{Im } S_{GB}^{p_i}) \tag{15b}$$

Another useful sensitivity measure is the magnitude-squared value of $S_{GB}^{p_i}$. This is especially useful in studies of the minimization of frequency-dependent sensitivity, which will be investigated later. The magnitude-squared value of $S_{GB}^{p_i}$ is given as

$$|S_{GB}^{p_i}|^2 = \frac{\omega_n^2}{k^2GB^2(4Q^2 - 1)}(x^2 - xy + Q^2y^2) \tag{16}$$

where x and y are defined in (10).

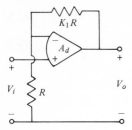

Figure 4.7-1 A noninverting amplifier.

The frequency dependence of single-amplifier realizations is found by determining the coefficients b_i and c_i as given in (10) of Sec. 4.1. The relationships are

$$b_2 s^2 + b_1 s + b_0 = Y_6(Y_1 + Y_2 + Y_3 + Y_5) + Y_2 Y_3 \qquad (17a)$$

$$c_2 s^2 + c_1 s + c_0 = (Y_3 + Y_4 + Y_6)(Y_1 + Y_2 + Y_5) + Y_3(Y_4 + Y_6) \qquad (17b)$$

The amplifier used in the single-amplifier finite-gain realizations can be represented as shown in Fig. 4.7-1. The amplifier gain $A(s)$ is easily found to be

$$A(s) = \frac{V_o(s)}{V_i(s)} = \frac{1}{\dfrac{1}{A_d(s)} + \dfrac{1}{1 + K_1}} \qquad (18)$$

where K_1 gives the ratio of the resistors as shown in Fig. 4.7-1. The frequency response of the operational amplifier can be incorporated into this model by using (1) for $A_d(s)$. Thus we obtain

$$A(s) = \frac{1}{\dfrac{s}{GB} + \sigma_A} \qquad (19)$$

where
$$\sigma_A = \frac{1}{A_0} + \frac{1}{1 + K_1} \qquad (20)$$

The general voltage transfer function for the single-amplifier realization of (11) of Sec. 4.1 can be written as

$$\frac{V_2(s)}{V_1(s)} = \frac{Y_1 Y_3}{[(c_2 s^2 + c_1 s + c_0)/A(s)] - (b_2 s^2 + b_1 s + b_0)} \qquad (21)$$

Substituting (19) into this and defining $q_i = c_i$ ($i = 0, 1, 2$) gives

$$\frac{V_2(s)}{V_1(s)} = \frac{Y_1 Y_3}{\sigma_A(q_2 s^2 + q_1 s + q_0) + \dfrac{s}{GB}(q_2 s^2 + q_1 s + q_0) - (b_2 s^2 + b_1 s + b_0)} \qquad (22)$$

which can be rewritten as

$$\frac{V_2(s)}{V_1(s)} = \frac{Y_1 Y_3}{(\sigma_A q_2 - b_2)\left[s^2 + \left(\frac{\sigma_A q_1 - b_1}{\sigma_A q_2 - b_2}\right)s + \frac{\sigma_A q_0 - b_0}{\sigma_A q_2 - b_2}\right] + \frac{s}{GB}(q_2 s^2 + q_1 s + q_0)} \tag{23}$$

As GB approaches infinity we obtain

$$\lim_{GB \to \infty} \frac{V_2(s)}{V_1(s)} = \frac{Y_1 Y_3/(\sigma_A q_2 - b_2)}{s^2 + \left(\frac{\sigma_A q_1 - b_1}{\sigma_A q_2 - b_2}\right)s + \frac{\sigma_A q_0 - b_0}{\sigma_A q_2 - b_2}} \tag{24}$$

Comparing this with the standard form of a second-order network function we see that

$$\omega_n^2 = \frac{\sigma_A q_0 - b_0}{\sigma_A q_2 - b_2} \tag{25a}$$

$$\frac{\omega_n}{Q} = \frac{\sigma_A q_1 - b_1}{\sigma_A q_2 - b_2} \tag{25b}$$

The numerator of such a standard function will have the form

Low-pass: $\qquad\qquad\qquad H_0 = \left|\dfrac{Y_1 Y_3}{\sigma_A q_0 - b_0}\right| \tag{26a}$

Bandpass: $\qquad\qquad\qquad sH_0 = \left|\dfrac{Y_1 Y_3}{\sigma_A q_1 - b_1}\right| \tag{26b}$

High-pass: $\qquad\qquad\qquad s^2 H_0 = \left|\dfrac{Y_1 Y_3}{\sigma_A q_2 - b_2}\right| \tag{26c}$

To simplify the above we define

$$k = \sigma_A q_2 - b_2 \tag{27}$$

Then the denominator of (23) becomes

$$D(s) = k\left(s^2 + \frac{\omega_n}{Q}s + \omega_n^2\right) + \frac{s}{GB}(q_2 s^2 + q_1 s + q_0) \tag{28}$$

which is of the form given in (6).

As an illustration of the application of the development given above, consider the low-pass realization defined by (10) of Sec. 4.2. For this using (17) we find

$$b_2 s^2 + b_1 s + b_0 = sC_2 G_3 \tag{29a}$$

$$q_2 s^2 + q_1 s + q_0 = s^2(C_2 C_4) + s(C_2 G_3 + C_4 G_1 + C_4 G_3) + G_1 G_3 \tag{29b}$$

From (20) and (27), since $b_2 = 0$ and $A_0 \gg 1 + K_1$,

$$k = \sigma_A q_2 - b_2 = \sigma_A q_2 \qquad \sigma_A \approx \frac{1}{1 + K_1} \tag{30}$$

and from (25), since b_0 is also zero,

$$\omega_n^2 = \frac{q_0}{q_2} \qquad \frac{\omega_n}{Q} = \frac{q_1}{q_2} - \frac{b_1}{\sigma_A q_2} \tag{31}$$

Rearranging these, we obtain

$$\frac{q_0}{\omega_n^2} = q_2 \qquad q_2 - \frac{Q}{\omega_n} q_1 = -\frac{Q}{\omega_n}\left(\frac{b_1}{\sigma_A}\right) \tag{32}$$

Substituting these in (10) gives

$$x = -\frac{Q b_1}{\omega_n \sigma_A} \qquad y = 0 \tag{33}$$

From (16), the magnitude-squared root sensitivity function with respect to GB is now found to be

$$|S_{GB}^{p_i}|^2 = \frac{Q^2}{\sigma_A^4 GB^2 (4Q^2 - 1)}\left(\frac{b_1}{q_2}\right)^2 \tag{34}$$

Taking the square root of both members of this we find

$$|S_{GB}^{p_i}| = \frac{Q(1 + K_1)^2}{GB\sqrt{4Q^2 - 1}}\left(\frac{1}{R_3 C_4}\right) = \frac{Q}{\sqrt{4Q^2 - 1}}\left(\frac{H_0^2 \omega_n}{GB}\right)\sqrt{\frac{R_1 C_2}{R_3 C_4}} \tag{35}$$

where $H_0 = 1 + K_1$. For the equal-resistance equal-capacitance design procedure (design 1) of Sec. 4.2, this becomes

$$|S_{GB}^{p_i}| = \frac{Q}{\sqrt{4Q^2 - 1}}\left(\frac{H_0^2 \omega_n}{GB}\right) \tag{36}$$

It is of interest at this point to illustrate the manner in which p_i depends on the frequency characteristics of the operational amplifier using graphical means. To do this we find the locus of the network function poles as GB is varied. First we write (28) in the form of

$$D(s) = q_2\left\{\frac{s^3}{GB} + s^2\left[\sigma_A + \frac{1}{GB}\left(\frac{\omega_n}{Q} + \frac{b_1}{\sigma_A q_2}\right)\right] + s\left(\frac{\sigma_A \omega_n}{Q} + \frac{\omega_n^2}{GB}\right) + \sigma_A \omega_n^2\right\} \tag{37}$$

For design 1 as defined in (14) of Sec. 6.2

$$\sigma_A = \frac{1}{K} = \frac{Q}{3Q - 1} = \frac{1}{1 + K_1} \qquad \frac{b_1}{\sigma_A q_2} = 3\omega_n - \frac{\omega_n}{Q} \tag{38}$$

Thus (37) becomes

$$D(s) = \frac{q_2}{GB}\left[s^3 + s^2\left(3\omega_n + \frac{GBQ}{3Q - 1}\right) + s\left(\omega_n^2 + \frac{GB\omega_n}{3Q - 1}\right) + \frac{GBQ\omega_n^2}{3Q - 1}\right] \tag{39}$$

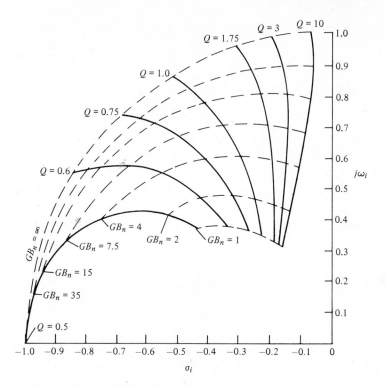

Figure 4.7-2 Effect of GB on the poles of the second-order low-pass or high-pass Sallen and Key equal-R equal-C (design 1) filters of Secs. 4.2 and 4.3.

Normalizing by ω_n such that $s_n = s/\omega_n$ and $GB_n = GB/\omega_n$, this may be rewritten

$$D(s) = \frac{q_2 \omega_n^2}{GB_n}\left[s_n^3 + s_n^2\left(3 + \frac{GB_n Q}{3Q - 1}\right) + s_n\left(1 + \frac{GB_n}{3Q - 1}\right) + \frac{GB_n Q}{3Q - 1}\right] \quad (40)$$

Plotting the upper complex pole as GB_n varies, and using Q as a parameter, we obtain the loci shown in Fig. 4.7-2. These curves vividly portray the effect of GB on p_i.[1]

Example 4.7-1 *The Actual ω_n and Q of a Finite-Gain Low-Pass Filter* It is desired to find the actual ω_n and Q of a low-pass finite-gain filter realization which uses design 1 of Sec. 4.2. The desired ω_n and Q are respectively $100 \times 2\pi$ krad/s and 1. The gain bandwidth of the amplifier is 1.5 MHz, and its frequency response has the form given in (1). The normalized gain bandwidth $GB_n = 15$. From Fig. 4.7-2 the actual normalized pole locations are found to

[1] A. Budak and D. M. Petrela, "Frequency Limitations of Active Filters Using Operational Amplifiers," *IEEE Trans. Circuit Theory*, vol. CT-19, no. 4, 1972, pp. 322–328.

be $\alpha_n \pm j\beta_n = -0.4 \pm j0.78$. From this result we calculate that

$$\omega_{nn} = \sqrt{\alpha_n^2 + \beta_n^2} = \sqrt{0.4^2 + 0.78^2} = 0.876 \text{ rad/s}$$

$$Q = \frac{\omega_{nn}}{2\alpha_n} = \frac{0.876}{2(.4)} = 1.096$$

Therefore the actual ω_n and Q are $87.6 \times 2\pi$ krad/s and 1.096. In this case the deviation is small enough so that the realization is readily tuned to the proper values. For some realizations, however, such a tuning procedure may be difficult to accomplish. $\qquad\square$

An analysis of the various other positive-gain filter realizations is readily accomplished following the method described above. The results of a series of such analyses is given in Table 4.7-1. The shifts of the pole p_i for these analyses are illustrated in Figs. 4.7-2 through 4.7-4. These shifts can be directly related to the phase lag introduced by the noninverting amplifier. This phase lag was analyzed in Sec. 4.6. Specifically, (4b) of Sec. 4.6 shows that as GB_n approaches a value of unity, the phase lag becomes large. However, since the magnitude of the open-loop transfer function of the positive-gain realization never exceeds unity, this does not cause the realization to become unstable.

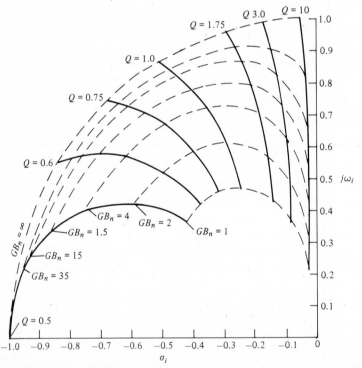

Figure 4.7-3 Effect of GB on the poles of the second-order low-pass or high-pass Sallen and Key unity-gain (design 2) filters of Secs. 4.2 and 4.3.

Table 4.7-1 Summary of the frequency-dependent sensitivities for positive-gain realizations

Positive-gain realization	Constraints	$\left|S^{p_a}_{GB}\right|$ and $D(s_n)$
Low-pass (Fig. 4.2-3)	None	$\left|S^{p_a}_{GB}\right| = \dfrac{Q}{\sqrt{4Q^2-1}}\left(\dfrac{H_0^2\,\omega_n}{GB}\right)\sqrt{\dfrac{R_1 C_2}{R_3 C_4}}$
	$R_1 = R_2 = R$, $C_2 = C_4 = C$ (Fig. 4.7-2)	$D(s_n) = \dfrac{q_2\,\omega_n^2}{GB_n}\left| s_n^3 + \left(3 + \dfrac{GB_n Q}{3Q-1}\right)s_n^2 + \left(1 + \dfrac{GB_n}{3Q-1}\right)s_n + \dfrac{GB_n Q}{3Q-1}\right|$
	$K = H_0 = 1$, $m = C_4/C_2$, $n = R_3/R_1 = 1$ (Fig. 4.7-3)	$D(s_n) = \dfrac{q_2\,\omega_n^2}{GB_n}\left| s_n^3 + \left(GB_n + \dfrac{1}{Q} + 2Q\right)s_n^2 + \left(1 + \dfrac{GB_n}{Q}\right)s_n + GB_n\right|$
Bandpass (Fig. 4.3-1)	None	$\left|S^{p_a}_{GB}\right| = \dfrac{(1+K_1)H_0\,\omega_n}{GB\sqrt{4Q^2-1}}\left(\dfrac{R_1}{R_2}\right)$
	$R_1 = R_2 = R_4 = R$, $C_3 = C_5 = C$ (Fig. 4.7-4)	$D(s_n) = \dfrac{q_2\,\omega_n^2}{GB_n}\left| s_n^3 + \left(2\sqrt{2} + \dfrac{QGB_n}{4Q-\sqrt{2}}\right)s_n^2 + \left(1 + \dfrac{GB_n}{4Q-\sqrt{2}}\right)s_n + \dfrac{QGB_n}{4Q-\sqrt{2}}\right|$
High-pass (Fig. 4.3-2)	None	$\left|S^{p_a}_{GB}\right| = \dfrac{Q}{4Q^2-1}\left(\dfrac{H_0^2\,\omega_n}{GB}\right)\sqrt{\dfrac{R_4 C_3}{R_2 C_1}}$
	$R_2 = R_4 = R$, $C_1 = C_3 = C$ (Fig. 4.7-2)	$D(s_n)$ is the same as the low-pass case for equal resistors and capacitors
	$A = H_0 = 1$, $n = R_4/R_2 = 1$, $m = C_3/C_1$ (Fig. 4.7-3)	$D(s_n)$ is the same as the low-pass case for unity gain and $n = 1$

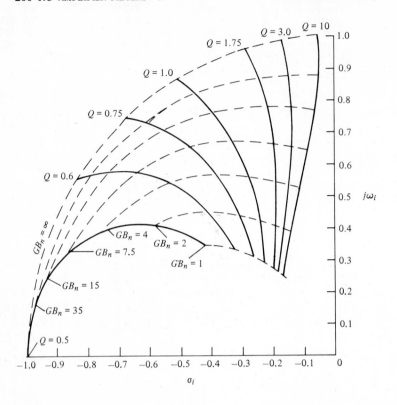

Figure 4.7-4 Effect of GB on the poles of the second-order bandpass Sallen and Key equal-*R* equal-*C* (design 1) filter of Sec. 4.3.

4.8 SUMMARY

In this chapter we have introduced the *RC* amplifier filter. In Sec. 4.1 the general configuration for the single-amplifier filter was presented. This configuration was applied to the realization of low-pass filter functions in Sec. 4.2, and to the realization of bandpass and high-pass ones in Sec. 4.3. Several different design procedures for each of these filter types were given. The sensitivity of these realizations were shown to be relatively high; thus they are, in general, useful only for the realization of low-*Q* functions. In Sec. 4.4 the realization of network functions with complex-conjugate zeros as well as poles was discussed. Realizations using one and two amplifiers with finite and infinite gain were presented. The realization of functions of higher than second order was discussed in Sec. 4.5. Finally, in Secs. 4.6 and 4.7 the effect of the limitations of operational amplifiers on the high-frequency performance of *RC* amplifier filters was discussed. In the following chapter some additional *RC* amplifier filter configurations will be presented.

PROBLEMS

4-1 (*Sec. 4.1*)   (*a*) Find the polynomials $N_{31}(s)$, $N_{32}(s)$, and $N_{33}(s)$ defined in (10) of Sec. 4.1 for the network shown in Fig. P4-1. Assume all the passive elements have unity value.

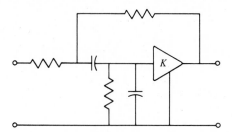

Figure P4-1

(*b*) In order for this circuit to realize complex-conjugate poles in a voltage transfer function, should the VCVS be inverting or noninverting?

4-2 (*Sec. 4.1*)   Repeat Prob. 4-1 for the network shown in Fig. P4-2.

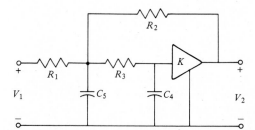

Figure P4-2

4-3 (*Sec. 4.2*)   Repeat the derivation determining the configuration for a positive-gain low-pass filter given in Sec. 4.2, but use the assumption that in (10) of Sec. 4.1, $b_0 \neq 0$, $b_1 \neq 0$, and $b_2 = 0$.

4-4 (*Sec. 4.2*)   Use the design relations of (13) of Sec. 4.2 to realize a second-order Bessel filter with a group delay of 1 ms.

4-5 (*Sec. 4.2*)   An alternate design procedure for the low-pass filter shown in Fig. 4.2-3 which allows the H_0 specifications as well as the ω_n and Q ones to be met is as follows: Choose $K = H_0$, and $C_2 = C_4 = C$, then

$$R_3 = \frac{1}{2\omega_n QC}[1 + \sqrt{1 + 4Q^2(H_0 - 2)}]$$

$$R_1 = \frac{1}{\omega_n^2 C^2 R_3}$$

where $H_0 > 2$.
Apply these relations to design a filter with $f_n = 30$ Hz, $Q = 1/\sqrt{2}$, $H_0 = 10$, and $C = 0.1$ μF.

4-6 (*Sec. 4.2*)   Derive the design relations given in Prob. 4-5.

4-7 (*Sec. 4.2*)   Determine the various sensitivities of (27) in Sec. 4.2 for the realization of the low-pass Sallen and Key filter shown in Fig. 4.2-4, in which $R_3 = 10R_1$ and $C_4 = 0.1C_2$.

4-8 (*Sec. 4.2*) Determine the various sensitivities of (27) of Sec. 4.2 for the realization of the low-pass Sallen and Key filter shown in Fig. 4.2-4, in which $C_2 = \sqrt{3} Q$, $C_4 = 1$, $R_1 = 1/Q\omega_n$, $R_3 = 1/(\sqrt{3}\,\omega_n)$. $C_4 = 1/2Q\omega_n$.

4-9 (*Sec. 4.2*) Determine the various sensitivities of (27) of Sec. 4.2 for the realization of the low-pass Sallen and Key filter shown in Fig. 4.2-4, in which $C_2 = \sqrt{3} Q$, $C_4 = 1$, $R_1 = 1/Q\omega_n$, $R_3 = 1/(\sqrt{3}\,\omega_n)$, and $K = \frac{4}{3}$. This design has been proposed as the best compromise between the active and passive sensitivities.[1]

4-10 (*Sec. 4.3*) Prove whether or not the positive-gain *RC*-amplifier filter circuit shown in Fig. 4.3-1 can be used to realize complex-conjugate poles if the gain K is set to unity.

4-11 (*Sec. 4.3*) Develop a bandpass *RC*-amplifier filter using the method outlined in Sec. 4.3, but with the initial choice of elements $Y_1 = sC_1$ and $Y_3 = G_3$. Obtain design relations similar to those given in (8) and (9) of that section.

4-12 (*Sec. 4.3*) An alternate design procedure for the high-pass filter shown in Fig. 4.3-2 which allows the H_0 specification as well as the ω_n and Q ones to be met is as follows: Choose $C_1 = C_3 = C$, then

$$R_2 = \frac{1/2Q + \sqrt{2(H_0 - 1) + 1/4Q^2}}{2\omega_n C}$$

$$R_4 = \frac{2/\omega_n C}{1/2Q + \sqrt{2(H_0 - 1) + 1/4Q^2}}$$

Apply these relations to design a filter with $Q = 0.707$, $f_n = 300$ Hz, and $H_0 = 100$.

4-13 (*Sec. 4.3*) Derive the design relations given in Prob. 4-12.

4-14 (*Sec. 4.3*) (*a*) For the Sallen and Key high-pass *RC* amplifier filter circuit whose voltage transfer function is given in (20) of Sec. 4.3, find expressions for the sensitivity similar to those given in (27) of Sec. 4.2 for the low-pass filter. Use an operational amplifier realization for the VCVS as shown in Fig. 4.2-4.

(*b*) Use the results obtained above to make a table of sensitivities for the first two designs identified in Sec. 4.3.

4-15 (*Sec. 4.3*) Repeat the preceding problem for the bandpass filter whose voltage transfer function is given in (8) of Sec. 4.3, and for the first design identified in that section.

4-16 (*Sec. 4.3*) (*a*) Find the voltage transfer function of the negative-gain low-pass filter configuration shown in Fig. P4-2 (where $K < 0$).

(*b*) Assume $R_1 = R_2 = R_3 = R$, and $C_4 = C_5 = C$, and find the design relations for RC and $|K|$ in terms of ω_n and Q of (1) of Sec. 4.2. Also find the expression for $|H_0|$.

4-17 (*Sec. 4.4*) (*a*) Using the filter circuit shown in Fig. 4.4-1, find a realization for the network function

$$\frac{V_2(s)}{V_1(s)} = \frac{H(s^2 + 1.2)}{s^2 + 0.1s + 1}$$

Set the parameter m to unity.

(*b*) What is the resulting value of the constant H_0?

4-18 (*Sec. 4.4*) Repeat Prob. 4-17 for the network function

$$\frac{V_2(s)}{V_1(s)} = \frac{H(s^2 + 0.9)}{s^2 + 0.1s + 1}$$

4-19 (*Sec. 4.4*) (*a*) Repeat Prob. 4-17 for the filter circuit shown in Fig. 4.4-3. Use the factor $(s + 1)$ in the partial-fraction expansion.

(*b*) Repeat Prob. 4-18 for this circuit.

[1] W. Saraga, "Sensitivity of 2nd-Order Sallen-Key-type Active RC Filters," *Electronics Letters*, vol. 3, no. 10, October 1967, pp. 442–444.

4-20 (*Sec. 4.4*) (*a*) Repeat Prob. 4-17 for the filter circuit shown in Fig. 4.4-7. Use the factor $(s + 1)$ in the partial fraction expansion.

(*b*) Repeat Prob. 4-18 for this circuit.

4-21 (*Sec. 4.4*) (*a*) Repeat Prob. 4-17 for the filter circuit shown in Fig. 4.4-8. Use the factor $(s + 1)$ in the partial-fraction expansion.

(*b*) Repeat Prob. 4-18 for this circuit.

4-22 (*Sec. 4.5*) (*a*) Design an eighth-order active *RC* low-pass filter with a 1-dB ripple in the passband and a cutoff frequency of 20 kHz. Use the filter shown in Fig. 4.2-3 for each of the component second-order realizations. Use design 1 of Sec. 4.2 to determine the element values.

(*b*) Repeat the problem using design 2.

(*c*) Repeat the problem using design 3 and verify the corresponding entries in Table 4.5-3.

4-23 (*Sec. 4.5*) (*a*) Sketch the overall magnitude characteristic of a fourth-order Butterworth bandpass filter function with a center frequency of 1 rad/s and a bandwidth of 1 rad/s realized by a cascade of second-order bandpass filters in which the constant H_0 is set to unity.

(*b*) Repeat for a cascade of a second-order low-pass filter and a second-order high-pass filter in which the low-pass one realizes the poles closest to the origin, and in which the H_0 constants of both filters are unity.

(*c*) Repeat for the case where the high-pass filter realizes the poles closest to the origin.

4-24 (*Sec. 4.6*) Find the frequency at which a noninverting operational amplifier configuration with a gain of 10 and GB of 1 MHz will have an excess phase lag of 10°.

4-25 (*Sec. 4.6*) (*a*) Find the value of the excess phase for a noninverting amplifier with a gain of 10 when $\omega = \text{GB}/100$.

(*b*) Repeat for $\omega = \text{GB}/10$.

(*c*) Repeat parts *a* and *b* for an inverting amplifier with a gain of -10.

4-26 (*Sec. 4.6*) Find the maximum frequency that will keep the phase lag caused by the amplifier in the negative-gain low-pass configuration of Prob. 4-16 to be less than 10° if GB = 1 MHz and $Q = 4$.

4-27 (*Sec. 4.6*) (*a*) Find the amount of excess phase that will be introduced by a unity-gain inverter at $\omega = 0.01$ GB.

(*b*) Repeat for $\omega = 0.1$ GB.

4-28 (*Sec. 4.7*) Derive (9) of Sec. 4.7.

4-29 (*Sec. 4.7*) (*a*) Find $|S_{\text{GB}}^{p_i}|$ for a low-pass positive-gain *RC*-amplifier realization in which $Q = 10$, $f_n = 10$ kHz, and the operational amplifier's gain bandwidth is 1 MHz. Assume that the equal-*R* equal-*C* (design 1) procedure of Sec. 4.2 is used.

(*b*) Find the actual pole positions of the filter and calculate the actual values of Q and f_n.

(*c*) Repeat the above if $f_n = 50$ kHz.

4-30 (*Sec. 4.7*) (*a*) Use Table 4.7-1 to find $|S_{\text{GB}}^{p_i}|$ for a bandpass positive-gain *RC*-amplifier realization in which $Q = 10$, $f_n = 10$ kHz, and the operational amplifier's gain bandwidth is 1 MHz. Assume that the equal-*R* equal-*C* (design 1) procedure of Sec. 4.3 is used.

(*b*) Verify the expressions given in Table 4.7-1 for $|S_{\text{GB}}^{p_i}|$ and $D(s_n)$ for this filter.

FIVE

RC-AMPLIFIER FILTERS—II

In the preceding chapter we introduced the *RC*-amplifier filter. Our primary concern through most of the chapter was with filters in which the amplifier was a VCVS with a relatively low gain, usually in the range of 1 to 5. In this chapter we shall investigate a somewhat different type of *RC*-amplifier filter, namely one in which the amplifier is a VCVS of (ideally) infinite gain, i.e., an operational amplifier. A filter using such an amplifier as its gain element is called an *infinite-gain filter*, where the word *infinite*, of course, refers to the gain of the active element and not the gain of the overall circuit. We shall see that this type of filter has both advantages and disadvantages when compared with the types discussed in the last chapter.

5.1 SINGLE-AMPLIFIER INFINITE-GAIN FILTERS

In this section we shall consider the first of two types of infinite-gain filter configurations. To derive the first configuration, we begin with the circuit shown in Fig. 4.1-3. Its voltage transfer function is given by (5) of Sec. 4.1. From this relation, letting K approach infinity we obtain

$$\frac{V_2(s)}{V_1(s)} = -\frac{y_{31}(s)}{y_{32}(s)} \tag{1}$$

We now define our first single-amplifier infinite-gain filter configuration by restricting the general passive network to the form shown in Fig. 4.1-4. When the VCVS is replaced with an operational amplifier, the configuration appears as

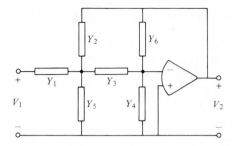

Figure 5.1-1 A general infinite-gain multiple-feedback filter configuration.

shown in Fig. 5.1-1. For this circuit, using (10) and (11) of Sec. 4.1, we obtain the general voltage transfer function

$$\frac{V_2(s)}{V_1(s)} = -\frac{N_{31}(s)}{N_{32}(s)} = \frac{-Y_1 Y_3}{Y_6(Y_1 + Y_2 + Y_3 + Y_5) + Y_2 Y_3} \tag{2}$$

Note that Y_4 does not appear in the above expression. The reason for this is readily seen from Fig. 5.1-1. Since the voltage between the input terminals of an operational amplifier is (ideally) zero, there is no voltage present across the element Y_4, and thus this element can be removed without any effect on the transfer function of the network. Since there are two feedback paths (through the elements Y_2 and Y_6) from the operational amplifier to the passive network, this type of filter is usually referred to as a *multiple-feedback filter*. By comparing (2) with the general form of the various second-order transfer functions discussed in Secs. 4.2 and 4.3 we can now develop the different types of filter realizations.

As a first filter type let us consider a low-pass one having a network function of the form given in (1) of Sec. 4.2. Comparing this to (2) we see it is necessary that $Y_1 = G_1$ and $Y_3 = G_3$. The denominator of (2) can thus be written as

$$D(s) = Y_6(G_1 + Y_2 + G_3 + Y_5) + Y_2 G_3 \tag{3}$$

For this polynomial to be of second order, it is necessary that $Y_2 = G_2$, $Y_5 = sC_5$, and $Y_6 = sC_6$. The resulting realization for a low-pass filter is shown in Fig. 5.1-2.

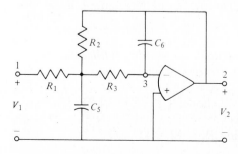

Figure 5.1-2 A low-pass infinite-gain multiple-feedback filter.

Its transfer function is

$$\frac{V_2(s)}{V_1(s)} = \frac{-G_1 G_3}{s^2 C_5 C_6 + s C_6 (G_1 + G_2 + G_3) + G_2 G_3} \tag{4}$$

This may also be put in the form

$$\frac{V_2(s)}{V_1(s)} = \frac{-1/R_1 R_3 C_5 C_6}{s^2 + s\dfrac{1}{C_5}\left(\dfrac{1}{R_1} + \dfrac{1}{R_2} + \dfrac{1}{R_3}\right) + \dfrac{1}{R_2 R_3 C_5 C_6}} \tag{5}$$

Comparing (5) to (1) of Sec. 4.2 gives the following equations

$$\omega_n = \frac{1}{\sqrt{R_2 R_3 C_5 C_6}} \tag{6a}$$

$$\frac{1}{Q} = \sqrt{\frac{C_6}{C_5}}\left(\sqrt{\frac{R_2 R_3}{R_1}} + \sqrt{\frac{R_3}{R_2}} + \sqrt{\frac{R_2}{R_3}}\right) \tag{6b}$$

$$|H_0| = \frac{R_2}{R_1} \tag{6c}$$

A design procedure which allows the selection of standard values of capacitors is given as follows:

Given: H_0, Q, and ω_n.
Choose: $C_5 = C$ (a convenient value).
Calculate:

$$C_6 = mC \qquad m \le \frac{1}{4Q^2(1 + |H_0|)} \tag{7a}$$

$$R_2 = \frac{1}{2\omega_n C m Q}\left[1 \pm \sqrt{1 - 4mQ^2(1 + |H_0|)}\right] \tag{7b}$$

$$R_1 = \frac{R_2}{|H_0|} \tag{7c}$$

$$R_3 = \frac{1}{\omega_n^2 C^2 R_2 m} \tag{7d}$$

Another approach which may be used for the design of the infinite-gain low-pass active filter is given in the Problems.

Example 5.1-1 *A Single-Amplifier Infinite-Gain Multiple-Feedback Low-Pass Filter* It is desired to realize a second-order low-pass Butterworth ($Q = 0.7071$) filter function in which $f_n = \omega_n/2\pi = 100$ Hz and $|H_0| = 1$. Using the design procedure of (7), we choose $C_5 = C = 0.1$ μF. Since from (7a), m must be less than 1/4, we select $m = 0.1$ so that $C_6 = C/10 = 0.01$ μF. We then find $R_1 = R_2 = 199.7$ kΩ and $R_3 = 12.68$ kΩ. $\qquad\square$

One of the major advantages of the single-amplifier infinite-gain multiple-feedback filter is the low sensitivities that this circuit possesses. Using the definitions of Sec. 3.5, from (6) we obtain

$$S_{R_1}^Q = Q\left(\frac{1}{R_1}\sqrt{\frac{R_2 R_3 C_6}{C_5}}\right) \tag{8a}$$

$$S_{R_2}^Q = -\frac{Q}{2}\left(\frac{1}{R_1}\sqrt{\frac{R_2 R_3 C_6}{C_5}} - \sqrt{\frac{R_3 C_6}{R_2 C_5}} + \sqrt{\frac{R_2 C_6}{R_3 C_5}}\right) \tag{8b}$$

$$S_{R_3}^Q = -\frac{Q}{2}\left(\frac{1}{R_1}\sqrt{\frac{R_2 R_3 C_6}{C_5}} + \sqrt{\frac{R_3 C_6}{R_2 C_5}} - \sqrt{\frac{R_2 C_6}{R_3 C_5}}\right) \tag{8c}$$

$$S_{C_5}^Q = -S_{C_6}^Q = \tfrac{1}{2} \tag{8d}$$

$$S_{R_1}^{\omega_n} = 0 \tag{8e}$$

$$S_{R_2,\, R_3,\, C_5,\, C_6}^{\omega_n} = -\tfrac{1}{2} \tag{8f}$$

The various terms enclosed in parentheses for the Q sensitivities of R_1, R_2, and R_3, however, are all less in magnitude than the expression for $1/Q$ given in (6b). Thus the product of Q and any of these bracketed terms must be less than unity in magnitude. We conclude that

$$|S_{R_1}^Q| < 1 \qquad |S_{R_2}^Q| < \tfrac{1}{2} \qquad |S_{R_3}^Q| < \tfrac{1}{2} \tag{9}$$

The sensitivity of this circuit to the operational amplifier gain is also readily shown to be even lower than the values given above.[1] Thus this circuit provides extremely low sensitivities with bounds on the sensitivities that are independent of Q. Similar observations hold for the bandpass and high-pass realizations which use this circuit. These will be discussed later in this section.

It is of interest to examine the circuit of Fig. 5.1-2 in more detail. From (1) we know that the zeros of the transfer admittance $y_{32}(s)$ create the complex pole pair of the overall voltage transfer function. To determine $y_{32}(s)$, it is necessary that terminal 1 be shorted to ground. The resulting passive RC network is shown in Fig. 5.1-3. Note that it is *not* a ladder network. As such the zeros of the transfer

[1] L. P. Huelsman, *Theory and Design of Active RC Circuits*, McGraw-Hill Book Company, New York, 1968, sec. 6.3.

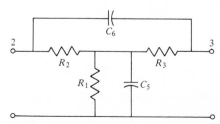

Figure 5.1-3 Determining $y_{32}(s)$ for the passive network in Fig. 5.1-2.

admittance $y_{32}(s)$ can be located anywhere in the left half of the complex frequency plane. Actually the network is simply the well-known bridged-T circuit, with an added shunt resistor. This same configuration will be the one which produces the complex natural frequencies of all the infinite-gain multiple-feedback filter types discussed in this section.

The second infinite-gain multiple-feedback filter we shall consider here is a bandpass one having a network function of the form given in (1) of Sec. 4.3. From (2) we note that there are two possible choices for the realization of this function. Let us first consider the one where $Y_1 = G_1$ and $Y_3 = sC_3$. Thus the denominator of (2) becomes

$$D(s) = Y_6(G_1 + Y_2 + sC_3 + Y_5) + Y_2 sC_3 \tag{10}$$

In order for $D(s)$ to be second order, $Y_2 = sC_2$, $Y_5 = G_5$, and $Y_6 = G_6$. The resulting realization is shown in Fig. 5.1-4. The transfer function is

$$\frac{V_2(s)}{V_1(s)} = \frac{-sG_1 C_3}{s^2 C_2 C_3 + s(G_6 C_2 + G_6 C_3) + G_6(G_1 + G_5)} \tag{11}$$

This can also be written as

$$\frac{V_2(s)}{V_1(s)} = \frac{-\dfrac{s}{R_1 C_2}}{s^2 + s\left(\dfrac{1}{R_6 C_3} + \dfrac{1}{R_6 C_2}\right) + \dfrac{1}{R_6 C_2 C_3}\left(\dfrac{1}{R_1} + \dfrac{1}{R_5}\right)} \tag{12}$$

Comparing this with (1) of Sec. 4.3 results in the following equations

$$\omega_n = \frac{\sqrt{1 + R_5/R_1}}{\sqrt{R_5 R_6 C_2 C_3}} \tag{13a}$$

$$\frac{1}{Q} = \frac{\sqrt{R_5 C_2/R_6 C_3} + \sqrt{R_5 C_3/R_6 C_2}}{\sqrt{1 + R_5/R_1}} \tag{13b}$$

$$|H_0| = \frac{R_6/R_1}{1 + C_2/C_3} \tag{13c}$$

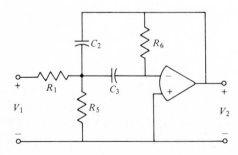

Figure 5.1-4 A bandpass infinite-gain multiple-feedback filter.

A design procedure can be obtained by letting $C_2 = C_3 = C$. Thus we can solve for R_1, R_5, and R_6 as follows:

$$R_1 = \frac{Q}{\omega_n C |H_0|} \tag{14a}$$

$$R_5 = \frac{Q}{(2Q^2 - |H_0|)\omega_n C} \tag{14b}$$

$$R_6 = \frac{2Q}{\omega_n C} \tag{14c}$$

A set of nomographs that allow implementation of (14) has appeared in the literature.[1]

Example 5.1-2 *A Single-Amplifier Infinite-Gain Multiple-Feedback Bandpass Filter* It is desired to realize a second-order bandpass filter in which $|H_0| = 2$, $Q = 2$, and $\omega_n = 10$ krad/s. If we choose $C_2 = C_3 = C = 0.01$ μF, then from (14), $R_1 = 10$ kΩ, $R_5 = 3.33$ kΩ, and $R_6 = 40$ kΩ. $\square$

It is possible to eliminate R_5 from the circuit shown in Fig. 5.1-4 and still achieve a bandpass realization. However, if this is done $|H_0|$ is no longer a free parameter. For example, if $C_2 = C_3$ and $R_5 = \infty$, then (13b) and (13c) become

$$\frac{1}{Q} = 2\sqrt{R_1/R_6} \tag{15a}$$

$$|H_0| = \frac{R_6}{2R_1} \tag{15b}$$

Thus $|H_0|$ is equal to $2Q^2$. For large values of Q, $|H_0|$ will be very large. This is undesirable since if similar stages were cascaded, the overall gain would become too large, and the allowable signal level and the stability of the resulting filter would be seriously degraded. Another bandpass realization is possible if we choose $Y_1 = sC_1$ and $Y_3 = G_3$. The resulting circuit requires three capacitors rather than the two needed in the one described above. A treatment of it is left to the reader as an exercise.

The final multiple-feedback filter realization to be considered in this section is the high-pass one having the form of (16) of Sec. 4.3. Using Table 4.3-1 to transform the low-pass circuit of Fig. 5.1-2 results in the filter shown in Fig. 5.1-5, in which $Y_1 = sC_1$, $Y_2 = sC_2$, $Y_3 = sC_3$, $Y_5 = G_5$, and $Y_6 = G_6$. The transfer function is

$$\frac{V_2(s)}{V_1(s)} = \frac{-s^2 C_1 C_3}{s^2 C_2 C_3 + s(G_6 C_1 + G_6 C_2 + G_6 C_3) + G_5 G_6} \tag{16}$$

[1] N. Doyle, "Swift, Sure Design of Active and Bandpass Filters," *Electronic Design News*, January 15, 1970, pp. 43–47.

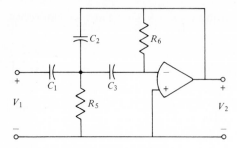

Figure 5.1-5 A high-pass infinite-gain multiple-feedback filter.

This may be put in the form

$$\frac{V_2(s)}{V_1(s)} = \frac{-s^2 \dfrac{C_1}{C_2}}{s^2 + s\dfrac{1}{R_6}\left(\dfrac{C_1}{C_2 C_3} + \dfrac{1}{C_2} + \dfrac{1}{C_3}\right) + \dfrac{1}{R_5 R_6 C_2 C_3}} \tag{17}$$

Note that three capacitors are required; i.e., the realization is not canonic. By equating (17) to (16) of Sec. 4.3, we obtain

$$\omega_n = \frac{1}{\sqrt{R_5 R_6 C_2 C_3}} \tag{18a}$$

$$\frac{1}{Q} = \sqrt{\frac{R_5}{R_6}}\left(\frac{C_1}{\sqrt{C_2 C_3}} + \sqrt{\frac{C_3}{C_2}} + \sqrt{\frac{C_2}{C_3}}\right) \tag{18b}$$

$$|H_0| = \frac{C_1}{C_2} \tag{18c}$$

A set of design equations can be developed if $C_1 = C_3 = C$, where C is selected as a convenient value. The equations are

$$R_5 = \frac{|H_0|}{\omega_n Q C(2|H_0| + 1)} \tag{19a}$$

$$R_6 = \frac{(2|H_0| + 1)Q}{\omega_n C} \tag{19b}$$

$$C_2 = \frac{C}{|H_0|} \tag{19c}$$

Example 5.1-3 *Single-Amplifier Infinite-Gain Multiple-Feedback High-Pass Filter* It is desired to realize a second-order high-pass Butterworth ($Q = 0.707$) transfer function where $f_n = \omega_n/2\pi = 100$ Hz and $|H_0| = 1$. Choosing $C = C_1 = C_3 = 0.1\ \mu$F, we find $C_2 = 0.1\ \mu$F, $R_5 = 7.503$ kΩ, and $R_6 = 33.757$ kΩ. $\square$

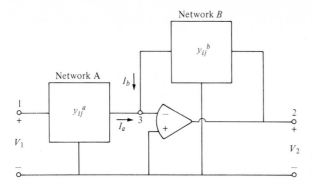

Figure 5.1-6 A general infinite-gain single-feedback filter configuration.

A second single-amplifier infinite-gain filter configuration is the *single-feedback* type.[1] Its basic form is shown in Fig. 5.1-6. The networks A and B are passive *RC* networks with y parameters $y_{ij}^a(s)$ and $y_{ij}^b(s)$. Analysis of this circuit shows that since the voltage at node 3 is zero and no current flows into the operational amplifier,

$$I_a(s) = -y_{12}^a(s)V_1(s) \qquad I_b(s) = -y_{12}^b(s)V_2(s) \tag{20}$$

Since $I_a(s) = -I_b(s)$, the transfer function is readily found to be

$$\frac{V_2(s)}{V_1(s)} = -\frac{y_{12}^a(s)}{y_{12}^b(s)} \tag{21}$$

The transfer admittances of (21) may be expressed in terms of their numerator and denominator polynomials as

$$y_{12}^a(s) = \frac{N_{12}^a(s)}{D_{12}^a(s)} \qquad y_{12}^b(s) = \frac{N_{12}^b(s)}{D_{12}^b(s)} \tag{22}$$

Substituting these into (21) we obtain

$$\frac{V_2(s)}{V_1(s)} = \frac{-N_{12}^a(s)D_{12}^b(s)}{N_{12}^b(s)D_{12}^a(s)} \tag{23}$$

We now assume that the passive networks are so chosen that the poles of $y_{12}^a(s)$ are the same as the poles of $y_{12}^b(s)$ so that $D_{12}^b(s)$ cancels with $D_{12}^a(s)$, giving

$$\frac{V_2(s)}{V_1(s)} = -\frac{N_{12}^a(s)}{N_{12}^b(s)} \tag{24}$$

Since they are the numerators of *RC* network transfer functions, $N_{12}^a(s)$ and $N_{12}^b(s)$ may have complex conjugate roots. Thus in (24) we may realize both complex poles and complex zeros.

[1] F. R. Bradley and R. McCoy, "Driftless D-C Amplifiers," *Electronics*, vol. 25, no. 4, April 1952, pp. 144–148. Also, R. Brennan and A. Bridgman, "Simulation of Transfer Functions Using Only One Operational Amplifier," *IRE WESCON Conv. Rec.*, vol. 1, 1957, pp. 273–277.

Table 5.1-1 RC passive networks suitable for y_{ij}^a and y_{ij}^b

Type	Network	Transfer admittance	Element values, ohms and farads
1		$y_{12} = \dfrac{-ks}{s+a}$	$k = \dfrac{1}{R}$ $a = \dfrac{1}{RC}$
2		$y_{12} = \dfrac{-k}{s+a}$	$k = \dfrac{1}{R_1 R_2 C}$ $a = \dfrac{1}{C}\left(\dfrac{1}{R_1}+\dfrac{1}{R_2}\right)$
3		$y_{12} = \dfrac{-ks^2}{s+a}$	$k = \dfrac{C_1 C_2}{C_1+C_2}$ $a = \dfrac{1}{R(C_1+C_2)}$
4		$y_{12} = \dfrac{-(s^2+as+1)}{s+a}$ $\dfrac{1}{2}<a<2$	$R_1 = \dfrac{1}{2.5-a}$ $R_2 = a - R_1$ $C_1 = 1$ $C_2 = \dfrac{1}{R_1 R_2}$
5		$y_{12} = \dfrac{-(s+1)(s^2+as+1)}{(s+\sigma_1)(s+\sigma_2)}$ $a<1$	$\sigma_1 = \dfrac{C_1+C_2}{C_3}$ $\sigma_2 = \dfrac{1}{\sigma_1}$ $\dfrac{1}{R_1} = (2.5-a)\dfrac{1+a}{2+a}$ $C_1 = \dfrac{1+a}{2+a}$ $\dfrac{1}{R_2} = \dfrac{C_1}{C_1-1}$ $C_3 = \dfrac{1}{R_3} = \dfrac{C_1 C_2}{1+a}$

A collection of useful networks realizing various $y_{12}^a(s)$ and $y_{12}^b(s)$ is given in Table 5.1-1.[1] The transfer admittance of each of these networks is also given in the table. The first three networks provide transmission zeros at the origin or infinity and are useful for realizing the numerators of low-pass, bandpass, and high-pass transfer functions. The transfer admittances for these networks each contain an $s + a$ denominator term which must be cancelled by a similar term in the denominator of the transfer admittance for the B network. The fourth and fifth networks are useful for realizing complex roots, either in the numerator or the denominator of the overall voltage transfer functions. The fourth network is a bridged-T which is useful for low-Q complex roots. If high-Q roots are desired, then the fifth network which is a twin-T network is more useful. The numerator of the transfer admittance for the twin-T network is third order while the denominator is second order. The design approach for this case assumes that one of the real poles, σ_1 or σ_2, of the denominator will be approximately cancelled by the numerator root at -1. Consequently, the remaining pole can be cancelled by the $s + a$ term in the denominator of the transfer admittance for the other network. This assumption turns out to be a good one if the Q of the complex poles, which are to be realized by the numerator of the twin-T network, are high.

Example 5.1-4 *A Single-Amplifier Infinite-Gain Single-Feedback Low-Pass Filter* It is desired to realize the following normalized transfer function

$$\frac{V_2(s)}{V_1(s)} = \frac{-10}{s^2 + \sqrt{2}\,s + 1} \tag{25}$$

Since this is a low-pass filter, from Table 5.1-1 we choose a type 2 circuit for network A and a type 4 circuit for network B. The constant a must be chosen as $\sqrt{2}$. Thus, from Table 5.1-1, for network B we obtain $R_1 = 0.921\ \Omega$, $R_2 = 0.493\ \Omega$, $C_1 = 1$ F, and $C_2 = 2.202$ F. If we let $R = R_1 = R_2$ for the type 2 circuit for network A, then we get $R = a/(2|H_0|)$. If $C = 1$ F, then $R = 0.07071\ \Omega$. The circuit is shown in Fig. 5.1-7. ☐

[1] J. V. Wait, L. P. Huelsman, and G. A. Korn, *Introduction to Operational Amplifier Theory and Applications*, McGraw-Hill Book Company, New York, 1975, chap. 4.

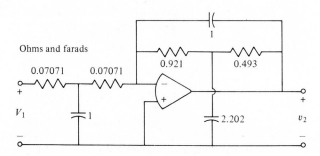

Figure 5.1-7 The low-pass infinite-gain single-feedback filter of Example 5.1-4.

The single-feedback infinite-gain realization described above has several disadvantages which are mostly caused by the fact that networks *A* and *B* must separately realize the same natural frequencies. Because of this requirement, more passive elements are required than are needed for other types of active filters. For example, the low-pass realization of Example 5.1-4 requires three capacitors and three resistors, compared to two capacitors and two or three resistors in other equivalent filters. In addition, the passive elements must be of high quality and low tolerance in order to achieve a sufficient degree of cancellation of the denominator polynomials. The tuning of this structure is thus complicated by the cancellation requirement. To offset these disadvantages, some of the advantages of this structure are the ability to satisfactorily realize high-*Q* circuits and to achieve a large value of H_0 without having large element spreads. In addition, the single-feedback infinite-gain realization can realize a completely general second-order network function, i.e., one which has both complex zeros and complex poles and is of the general form discussed in Sec. 4.4.

5.2 MULTIPLE-AMPLIFIER FILTERS

The technology of making active devices has advanced to the point where certain traditional guidelines are no longer valid. One of these guidelines was that one should minimize the number of active devices. This viewpoint is reflected in the single-amplifier filters discussed in the preceding sections. With respect to modern integrated circuit technology, however, there is frequently little advantage to minimizing the number of active elements. Therefore if multiple-amplifier realizations can provide better performance, such realizations may be preferable to single-amplifier ones. The purpose of this section is to introduce two multiple-amplifier realizations. These two realizations are called the *state variable* and the *resonator*. They use from two to four operational amplifiers depending upon the specific filter characteristics desired.

The *state-variable filter* realization, also called the *KHN filter* from the names of its originators,[1] has extreme flexibility, good performance, and low sensitivities. These characteristics have made it the workhorse of commercially available active filters. The name *state variable* is derived from the fact that state-variable methods of solving differential equations are used in the development of the realization. To see this, consider the inverting second-order bandpass transfer function

$$\frac{V_2(s)}{V_1(s)} = \frac{-|H|s}{s^2 + a_1 s + a_0} \tag{1}$$

---

[1] W. J. Kerwin, L. P. Huelsman, and R. W. Newcomb, "State-Variable Synthesis for Insensitive Integrated Circuit Transfer Functions," *IEEE J. Solid-State Circuits*, vol. SC-2, September 1967, pp. 87–92. An earlier reference to this configuration was given in W. H. Schussler, *On the Representation of Transfer Functions and Networks on Analog Computers*, Westdeutscher Verlag, Cologne, 1961.

In this equation we now introduce an arbitrary frequency-domain variable $X(s)$ and multiply both the numerator and denominator by $X(s)/s^2$ to achieve

$$\frac{V_2(s)}{V_1(s)} = \frac{-|H|\dfrac{X(s)}{s}}{X(s) + \dfrac{a_1 X(s)}{s} + \dfrac{a_0 X(s)}{s^2}} \tag{2}$$

If we separately equate the numerators and denominators of both members of (2), two equations result

$$X(s) = V_1(s) - \frac{a_1 X(s)}{s} - \frac{a_0 X(s)}{s^2} \tag{3a}$$

$$V_2(s) = -|H|\frac{X(s)}{s} \tag{3b}$$

If we now take the inverse Laplace transform of both sides of the equations of (3), we get the following integral (time-domain) equations

$$x(t) = v_1(t) - a_1 \int x(t)\,dt - a_0 \int \left[\int x(t)\,dt \right] dt \tag{4a}$$

$$v_2(t) = -|H| \int x(t)\,dt \tag{4b}$$

where $x(t) = L^{-1}[X(s)]$. The quantities $x(t)$, $\int x(t)\,dt$, and $\int [\int x(t)\,dt]\,dt$ are called *state variables*,[1] which accounts for the name of the filter. An analog-computer block diagram for solving (4) is shown in Fig. 5.2-1. This block diagram is readily converted to a filter circuit by using operational amplifiers to model the inverting

[1] P. M. DeRusso, R. J. Roy, and C. M. Close, *State Variables for Engineers*, John Wiley & Sons, Inc., New York, 1965.

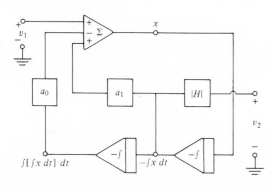

Figure 5.2-1 The general state-variable filter configuration.

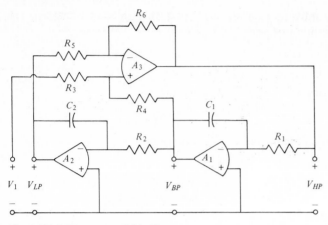

Figure 5.2-2 The state-variable filter.

integrators. The result is shown in Fig. 5.2-2. The bandpass transfer function for this circuit is developed by first observing that

$$V_{LP}(s) = \frac{-V_{BP}(s)}{sR_2C_2} \qquad V_{BP}(s) = \frac{-V_{HP}(s)}{sR_1C_1} \tag{5}$$

where $V_{LP}(s)$, $V_{BP}(s)$, and $V_{HP}(s)$ are respectively the low-pass, bandpass, and high-pass outputs. In addition one can express $V_{HP}(s)$ as

$$V_{HP}(s) = -\frac{R_6}{R_5} V_{LP}(s) + \frac{R_4}{R_3 + R_4} \frac{R_5 + R_6}{R_5} V_1(s) + \frac{R_3}{R_3 + R_4} \frac{R_5 + R_6}{R_5} V_{BP}(s) \tag{6}$$

Using (5) to eliminate $V_{LP}(s)$ and $V_{HP}(s)$ we obtain

$$V_{BP}(s) = -\frac{1}{sR_1C_1} \left[\frac{R_6}{R_5} \frac{V_{BP}(s)}{sR_2C_2} + \frac{1 + R_6/R_5}{1 + R_3/R_4} V_1(s) + \frac{1 + R_6/R_5}{1 + R_4/R_3} V_{BP}(s) \right] \tag{7}$$

Solving for $V_{BP}(s)/V_1(s)$ in the above and using (5) we obtain the bandpass, low-pass, and high-pass transfer functions

$$\frac{V_{BP}(s)}{V_1(s)} = \frac{- \left[\dfrac{1 + R_6/R_5}{1 + R_3/R_4} \dfrac{s}{R_1C_1} \right]}{D(s)} \tag{8a}$$

$$\frac{V_{LP}(s)}{V_1(s)} = \frac{\dfrac{1 + R_6/R_5}{1 + R_3/R_4} \dfrac{1}{R_1R_2C_1C_2}}{D(s)} \tag{8b}$$

$$\frac{V_{HP}(s)}{V_1(s)} = \frac{\dfrac{1 + R_6/R_5}{1 + R_3/R_4} s^2}{D(s)} \tag{8c}$$

$$D(s) = s^2 + \frac{s}{R_1C_1} \frac{1 + R_6/R_5}{1 + R_4/R_3} + \frac{R_6/R_5}{R_1R_2C_1C_2} \tag{8d}$$

Note that the low-pass and high-pass realizations are noninverting, whereas the bandpass is inverting. The expressions for ω_n and Q are given as

$$\omega_n = \sqrt{\frac{R_6/R_5}{R_1 R_2 C_1 C_2}} \tag{9a}$$

$$\frac{1}{Q} = \frac{1 + R_6/R_5}{1 + R_4/R_3} \sqrt{\frac{R_5 R_2 C_2}{R_6 R_1 C_1}} \tag{9b}$$

H_0 will be different for all three realizations. It is given as follows:

Low-pass: $$H_0 = \frac{1 + R_5/R_6}{1 + R_3/R_4} \tag{10a}$$

[See (1) of Sec. 4.2]

Bandpass: $$H_0 = -R_4/R_3 \tag{10b}$$
[See (1) of Sec. 4.3]

High-pass: $$H_0 = \frac{1 + R_6/R_5}{1 + R_3/R_4} \tag{10c}$$

[See (16) of Sec. 4.3]

If we choose $R_5 = R_6$, $R_1 = R_2 = R$, and $C_1 = C_2 = C$, (9) becomes

$$\omega_n = \frac{1}{RC} \tag{11a}$$

$$\frac{1}{Q} = \frac{2}{1 + R_4/R_3} \tag{11b}$$

For these choices the following design procedure may be used:

1. Assume ω_n and Q are specified.
2. Choose convenient values for $C_1 = C_2 = C$ and $R_3 = R_5 = R_6$.
3. Calculate:

$$R_1 = R_2 = \frac{1}{\omega_n C} \tag{12a}$$

$$R_4 = (2Q - 1)R_3 \tag{12b}$$

4. H_0 becomes:

Low-pass and high-pass: $$H_0 = \frac{2Q - 1}{Q} \tag{13a}$$

Bandpass: $$H_0 = 1 - 2Q \tag{13b}$$

Example 5.2-1 *A State-Variable Bandpass Filter* It is desired to design a bandpass state-variable filter with $Q = 20$ and $f_n = 1$ kHz. Let us choose $C_1 = C_2 = 0.01 \ \mu F$ and $R_3 = R_5 = R_6 = 10$ kΩ. Thus, from (12) we get

$R_1 = R_2 = 15.9$ kΩ and $R_4 = 390$ kΩ. From (10) or (13), $H_0 = -39.0$. This same realization yields low-pass and high-pass transfer functions with $Q = 20$ and $f_n = 1$ kHz at the appropriate output terminals of Fig. 5.2-2. H_0 for these realizations is 1.95. ☐

One of the reasons for the popularity of the state-variable filter is the low sensitivities that characterize its performance. Using the definitions of Sec. 3.5 and the relations of (9) we find

$$S^Q_{R_1, C_1} = -S^Q_{R_2, C_2} = -S^{\omega_n}_{R_1, R_2, R_5, C_1, C_2} = +S^{\omega_n}_{R_6} = \tfrac{1}{2} \tag{14a}$$

$$S^Q_{R_3} = \frac{-1}{1 + R_3/R_4} = -S^Q_{R_4} \tag{14b}$$

$$S^Q_{R_5} = -S^Q_{R_6} = -\frac{Q}{2} \frac{R_5 - R_6}{1 + R_4/R_3} \sqrt{\frac{R_2 C_2}{R_5 R_6 R_1 C_1}} \tag{14c}$$

An examination of these relations shows that the Q sensitivity for R_5 and R_6 can be set to zero by choosing $R_5 = R_6$. The remaining sensitivities are all less than unity in magnitude, and most of them have a magnitude of $\tfrac{1}{2}$. The Q sensitivities of the realization to the gains of the operational amplifiers are even smaller, being of the order of Q/K_0, where K_0 is the open-loop operational amplifier gain.[1] As a result of its low sensitivities, the state-variable filter has been successfully used to realize functions with Qs in the hundreds. The concept of such a filter has also been extended to the *n*th-order case.[2]

The second multiple-amplifier filter to be discussed in this section is the *resonator* active filter.[3] The basic approach used here is to take an *RC* oscillator and apply degenerative feedback. This technique has also been referred to as producing a "biquad" filter. We shall prefer the name resonator since biquad is also used to describe a class of transfer functions. The resonator can be developed by considering the general analog computer block diagram for a quadrature oscillator as shown in Fig. 5.2-3. Without the degenerate feedback, this circuit solves the differential equation

$$x''(t) + \omega_0^2 x(t) = 0 \tag{15}$$

which has the solution

$$x(t) = A \sin \omega_0 t \tag{16}$$

Implementing Fig. 5.2-3 with operational amplifiers results in the circuit shown in Fig. 5.2-4, which is called the *resonator active filter*. The resistor R_1 serves as the

---

[1] Huelsman, op. cit., p. 209.

[2] Huelsman, op. cit., p. 205.

[3] J. Tow, "Design Formulas for Active RC Filters Using Operational Amplifier Biquad," *Electronic Letters*, July 24, 1969, pp. 339–341.

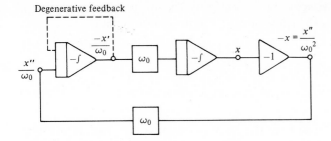

Figure 5.2-3 The general resonator filter configuration.

degenerate feedback path. The input signal V_1 is injected into the circuit through R_4. To analyze this filter we begin by finding $V_{BP}(s)/V_1(s)$. $V_{BP}(s)$ can be expressed as

$$V_{BP}(s) = \frac{-\dfrac{1}{R_4 C_1}}{s + \dfrac{1}{R_1 C_1}} V_1(s) - \frac{\dfrac{1}{R_3 C_1}}{s + \dfrac{1}{R_1 C_1}} V_x(s) \tag{17}$$

However, we see that

$$V_x(s) = -V_{LP}(s) = \frac{V_{BP}(s)}{s R_2 C_2} \tag{18}$$

so that substituting (18) into (17) results in the bandpass transfer function of the resonator active filter.

$$\frac{V_{BP}(s)}{V_1(s)} = \frac{-\dfrac{s}{R_4 C_1}}{s^2 + \dfrac{s}{R_1 C_1} + \dfrac{1}{R_2 R_3 C_1 C_2}} \tag{19}$$

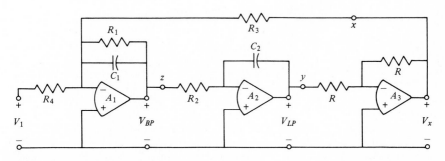

Figure 5.2-4 The resonator filter.

This transfer function is attractive in its simplicity. The low-pass transfer function can now be readily derived by substituting (19) into (18) to obtain

$$\frac{V_{LP}(s)}{V_1(s)} = \frac{\dfrac{1}{R_2 R_4 C_1 C_2}}{s^2 + \dfrac{s}{R_1 C_1} + \dfrac{1}{R_2 R_3 C_1 C_2}} \tag{20}$$

If the low-pass output is taken from the output of the inverter, then an inverting low-pass realization may also be obtained. The ideal integrator between x and y may be interchanged with the unity gain inverter between y and z to provide a noninverting bandpass transfer function at V_x. The three-amplifier circuit of Fig. 5.2-4 cannot be used to provide a high-pass transfer function.

Equating the denominator of (19) to that of the standard second-order bandpass function of (1) of Sec. 4.2, we obtain

$$\omega_n = \frac{1}{\sqrt{R_2 R_3 C_1 C_2}} \tag{21a}$$

$$\frac{1}{Q} = \frac{1}{R_1} \sqrt{\frac{R_2 R_3 C_2}{C_1}} \tag{21b}$$

From these relations we readily find that the Q and ω_n sensitivities are very low, namely, they all have magnitudes of either one or $\frac{1}{2}$. The expressions for H_0 are

Low-pass:
$$H_0 = \frac{R_3}{R_4} \tag{22a}$$

Bandpass:
$$|H_0| = \frac{R_1}{R_4} \tag{22b}$$

A design procedure can now be developed by assuming that $R_2 = R_3 = R$ and $C_1 = C_2 = C$, and proceeding as follows:

1. Assume ω_n, Q, and H_0 are specified.
2. Let $R_2 = R_3 = R$ and $C_1 = C_2 = C$.
3. Select either R or C and solve for the other, using

$$\omega_n = \frac{1}{RC} \tag{23a}$$

4. Calculate:
$$R_1 = QR \tag{23b}$$

Low-pass:
$$R_4 = \frac{R}{H_0} \tag{23c}$$

Bandpass:
$$R_4 = \frac{R_1}{|H_0|} \tag{23d}$$

Example 5.2-2 *A Resonator Filter* It is desired to design a low-pass and a bandpass second-order resonator active filter with $Q = 20$, $f_n = 1$ kHz, and $|H_0| = 1$. Let us choose $C_1 = C_2 = C = 0.01$ μF. Thus $R = 15.915$ kΩ, and $R_1 = 318.31$ kΩ. For the low-pass realization $R_4 = 15.915$ kΩ and for the bandpass realization $R_4 = 318.31$ kΩ. $\square$

In this section we have discussed two of the most useful realizations for second-order active filters that will be considered in this text. Of the two, the state-variable filter has the advantage of being more versatile, in that it can be used to realize a high-pass filter as well as a low-pass or bandpass one. The state-variable configuration of Fig. 5.2-2 is also capable of realizing (1) of Sec. 4.4 with the use of one additional amplifier. The amplifier is used to sum the low-pass, bandpass, and high-pass outputs to yield a biquadratic realization. If we wish to realize (1) with H, b_1, and b_0 all greater than 0, then we may use the configuration shown in Fig. 5.2-5. Since the low-pass and high-pass realizations are already positive they are applied to the noninverting input of A_4. The sign of the bandpass transfer function is changed by applying it to the inverting input of A_4. We may write V_2 as

$$V_2(s) = \frac{1 + R_9/R_{10}}{1 + R_7/R_8} V_{\text{HP}}(s) - \frac{R_9}{R_{10}} V_{\text{BP}}(s) + \frac{1 + R_9/R_{10}}{1 + R_8/R_7} V_{\text{LP}}(s) \qquad (24)$$

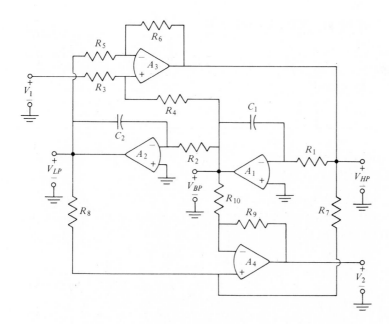

Figure 5.2-5 A state-variable realization for biquadratic network functions.

Using (8) and (24) we obtain

$$\frac{V_2(s)}{V_1(s)} = \frac{1 + R_9/R_{10}}{1 + R_7/R_8} \frac{1 + R_6/R_5}{1 + R_3/R_4} \frac{s^2 + \dfrac{s}{R_1 C_1} \dfrac{1 + R_7/R_8}{1 + R_{10}/R_9} + \dfrac{R_7/R_8}{R_1 R_2 C_1 C_2}}{s^2 + \dfrac{s}{R_1 C_1} \dfrac{1 + R_6/R_5}{1 + R_4/R_3} + \dfrac{R_6/R_5}{R_1 R_2 C_1 C_2}} \tag{25}$$

One approach to simplifying these equations is to let $R_3 = R_5 = R_8 = R_9 = 1$ and $R_1 C_1 = R_2 C_2 = 1$ s. Thus

$$\frac{V_2(s)}{V_1(s)} = \frac{R_4}{R_{10}} \frac{(1 + R_6)(1 + R_{10})}{(1 + R_4)(1 + R_7)} \frac{s^2 + s\dfrac{1 + R_7}{1 + R_{10}} + R_7}{s^2 + s\dfrac{1 + R_6}{1 + R_4} + R_6} \tag{26}$$

Equating (1) of Sec. 4.4 to (26) results in

$$\omega_z = \sqrt{R_7} \qquad Q_z = \sqrt{R_7} \frac{1 + R_{10}}{1 + R_7} \tag{27a}$$

$$\omega_p = \sqrt{R_6} \qquad Q_p = \sqrt{R_6} \frac{1 + R_4}{1 + R_6} \tag{27b}$$

From these equations we see that the state-variable biquad may be designed by using R_6 and R_7 to control ω_p and ω_z and R_4 and R_{10} to control Q_p and Q_z. In addition, zeros can be shifted to the right half of the complex plane by taking R_{10} to the noninverting side of the output summer A_4. In this case another resistor must be connected from the inverting terminal of A_4 to ground and the element values for R_7, R_8, and R_{10} must be recalculated (see Problems).

The resonator can also be used to develop a realization of a biquadratic transfer function. To modify this circuit we first rotate it counterclockwise until the output of operational amplifier A_1 is at the far left. An additional capacitor C_3 is added and the input signal is then applied simultaneously to each amplifier as shown in Fig. 5.2-6. The transfer function of this circuit can be shown to be

$$\frac{V_2(s)}{V_1(s)} = \frac{-C_3}{C_1} \frac{s^2 + s\left(\dfrac{1}{R_4} - \dfrac{R_6}{R_8 R_3}\right)\dfrac{1}{C_3} + \dfrac{R_6}{R_3 R_5 R_7 C_2 C_3}}{s^2 + \dfrac{s}{R_1 C_1} + \dfrac{R_6}{R_2 R_3 R_5 C_1 C_2}} \tag{28}$$

The generality of this realization is shown in Table 5.2-1 where the type of second-order function and the corresponding elements are given. Equating (1) of Sec. 4.4 and (28) we obtain

$$\omega_z = \sqrt{\frac{R_6}{R_3 R_5 R_7 C_2 C_3}} \qquad \frac{1}{Q_z} = \left(\frac{1}{R_4} - \frac{R_6}{R_3 R_8}\right)\sqrt{\frac{R_3 R_5 R_7 C_2}{R_6 C_3}} \tag{29a}$$

$$\omega_p = \sqrt{\frac{R_6}{R_2 R_3 R_5 C_1 C_2}} \qquad \frac{1}{Q_p} = \frac{1}{R_1}\sqrt{\frac{R_2 R_3 R_5 C_2}{R_6 C_1}} \tag{29b}$$

Example 5.2-3 It is desired to design a second-order realization using the circuit of Fig. 5.2-6. The specifications are $f_z = 1.6$ kHz, $f_p = 1.5$ kHz,

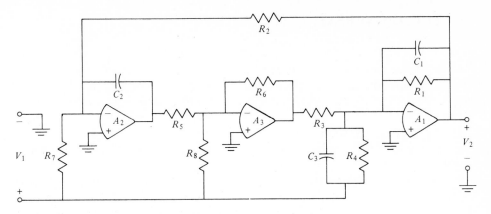

Figure 5.2-6 A resonator realization for biquadratic network functions.

$Q_z = \infty$, and $Q_p = 10$. It is also desired to have a low-frequency gain of -1. From Table 5.2-1 we see that $R_4 = R_8 = \infty$. If we select $R_3 = R_5 = R_6 = 10 \text{ k}\Omega$, and $C_1 = C_2 = 0.01 \ \mu\text{F}$, then from (29b) we have $R_1 = 106.1 \text{ k}\Omega$ and $R_2 = 11.26 \text{ k}\Omega$. The low-frequency gain is given as $-(R_2/R_7)$. Thus to achieve the gain specification, $R_7 = 11.3 \text{ k}\Omega$. Since all elements but C_3 are specified, we use (29a) and ω_z to find that $C_3 = 8.76 \times 10^{-9}$ F.   □

It should be observed that Fig. 5.2-6 is also capable of a high-pass, second-order realization.

Table 5.2-1 Design and tuning relationships for Fig. 5.2-6

Type of second-order transfer function	Element constraints	Tuning elements
Low-pass	$R_4 = R_8 = \infty$ $C_3 = 0$	R_1, R_2, R_7
Bandpass	R_4 or $R_8 = \infty$ $R_7 = \infty$ $C_3 = 0$	$R_1, R_2,$ and R_4 or R_8, whichever is not infinite
High-pass	$R_4 = R_7 = R_8 = \infty$	$R_1, R_2,$ and $C_3{}^*$
Biquad	$R_4 = \infty$	$R_1, R_2, R_7,$ $R_8,$ and $C_3{}^*$
$j\omega$-axis zeros	$R_4 = R_8 = \infty$	$R_1, R_2, R_7,$ and $C_3{}^*$

* Note that C_3 may be fixed if the passband gain is a free parameter.

5.3 THE UNIVERSAL ACTIVE FILTER

One of the limitations of the state-variable filter described in the preceding section is that the bandpass network function can only be realized in an inverting form, while the low-pass and high-pass functions can only be realized as noninverting ones. In addition, for the specified design procedure, the value of the multiplicative constant H_0 is not free to be chosen. In this section we present a *modified state-variable filter* realization which overcomes these difficulties. This circuit is commonly called the *universal active filter*, and in packaged form it is commercially available from several sources.[1]

The modified state-variable filter is shown in Fig. 5.3-1. It differs from the original circuit given in Fig. 5.2-2 in that two resistors, R_7 and R_8, have been added, thus forming a new input V_{inB}. The original input has been relabeled as V_{inA}. Analysis of Fig. 5.3-1 begins by assuming $V_{inB} = 0$, and expressing $V_{HP}(s)$ as a function of all the inputs to the amplifier A_3. We get

$$V_{HP}(s) = \left[V_{inA}(s) \frac{R_4 \| R_7}{R_3 + R_4 \| R_7} + V_{BP}(s) \frac{R_3 \| R_7}{R_4 + R_3 \| R_7} \right] \left(1 + \frac{R_6}{R_5 \| R_8} \right) - \frac{R_6}{R_5} V_{LP}(s) \tag{1}$$

where the symbol $\|$ stands for *the parallel connection of.* To simplify the algebra, let us define

$$K_1 = \frac{R_4 \| R_7}{R_3 + R_4 \| R_7} = \frac{R_4 R_7}{R_3 R_7 + R_3 R_4 + R_4 R_7} = \frac{1}{1 + R_3/R_4 + R_3/R_7} \tag{2a}$$

$$K_2 = \frac{R_3 \| R_7}{R_4 + R_3 \| R_7} = \frac{R_3 R_7}{R_3 R_4 + R_3 R_7 + R_4 R_7} = \frac{1}{1 + R_4/R_3 + R_4/R_7} \tag{2b}$$

$$K_3 = \frac{R_6}{R_5} \tag{2c}$$

$$K_4 = \frac{R_6}{R_8} \tag{2d}$$

In terms of these quantities, (1) may be rewritten as

$$V_{HP}(s) = [K_1 V_{inA}(s) + K_2 V_{BP}(s)](1 + K_3 + K_4) - K_3 V_{LP}(s) \tag{3}$$

If we now define

$$\omega_1 = \frac{1}{R_1 C_1} \quad \text{and} \quad \omega_2 = \frac{1}{R_2 C_2} \tag{4}$$

---

[1] Beckman model 881, Baldwin Electronics, Inc. model FS-50 (FS-51), Integrated Microsystem model μAR-2000, National Semiconductor model AF100, Burr Brown Research Corp. model UAF41, and General Instrument Corp. model ACF 7092C.

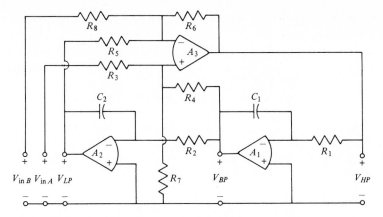

Figure 5.3-1 The modified state-variable (universal) active filter.

and observe that

$$V_{BP}(s) = -\frac{\omega_1}{s} V_{HP}(s) \tag{5}$$

and

$$V_{LP}(s) = -\frac{\omega_2}{s} V_{BP}(s) = \frac{\omega_1 \omega_2}{s^2} V_{HP}(s) \tag{6}$$

then (3) can be written as follows:

$$V_{HP}(s) = -(1 + K_3 + K_4)\frac{\omega_1 K_2}{s} V_{HP}(s)$$

$$- \frac{K_3 \omega_1 \omega_2}{s^2} V_{HP}(s) + (1 + K_3 + K_4)K_1 V_{inA}(s) \tag{7}$$

Solving this, the high-pass transfer function for the circuit in Fig. 5.3-1 with $V_{inB} = 0$ is

$$\frac{V_{HP}(s)}{V_{inA}(s)} = \frac{(1 + K_3 + K_4)K_1 s^2}{s^2 + (1 + K_3 + K_4)K_2 \omega_1 s + K_3 \omega_1 \omega_2} \tag{8}$$

If $V_{inB} = 0$, then we may set $R_8 = \infty$ so that $K_4 = 0$. In this case (8) can be written as

$$\frac{V_{HP}(s)}{V_{inA}(s)} = \frac{(1 + K_3)K_1 s^2}{s^2 + (1 + K_3)K_2 \omega_1 s + \omega_1 \omega_2 K_3} \tag{9}$$

The bandpass and low-pass transfer functions for Fig. 5.3-1 with $V_{inB} = 0$ and $R_8 = \infty$ may be found from the above using (5) and (6). Thus we obtain

$$\frac{V_{BP}(s)}{V_{inA}(s)} = \frac{-(1 + K_3)K_1 \omega_1 s}{s^2 + (1 + K_3)K_2 \omega_1 s + \omega_1 \omega_2 K_3} \tag{10}$$

and

$$\frac{V_{LP}(s)}{V_{inA}(s)} = \frac{(1 + K_3)K_1 \omega_1 \omega_2}{s^2 + (1 + K_3)K_2 \omega_1 s + \omega_1 \omega_2 K_3} \tag{11}$$

Equating the denominator of (9), (10), or (11) to the standard second-order polynomial gives the equations

$$\omega_n = \sqrt{\omega_1 \omega_2 K_3} = \sqrt{\frac{R_6/R_5}{R_1 R_2 C_1 C_2}} \tag{12a}$$

$$\frac{1}{Q} = (1 + K_3)K_2 \sqrt{\frac{\omega_1}{\omega_2 K_3}} = \frac{1 + R_6/R_5}{1 + R_4/R_3 + R_4/R_7} \sqrt{\frac{R_2 R_5 C_2}{R_1 R_6 C_1}} \tag{12b}$$

Comparing (9), (10), and (11) with the respective low-, band-, and high-pass general second-order transfer functions yields the expressions for H_0 which are given below:

Low-pass:

$$H_0 = (1 + K_3)\frac{K_1}{K_3} = \frac{1 + R_5/R_6}{1 + R_3/R_4 + R_3/R_7} \tag{13a}$$

[See (1) of Sec. 4.2]

Bandpass:

$$H_0 = -\frac{K_1}{K_2} = -\frac{R_4}{R_3} \tag{13b}$$

[See (1) of Sec. 4.3]

High-pass:

$$H_0 = (1 + K_3)K_1 = \frac{1 + R_6/R_5}{1 + R_3/R_4 + R_3/R_7} \tag{13c}$$

[See (16) of Sec 4.3]

Typically $K_3 = \frac{1}{10}$, $C_1 = C_2 = 1000$ pF, $R_4 = R_5 = 100$ kΩ, and $R_6 = 10$ kΩ. For these values (12) and (13) reduce to the following expressions:

$$\omega_n = \sqrt{\frac{\omega_1 \omega_2}{10}} = \frac{3.162 \times 10^8}{\sqrt{R_1 R_2}} \tag{14a}$$

$$Q = 28{,}748 \left(\frac{1}{100 \text{ k}\Omega} + \frac{1}{R_3} + \frac{1}{R_7} \right) \sqrt{\frac{R_1}{R_2}} \tag{14b}$$

Low-pass:

$$H_0 = \frac{11}{1 + R_3 \left(\dfrac{1}{100 \text{ k}\Omega} + \dfrac{1}{R_7} \right)} \tag{14c}$$

Bandpass:

$$H_0 = -\frac{100 \text{ k}\Omega}{R_3} \tag{14d}$$

High-pass:

$$H_0 = \frac{1.1}{1 + R_3 \left(\dfrac{1}{100 \text{ k}\Omega} + \dfrac{1}{R_7} \right)} \tag{14e}$$

A design procedure which assumes these values, namely, $C_1 = C_2 = 1000$ pF, $R_4 = R_5 = 100$ kΩ, $R_6 = 10$ kΩ, $R_8 = \infty$, and which in addition uses $R_1 = R_2$, can be developed for the circuit of Fig. 5.3-1 with $V_{inB} = 0$. From (14a) we have

$$R_1 = R_2 = \frac{5.0329 \times 10^4}{f_n} \text{ k}\Omega \tag{15}$$

To complete the design, it remains to find the values of R_3 and R_7 from (14b) through (14e). If we assume that $|H_0| = 1$, then we find

Low-pass:
$$R_3 = \frac{316.2 \text{ k}\Omega}{Q} \tag{16a}$$

Bandpass:
$$R_3 = 100 \text{ k}\Omega \tag{16b}$$

High-pass:
$$R_3 = \frac{31.62 \text{ k}\Omega}{Q} \tag{16c}$$

Solving for R_7 in (14b) gives

$$R_7 = \frac{100 \text{ k}\Omega}{3.4785Q - 1 - \dfrac{100 \text{ k}\Omega}{R_3}} \tag{17}$$

Substitution of (16) into (17) gives:

Low-pass:
$$R_7 = \frac{100 \text{ k}\Omega}{3.162Q - 1} \tag{18a}$$

Bandpass:
$$R_7 = \frac{100 \text{ k}\Omega}{3.4785Q - 2} \tag{18b}$$

High-pass:
$$R_7 = \frac{100 \text{ k}\Omega}{0.3162Q - 1} \tag{18c}$$

We note that Q must be greater than $\sqrt{10}$ for the high-pass case. If this lower limit of Q is unacceptable, then the value of R_2/R_1 may be adjusted to lower this limitation. If the product of R_1 and R_2 is equal to the original product when R_1 and R_2 were equal, then R_2/R_1 modifies only Q. This concludes the design procedure for network functions in which $|H_0|$ is unity.

Example 5.3-1 *A Unity-Gain Inverting State-Variable Bandpass Filter* It is desired to design a bandpass filter with $Q = 20$, $f_n = 1$ kHz, and $|H_0| = 1$. From the design procedure given above ($R_8 = \infty$, $V_{\text{inB}} = 0$) we have $R_4 = R_5 = 100$ kΩ, $R_6 = 10$ kΩ, and $C_1 = C_2 = 1000$ pF. From (15) we get $R_1 = R_2 = 50.329$ kΩ, from (16b) $R_3 = 100$ kΩ, and from (18b) $R_7 = 1480$ Ω. (See Example 5.2-1 for a state-variable realization in which $|H_0| \neq 1$.) $\qquad\square$

It is of interest to examine the relative magnitudes of $V_{\text{LP}}(j\omega)$, $V_{\text{BP}}(j\omega)$, and $V_{\text{HP}}(j\omega)$ for the filter of Fig. 5.3-1. If we assume that $\omega_1 = \omega_2$ and $R_6/R_5 = 0.1$, then (12a) gives $\omega_1 = \omega_2 = \sqrt{10}\,\omega_n = 3.162\omega_n$. Equation (6) shows that at a constant frequency ω_x

$$|V_{\text{LP}}(j\omega_x)| = 3.162\left|\frac{\omega_n}{\omega_x}\right|\,|V_{\text{BP}}(j\omega_x)| = 10\left|\frac{\omega_n^2}{\omega_x^2}\right|\,|V_{\text{HP}}(j\omega_x)| \tag{19}$$

Thus the signal level at the various output points in the circuit can differ by a level of as much as 10. Note that this could cause saturation even though it was not indicated by calculations for the magnitude of the output voltage at the particular output point used.

To complete the introduction of the modified state-variable realization we need to consider the use of the V_{inB} input. In this configuration we have $V_{inA} = 0$ and $R_3 = \infty$ in Fig. 5.3-1. Following a development similar to that used for the V_{inA} configuration, we find

$$V_{HP}(s) = \frac{R_7}{R_4 + R_7}\left(1 + \frac{R_6}{R_5 \| R_8}\right)V_{BP}(s) - \frac{R_6}{R_8}V_{inB}(s) - \frac{R_6}{R_5}V_{LP}(s) \qquad (20)$$

This can be simplified to

$$V_{HP}(s) = -(1 + K_3 + K_4)K_2\frac{\omega_1 V_{HP}(s)}{s} - \frac{\omega_1 \omega_2 K_3}{s^2}V_{HP}(s) - K_4 V_{inB}(s) \quad (21)$$

through the use of (2) and (6), and by noting that, since $R_3 = \infty$, $K_2 = 1/(1 + R_4/R_7)$. Solving for the high-pass transfer function we get

$$\frac{V_{HP}(s)}{V_{inB}(s)} = \frac{-K_4 s^2}{s^2 + (1 + K_3 + K_4)K_2\omega_1 s + \omega_1\omega_2 K_3} \qquad (22)$$

Equation (6) may be used to obtain the bandpass and low-pass transfer functions for this case. These are

$$\frac{V_{BP}(s)}{V_{inB}(s)} = \frac{\omega_1 K_4 s}{s^2 + (1 + K_3 + K_4)K_2\omega_1 s + \omega_1\omega_2 K_3} \qquad (23)$$

and

$$\frac{V_{LP}(s)}{V_{inB}(s)} = \frac{-\omega_1\omega_2 K_4}{s^2 + (1 + K_3 + K_4)K_2\omega_1 s + \omega_1\omega_2 K_3} \qquad (24)$$

Comparing (9), (10), and (11) with the above three equations shows that the universal active filter has the capability to realize second-order low-pass, bandpass, and high-pass filters with either 0 or 180° phase shift at the frequency where H_0 is specified. Equating (22), (23), and (24) to the standard second-order denominator results in

$$\omega_n = \sqrt{\omega_1\omega_2 K_3} = \sqrt{\frac{R_6/R_5}{R_1 R_2 C_1 C_2}} \qquad (25a)$$

$$\frac{1}{Q} = (1 + K_3 + K_4)K_2\sqrt{\frac{\omega_1}{\omega_2 K_3}} = \frac{1 + R_6/R_5 + R_6/R_8}{1 + R_4/R_7}\sqrt{\frac{R_2 R_5 C_2}{R_1 R_6 C_1}} \qquad (25b)$$

The values of H_0 for the three cases are found to be

Low-pass: $\qquad H_0 = -\dfrac{K_4}{K_3} = -\dfrac{R_5}{R_8}$ $\qquad\qquad\qquad\qquad\qquad$ (26a)

Bandpass: $\qquad H_0 = \dfrac{K_4}{K_2(1 + K_3 + K_4)} = \dfrac{1 + R_4/R_7}{1 + R_8/R_6 + R_8/R_5}$ $\qquad$ (26b)

High-pass: $\qquad H_0 = -K_4 = -\dfrac{R_6}{R_8}$ $\qquad\qquad\qquad\qquad\qquad$ (26c)

Assuming that $R_4 = R_5 = 100$ kΩ, $R_6 = 10$ kΩ, and $C_1 = C_2 = 1000$ pF, as we did for the $V_{inA}(s)$ configuration, we obtain

$$\omega_n = \frac{3.162 \times 10^8}{\sqrt{R_1 R_2}} \tag{27a}$$

$$Q = 0.3162 \sqrt{\frac{R_1}{R_2} \frac{1 + \dfrac{100 \text{ k}\Omega}{R_7}}{1.1 + \dfrac{10 \text{ k}\Omega}{R_8}}} \tag{27b}$$

Low-pass: $\qquad H_0 = -\dfrac{100 \text{ k}\Omega}{R_8}$ $\qquad\qquad\qquad\qquad$ (27c)

Bandpass: $\qquad H_0 = \dfrac{1 + \dfrac{100 \text{ k}\Omega}{R_7}}{1 + \dfrac{R_8}{10 \text{ k}\Omega} + \dfrac{R_8}{100 \text{ k}\Omega}} = \dfrac{1 + \dfrac{100 \text{ k}\Omega}{R_7}}{1 + \dfrac{R_8}{9.091 \text{ k}\Omega}}$ $\qquad$ (27d)

High-pass: $\qquad H_0 = -\dfrac{10K}{R_8}$ $\qquad\qquad\qquad\qquad$ (27e)

A design procedure for the V_{inB} input can be developed using the values given and also letting $R_1 = R_2$. The values of R_1 and R_2 are then found from

$$R_1 = R_2 = \frac{5.0329 \times 10^4}{f_n} \text{ k}\Omega \tag{28}$$

Assuming $|H_0| = 1$ in (27), we get the following design formulas for R_8:

Low-pass: $\qquad\qquad\qquad\qquad R_8 = 100$ kΩ $\qquad\qquad\qquad\qquad$ (29a)

Bandpass: $\qquad\qquad\qquad\qquad R_8 = 31.62Q$ kΩ $\qquad\qquad\qquad\qquad$ (29b)

High-pass: $\qquad\qquad\qquad\qquad R_8 = 10$ kΩ $\qquad\qquad\qquad\qquad$ (29c)

Solving for R_7 in (27b) and noting that $R_1 = R_2$ give

$$R_7 = \frac{100 \text{ k}\Omega}{3.162Q\left(1.1 + \dfrac{10K}{R_8}\right) - 1} \tag{30}$$

Substitution of (29) into (30) results in

Low-pass: $\qquad\qquad\qquad R_7 = \dfrac{100 \text{ k}\Omega}{3.7947Q - 1}$ $\qquad\qquad$ (31a)

Bandpass: $\qquad\qquad\qquad R_7 = \dfrac{100 \text{ k}\Omega}{3.4785Q}$ $\qquad\qquad$ (31b)

High-pass: $\qquad\qquad\qquad R_7 = \dfrac{100 \text{ k}\Omega}{6.6402Q - 1}$ $\qquad\qquad$ (31c)

Example 5.3-2 *A Unity-Gain Noninverting State-Variable Bandpass Filter*   It is desired to design a noninverting bandpass filter with $Q = 20$ and $f_n = 1$ kHz using the V_{inB} input in Fig. 5.3-1. From the design procedure we have initially assumed that $R_4 = R_5 = 100$ kΩ, $R_6 = 10$ kΩ, and $C_1 = C_2 = 1000$ pF. Therefore from (15) we find that $R_1 = R_2 = 50.329$ kΩ. Equations (29b) and (31b) give $R_8 = 632.40$ kΩ, and $R_7 = 1.437$ kΩ. From (27d) we determine that $H_0 = 1$.   □

Table 5.3-1 summarizes the design procedures for both the V_{inA} and the V_{inB} low-pass, bandpass, and high-pass configurations of the modified state-variable active filter. The design equations of this table are satisfactory as long as f_n is much less than the gain bandwidths of the operational amplifiers. The versatility of this "universal" active filter realization is well illustrated by the table.

Now let us consider the tuning of the general-purpose universal active filter. A typical filter with external resistors R_1, R_2, R_3, R_7, and R_8 is shown in Fig. 5.3-2. Of these, R_1 and R_2 are used to tune ω_n. For the noninverting configuration, R_7 or R_3 (in the bandpass case) may also be used to tune Q. R_3 and R_7 (*except* for the bandpass case) may also be used to tune H_0. For the inverting configuration, R_7 and R_8 may be used to tune Q while R_8 and R_7 (only for the bandpass case) may be used to tune H_0.

Some final observations on the UAF may now be made. First let us consider the signal levels in the amplifier. These are important, since high signal levels

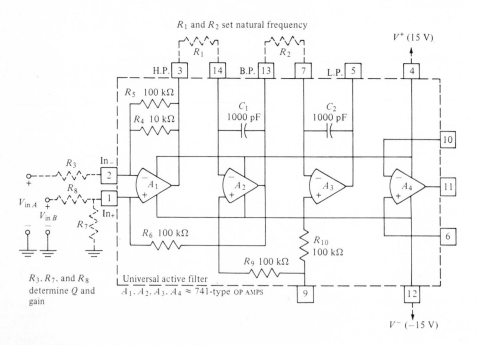

Figure 5.3-2 A typical universal active filter.

Table 5.3-1 Low-frequency design equations for the state-variable active filter with $H_0 = 1$*

	Noninverting Configuration ($R_8 = \infty$, $V_{inB} = 0$)			Inverting Configuration ($R_3 = \infty$, $V_{inA} = 0$)		
	Low-pass	Bandpass	High-pass	Low-pass	Bandpass	High-pass
Phase shift of transfer function at freq. where H_0 is specified	$0°$	$\pm 180°$	$0°$	$\pm 180°$	$0°$	$\pm 180°$
R_1, R_2 (kΩ)	$\dfrac{5.0329 \times 10^4}{f_n}$	$\dfrac{5.0329 \times 10^4}{f_n}$	$\dfrac{5.0329 \times 10^4}{f_n}$	$\dfrac{5.0329 \times 10^4}{f_n}$	$\dfrac{5.0329 \times 10^4}{f_n}$	$\dfrac{5.0329 \times 10^4}{f_n}$
R_3 (kΩ)	$\dfrac{316.2}{Q}$	100	$\dfrac{31.26}{Q}$	—	—	—
R_8 (kΩ)	—	—	—	100	$31.62Q$	10
R_7 (kΩ)	$\dfrac{100}{3.162Q - 1}$	$\dfrac{100}{3.4785Q - 2}$	$\dfrac{100}{0.3162Q - 1}$	$\dfrac{100}{3.4785Q - 1}$	$\dfrac{100}{3.4785Q}$	$\dfrac{100}{6.6402Q - 1}$

* This design procedure assumes that $R_1 = R_2$, $R_4 = R_5 = 100$ kΩ, $R_6 = 10$ kΩ, $C_1 = C_2 = 1000$ pF.

require a high slew rate (see Sec. 5.4), and too high a slew rate may cause instability. Typically, the maximum output voltage swing of a general-purpose UAF is 20, 8, and 2 V peak-to-peak for the low-pass, bandpass, and high-pass outputs, respectively. These limits hold over the frequency range of 10 Hz to 1 kHz. At 10 kHz, the output swings are reduced to 8, 3, and 0.8 V peak-to-peak. These values assume ± 15 V power supplies. Another observation on the UAF is the maximum Q. This is approximately 50 and is determined primarily by the frequency response of the amplifier and the dissipation of the capacitors. In designing a UAF, especially for high values of Q, it is usually wise to use the procedure given for $H_0 = 1$ in Table 5.3-1. If gains greater than one are desired, these may be obtained by using the uncommitted amplifier in the UAF package.

The general-purpose UAF is useful for frequencies up to about 10 kHz. Above this frequency, the Q-enhancement effect described in Sec. 5.5 becomes significant, making tuning more complex. While the lagging phase shift due to the gain bandwidth can be compensated with leading phase shift (i.e., by shunting R_1 and/or R_2 with a small capacitance) there are still problems of signal levels and the effects of slew rate. For frequencies greater than 10 kHz, high performance UAFs must be used. These are available in hybrid integrated circuit form for frequencies of 100 kHz and more. They use high-performance operational amplifiers and as a result are more expensive than the standard UAF.

5.4 INTEGRATORS

The use of an operational amplifier as an integrator has been shown to be of importance in the active filter realizations presented in Secs. 5.2 and 5.3. Here we present some of the details concerning such usage. First let us consider the inverting integrator. Its basic circuit is shown in Fig. 5.4-1. For this we obtain

$$A(s) = \frac{V_o(s)}{V_s(s)} = \frac{-A_d(s)\dfrac{\omega_{RC}}{s + \omega_{RC}}}{1 + \dfrac{A_d(s)s}{s + \omega_{RC}}} \tag{1}$$

where

$$\omega_{RC} = \frac{1}{RC} \tag{2}$$

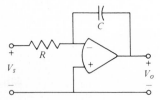

Figure 5.4-1 An inverting integrator.

Using a dominant pole model for $A_d(s)$ [see (17) of App. B] we obtain

$$A(s) = \frac{V_o(s)}{V_s(s)} = \frac{-\omega_{RC}\,\text{GB}}{s^2 + (\omega_a + \omega_{RC} + \text{GB})s + \omega_a\omega_{RC}} \tag{3}$$

If $\omega_a < \omega_{RC}$, for values of ω less than GB, the magnitude of (3) may be approximated as

$$|A(j\omega)| \approx \frac{\omega_{RC}\,\text{GB}}{\omega\sqrt{\omega^2 + \text{GB}^2}} \tag{4}$$

The phase shift of (3) may be written as

$$\arg[A(j\omega)] = \pi - \tan^{-1}\left[\frac{\omega(\omega_a + \omega_{RC} + \text{GB})}{\omega_a\omega_{RC} - \omega^2}\right] \tag{5}$$

If ω_a and ω_{RC} are both small compared to GB, then (5) reduces to

$$\arg[A(j\omega)] \approx \pi - \tan^{-1}\left(\frac{\omega\,\text{GB}}{\omega_a\omega_{RC} - \omega^2}\right) \tag{6}$$

In this relation, the argument of the arctangent function is much greater than unity. As a result (6) can be approximated as

$$\arg[A(j\omega)] \approx \frac{\pi}{2} + \frac{\omega_a\omega_{RC}}{\omega\,\text{GB}} - \frac{\omega}{\text{GB}} = \frac{\pi}{2} + \frac{\omega_{RC}}{\omega A_0} - \frac{\omega}{\text{GB}} \tag{7}$$

This expression illustrates several important points concerning the performance of the inverting integrator. From it we see that a decrease in A_0 causes phase lead, whereas an increase in ω/GB causes phase lag. Unfortunately, the second term of (7) is much smaller than the third term so that cancellation is not practical. If we construct a Bode plot of (3) as shown in Fig. 5.4-2, we see that there are two different regions in the frequency response. The first of these has the range defined by $\omega < \omega_{RC}$. For this range

$$A(s) = \frac{V_o(s)}{V_s(s)} \approx -\frac{\omega_a A_0}{s + \omega_a} \tag{8}$$

The second frequency range is defined by $\omega_a < \omega < \text{GB}$. In this frequency range the expressions given in (4) and (7) may be used to analyze the frequency characteristics of the inverting integrator.

The input impedance of the inverting integrator is essentially equal to the value of the input resistor R. It has little effect on the performance of active filters. The operational amplifier output impedance, however, can seriously influence the performance of the inverting integrator. The reason for this is that as the frequency increases, the magnitude of the closed-loop output impedance of the integrator also increases. This gives the integrator an apparent inductive output impedance which, in combination with the capacitive loading of a following stage, can produce complex poles. Complementary transmission zeros may also be formed. These can cause significant changes in the high-frequency response of the circuit. A model for this effect can be achieved by placing a resistor

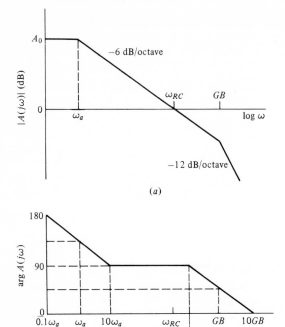

(a)

(b)

Figure 5.4-2 Bode plots for the inverting integrator of Fig. 5.4-1.

R_o in series with the output of the amplifier and by placing a capacitance to ground C_L in shunt with the output. The equivalent circuit is shown in Fig. 5.4-3a. A circuit suitable for calculation of $V_o(s)/V_s(s)$ is shown in Fig. 5.4-3b. If $\omega_{RC} <$ GB and $\omega_a \approx 0$ it can be shown that

$$A(s) = \frac{V_o(s)}{V_s(s)} \approx \frac{(1/RC_L)(s^2 - \text{GB}/R_o C)}{s[s^2 + s(1/R_o C_L) + \text{GB}(1/R_o C_L)]} \tag{9}$$

The poles of this function are found as

$$p_1 \approx 0 \qquad \text{(dominant integrator pole)} \tag{10}$$

and
$$p_2, p_3 \approx \frac{-\omega_o}{2}\left(1 \pm \sqrt{1 - 4\text{GB}/\omega_o}\right) \qquad \omega_o = \frac{1}{R_o C_L} \tag{11}$$

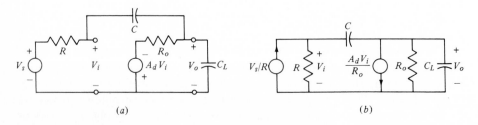

(a)   (b)

Figure 5.4-3 Effect of operational amplifier output resistance on an inverting integrator.

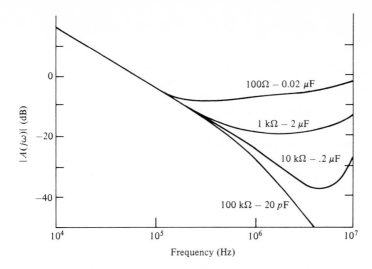

Figure 5.4-4 Frequency response characteristics for inverting integrators.

The zeros are

$$z_1, z_2 \approx \pm \sqrt{\frac{GB}{R_o C}} \tag{12}$$

The poles p_2 and p_3 become complex if $\omega_o < 4GB$. For an operational amplifier with a nominal 50-Ω open-loop output impedance and with GB = 1 MHz, complex poles will appear when C_L is greater than 800 pF. The two zeros defined by (12) have cancelling phase effects since they appear as approximately mirror images about the $j\omega$ axis. The magnitude response, however, is given a 12-dB/octave boost from these two zeros. These effects are illustrated in Fig. 5.4-4. This shows the magnitude characteristic for an inverting integrator in which $\omega_a < 4GB$. The operational amplifier is a 741 with a GB of approximately 1 MHz. The effects of various combinations of R and C are illustrated. The value of ω_{RC}, however, is kept constant at 500 krad/s. From the curves we readily see that the effect of output impedance is minimized as the value of C is lowered. Correspondingly the capacitive loading should be kept as small as possible.

The phase lag introduced by the gain-bandwidth limitation of the operational amplifier may be large enough to require compensation. One method for accomplishing this is to place a small resistance R_z in series with C as shown in the circuit of Fig. 5.4-5. This produces a pole in the feedback function which becomes

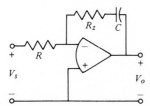

Figure 5.4-5 Phase lag compensation for an inverting integrator.

a zero in the overall integrator response. Analysis of the circuit of Fig. 5.4-5 gives the transfer function

$$A(s) = \frac{V_o(s)}{V_s(s)} = \frac{-GB[s(R_z/R) + \omega_{RC}]}{s^2 + (\omega_a + \omega_{RC} + GB)s + \omega_{RC}\omega_a} \tag{13}$$

where it has been assumed that $R_z < R$. Note that the denominators of (13) and (3) are identical; however, a zero, located at $\omega_z = 1/R_zC$, has been introduced into the numerator. We may approximate the phase shift of (13) as

$$\arg\left[A(j\omega)\right] \approx \pi + \tan^{-1}\frac{\omega}{\omega_z} - \tan^{-1}\frac{\omega GB}{\omega_a\omega_{RC} - \omega^2} \tag{14}$$

Assuming that $\omega_a\omega_{RC} \ll \omega^2$ gives

$$\arg\left[A(j\omega)\right] \approx \frac{\pi}{2} + \frac{\omega}{\omega_z} - \frac{\omega}{GB} \tag{15}$$

Thus, if $\omega_z = $ GB, the phase lag due to GB is eliminated until higher-order effects come into the picture. An example of what can be accomplished is shown in Fig. 5.4-6. In this example the values GB = 1 MHz, R = 200 kΩ, and C = 20 pF were used. One of the disadvantages of the method of compensation which uses R_z is that it depends on having $\omega_z = $ GB. The temperature coefficients of ω_z and GB, however, will probably not be equal. Thus the compensation actually obtained is dependent upon temperature. Another disadvantage of this method is that GB varies among operational amplifiers. As a result each operational amplifier must be individually compensated.

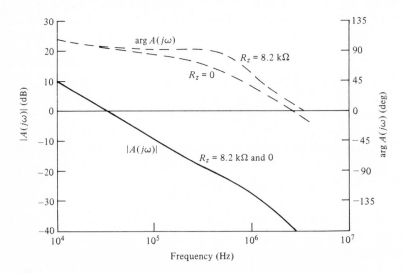

Figure 5.4-6 Frequency response of a compensated inverting integrator.

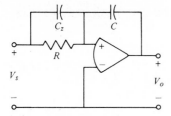

Figure 5.4-7 Phase lag compensation for an inverting integrator.

A second method for the compensation of phase lag is shown in Fig. 5.4-7. Here a capacitor C_z has been added in parallel with R. Analysis of this circuit results in

$$A(s) = \frac{V_o(s)}{V_s(s)} \approx -\frac{\omega_{RC}}{\omega_{RC} + \omega_z} \frac{\mathrm{GB}}{s} \frac{s + \omega_z}{s + \dfrac{\omega_z(\omega_{RC} + \mathrm{GB})}{\omega_z + \omega_{RC}}} \tag{16}$$

where $\omega_a \approx 0$ and $\omega_z = 1/RC_z$. To achieve the desired compensation the negative-real pole and zero of this expression must cancel. Thus we require

$$\omega_z = \mathrm{GB} \tag{17}$$

In this case, if $\mathrm{GB} > \omega_{RC}$, then (16) gives the expression of an ideal inverting integrator. Obviously the disadvantages of having $\omega_z = \mathrm{GB}$ described above also apply to this compensation method.

A third method for compensating for the phase lag of an inverting integrator is the use of another operational amplifier. The method is illustrated in Fig. 5.4-8 in which a unity-gain amplifier has been inserted as a buffer in the feedback path of the inverting integrator. The transfer function for this circuit is

$$A(s) = \frac{V_o(s)}{V_s(s)} = \frac{\left(1 + \dfrac{1}{A_{d2}(s)}\right)\left(\dfrac{\omega_{RC}}{s + \omega_{RC}}\right)}{\dfrac{1}{A_{d1}(s)} + \dfrac{1}{A_{d1}(s)A_{d2}(s)} + \dfrac{s}{s + \omega_{RC}}} \tag{18}$$

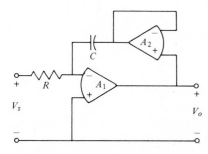

Figure 5.4-8 Using an operational amplifier to compensate for the phase lag of an inverting integrator.

where $A_{d1}(s)$ is the open-loop gain of amplifier A_1, and $A_{d2}(s)$ is that of A_2. Using the one-pole model for these quantities we obtain

$$A(s) = \frac{V_o(s)}{V_s(s)} = \frac{-\omega_{RC}\, GB_1(s + GB_2)}{s\left[s^2 + (GB_2 + \omega_{RC})s + (\omega_{RC}\, GB_2 + GB_1\, GB_2) + \dfrac{\omega_{RC}\,\omega_{a1}\, GB_1}{s}\right]}$$

(19)

The phase of (19) may be written as

$$\arg\,[A(j\omega)] = \frac{\pi}{2} + \tan^{-1}\frac{\omega}{GB_2} - \tan^{-1}\left(\frac{\omega}{GB_1} + \frac{\omega\omega_{RC}}{GB_1\, GB_2} - \frac{\omega_{RC}\,\omega_{a1}}{\omega GB_2}\right)$$

(20)

where we have assumed $GB_1\, GB_2 > \omega^2$ and $GB_1 > \omega_{RC}$. Taking account of the fact that the arguments of the arctangent functions in (20) are small, it may be simplified to

$$\arg\,[A(j\omega)] \approx \frac{\pi}{2} + \frac{\omega}{GB_2} - \frac{\omega}{GB_1} - \frac{\omega\omega_{RC}}{GB_1\, GB_2} + \frac{\omega_{RC}\,\omega_{a1}}{\omega GB_2}$$

(21)

Comparing this result with the one given in (7), we see that the buffer in the feedback loop has contributed phase lead which can be used to offset the phase lag of the integrator. If $\omega = \omega_{RC}$, $GB_i = GB$, $A_{0i} = A_0$, $\omega_{ai} = \omega_a$, then

$$\arg\,[A(j\omega_{RC})] \approx \frac{\pi}{2} - \left(\frac{\omega_{RC}}{GB}\right)^2 + \frac{\omega_{RC}}{\omega A_0}$$

(22)

Now let us consider the realization of a noninverting integrator. One approach is to cascade an inverting integrator with an inverter as shown in Fig. 5.4-9. The phase of this configuration is

$$\arg\,[A(j\omega)] \approx -\frac{\pi}{2} - \frac{\omega}{GB_1} - \frac{2\omega}{GB_2}$$

(23)

If the amplifiers are matched then the phase is given as

$$\arg\,[A(j\omega)] \approx -\frac{\pi}{2} - \frac{3\omega}{GB}$$

(24)

From this result we see that the circuit of Fig. 5.4-9 is in general not a satisfactory one because of its significant excess phase lag.

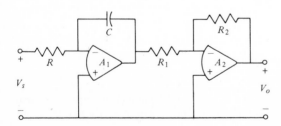

Figure 5.4-9 A realization for a non-inverting integrator.

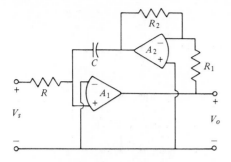

Figure 5.4-10 Using an operational amplifier to compensate for phase lag in a noninverting integrator.

Another noninverting integrator is shown in Fig. 5.4-10. In this circuit, an inverter is put in the feedback path of an inverting integrator both to achieve a noninverting integrator and also to provide phase lead which is used to cancel the phase lag of the inverting integrator. Note that the feedback must be returned to the + terminal of A_1 to achieve negative feedback. Analysis of Fig. 5.4-10 gives the following transfer function:

$$A(s) = \frac{V_o(s)}{V_s(s)} = \frac{\omega_{RC}\left(\dfrac{1}{A_{d2}(s)} + \dfrac{R_1}{R_1 + R_2}\right)}{(s + \omega_{RC})\left[\dfrac{1}{A_{d1}(s)A_{d2}(s)} + \dfrac{R_1}{A_{d1}(s)(R_1 + R_2)}\right] + \dfrac{R_2}{R_1 + R_2}s} \tag{25}$$

Substituting the dominant pole model for $A_{d1}(s)$ and $A_{d2}(s)$ gives

$$A(s)$$

$$\approx \frac{\omega_{RC}\,\mathrm{GB}_1\left(s + \mathrm{GB}_2\,\dfrac{R_1}{R_1 + R_2}\right)}{s\left[s^2 + \left(\mathrm{GB}_2\,\dfrac{R_1}{R_1 + R_2} + \omega_{RC}\right)s + \left(\omega_{RC}\,\dfrac{R_1}{R_1 + R_2}\,\mathrm{GB}_2 + \dfrac{R_2}{R_1 + R_2}\,\mathrm{GB}_1\,\mathrm{GB}_2\right.\right.}$$

$$\left.\left. + \dfrac{1}{s}\left(\omega_{a1}\,\omega_{RC}\,\mathrm{GB}_2\,\dfrac{R_1}{R_1 + R_2}\right)\right]}$$

$$\tag{26}$$

where it is assumed that $\omega_a < \mathrm{GB}$. The phase of $A(j\omega)$ is given as

$$\arg\left[A(j\omega)\right] \approx -\frac{\pi}{2} + \tan^{-1}\left(\frac{\omega}{\mathrm{GB}_2}\,\frac{R_1 + R_2}{R_1}\right)$$

$$- \tan^{-1}\left(\frac{\omega\omega_{RC}}{\mathrm{GB}_1\,\mathrm{GB}_2}\,\frac{R_1 + R_2}{R_2} + \frac{\omega R_1}{\mathrm{GB}_2 R_2} - \frac{\omega_{RC} R_1}{\omega A_{01} R_2}\right) \tag{27}$$

where $\omega < \mathrm{GB}_1$ (or GB_2) and $\omega_{RC} < \mathrm{GB}_1$. If the arguments of the arctangents are small and $\mathrm{GB}_1 = \mathrm{GB}_2 = \mathrm{GB}$, then (27) can be written as

$$\arg\left[A(j\omega)\right] \approx -\frac{\pi}{2} + \frac{\omega}{\mathrm{GB}} + \frac{RC}{A_{01}} - 2\left(\frac{\omega}{\mathrm{GB}}\right)^2 \approx -\frac{\pi}{2} + \frac{\omega}{\mathrm{GB}} \tag{28}$$

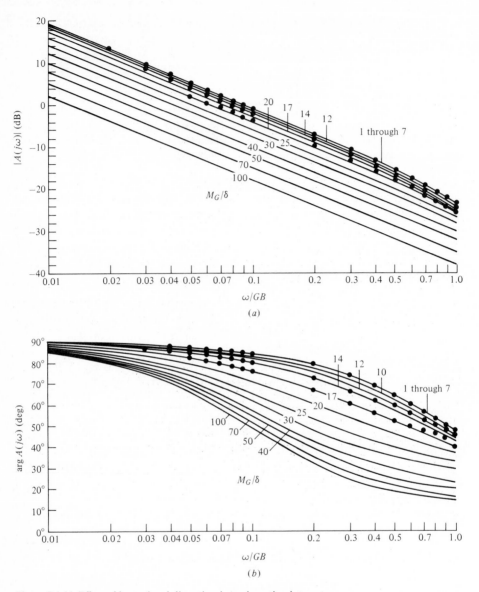

Figure 5.4-11 Effect of large-signal distortion in an inverting integrator.

Thus this noninverting integrator has an excess phase lead of approximately ω/GB. As such it will prove to be useful in cancelling out the phase lag introduced by the standard inverting integrator of Fig. 5.4-1 in feedback loops which contain both types of integrators.[1]

[1] See K. Martin and A. Sedra, "On the Stability of the Phase-Lead Integrator," *IEEE Trans. Circuits and Systems*, vol. CAS-24, no. 6, June 1977, pp. 321–324.

A final consideration in the treatment of inverting integrators is the subject of large-signal distortion caused by slew-rate limitations. To see this, we first express the phase shift of an inverting integrator as

$$\arg[A(j\omega)] = \frac{\pi}{2} - \tan^{-1}\left[\frac{1}{1 + \dfrac{GB}{\omega}N(A)}\right] \tag{29}$$

in which $N(A)$ is the system-describing function. $N(A) = 1$ corresponds to no large signal distortion whereas $N(A) < 1$ corresponds to slew-induced distortion. As $N(A)$ approaches zero, we note that the results in (29) and (10) of Sec. 4.6 are different, namely, the inverting integrator is less influenced by large signal distortion than the inverting amplifier is. The influence of large-signal distortion upon the small-signal frequency response of the inverting integrator is illustrated in Fig. 5.4-11 for $\omega_{RC} = GB/10$. The quantity M_g is the peak amplitude of the input signal, while δ is the threshold level (see Fig. B-13 in App. B). The results given above are for the inverting integrator. The noninverting integrator may be similarly treated. However, since slew creates phase lag, and the noninverting integrator characteristic produces phase lead, we may anticipate that its small signal characteristics would be less influenced than those of the inverting integrator.

5.5 Q ENHANCEMENT

When the state-variable or resonator circuits described in Secs. 5.2 and 5.3 are used to realize high-Q filter functions, the Q obtained is usually higher than that called for by the design procedure. This is called Q *enhancement*. It is primarily caused by the phase lag introduced by the operational amplifiers. If we let Q_o and f_{no} be the original design values of Q and f_n, then when $Q_o \times f_{no}$ is greater than 10^4, it is usually necessary to correct the design. We now consider how Q enhancement occurs in the state-variable filter. We first find the phase shift of the open-loop transfer function of the circuit with $V_{in}(s) = 0$ and R_7 and R_8 infinite as shown in Fig. 5.5-1. For this, the voltage $V'_{HP}(s)$ can be written as

$$V'_{HP}(s) = \frac{A_3(s)}{1 + \dfrac{A_3(s)}{1 + R_6/R_5}}\left[\frac{V_{BP}(s)}{1 + R_4/R_3} - \frac{V_{LP}(s)}{1 + R_5/R_6}\right] \tag{1}$$

If we assume that $A_3(s) \approx GB_3/s$ this becomes

$$V'_{HP}(s) = \left[\frac{GB_3}{s + GB_3/(1 + R_6/R_5)}\right]\left[\frac{V_{BP}(s)}{1 + R_4/R_3} - \frac{V_{LP}(s)}{1 + R_5/R_6}\right] \tag{2}$$

For sinusoidal steady-state conditions, if we define the phasors $\mathscr{V}_{HP}$, $\mathscr{V}_{BP}$, and $\mathscr{V}_{LP}$, this relation may be written as

$$\mathscr{V}'_{HP} = [M^+ \underline{/\theta^+}]\mathscr{V}_{BP} - [M^- \underline{/\theta^-}]\mathscr{V}_{LP} \tag{3}$$

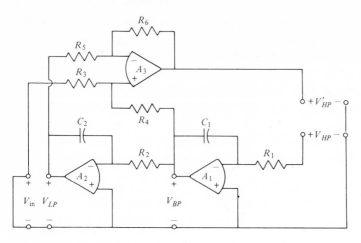

Figure 5.5-1 A state-variable filter.

where, assuming $\omega \approx \omega_{no} < GB_3$,

$$\theta^+ = \theta^- = -\tan^{-1}\left[\frac{\omega_{no}}{GB_3}\left(1 + \frac{R_6}{R_5}\right)\right] \approx -\frac{\omega_{no}}{GB_3}\left(1 + \frac{R_6}{R_5}\right) \tag{4a}$$

$$M^+ = \left|\frac{GB_3}{j\omega + \dfrac{GB_3}{1 + R_6/R_5}}\right| \frac{1}{1 + R_4/R_3} \tag{4b}$$

$$M^- = \left|\frac{GB_3}{j\omega + \dfrac{GB_3}{1 + R_6/R_5}}\right| \frac{1}{1 + R_5/R_6} \tag{4c}$$

We may express the transfer function from $\mathscr{V}_{BP}$ to $\mathscr{V}_{LP}$ as

$$\frac{\mathscr{V}_{LP}}{\mathscr{V}_{BP}} = \frac{-\omega_2 GB_2}{-\omega^2 + j\omega GB_2 + \omega_{a2}\omega_2} = M_2\underline{/\theta_2} \tag{5}$$

where $\omega_2 = 1/R_2C_2$ and ω_{a2} and GB_2 are respectively the bandwidth and gain bandwidth of operational amplifier A_2. If $GB_2 > \omega$, then

$$M_2 \approx \frac{\omega_2}{\omega} \tag{6a}$$

$$\theta_2 = \frac{\pi}{2} + \frac{\omega_2}{\omega_n A_0} - \frac{\omega_n}{GB_2} \tag{6b}$$

Thus (3) can be expressed as

$$\frac{\mathscr{V}'_{HP}}{\mathscr{V}_{BP}} = M^+\underline{/\theta^+} - M^- M_2\underline{/\theta^- + \theta_2} = M^+\left(1\underline{/\theta^+} - \frac{M^- M_2}{M^+}\underline{/\theta^- + \theta_2}\right) \tag{7}$$

We can show from (4), (6), and (9b) of Sec. 5.2 that if $\omega \approx \omega_{no}$, then

$$\frac{M^- M_2}{M^+} = Q_o \tag{8}$$

This relation shows that Q_o is equal to the ratio of two factors. The numerator factor is the product of the transmission through amplifier A_2 and the inverting section of amplifier A_3, while the denominator factor is the magnitude of the transmission through the noninverting section of amplifier A_3. We may now write (7) in the form

$$\frac{\mathscr{V}'_{HP}}{\mathscr{V}_{BP}} = M^+[\cos \theta^+ + j \sin \theta^+ - Q_o \cos (\theta^- + \theta_2) - jQ_o \sin (\theta^- + \theta_2)] \tag{9}$$

The phase shift from $\mathscr{V}_{BP}$ to $\mathscr{V}'_{HP}$, $\theta_{H'B}$, can now be written as

$$\theta_{H'B} = \tan^{-1} \frac{\sin \theta^+ - Q_o \sin (\theta^- + \theta_2)}{\cos \theta^+ - Q_o \cos (\theta^- + \theta_2)} \tag{10}$$

This may be simplified by noting that $\theta^+ \approx 0$ and that

$$\theta^- + \theta_2 = \frac{\pi}{2} + \left[\frac{\omega_2}{\omega_{no} A_0} - \frac{\omega_{no}}{GB_2} - \frac{\omega_{no}}{GB_3} \left(1 + \frac{R_6}{R_5}\right)\right] = \frac{\pi}{2} + \varepsilon \tag{11}$$

Using series approximations and the trigonometric relationships for cos and sin of the sum of angles allows (10) to be written as

$$\theta_{H'B} \approx \tan^{-1} \frac{\theta^+ - Q_o}{1 + Q_o \varepsilon}$$

$$= -\tan^{-1} \frac{Q_o + \dfrac{\omega_{no}}{GB_3}\left(1 + \dfrac{R_6}{R_5}\right)}{1 + Q_o\left[\dfrac{\omega_2}{\omega_{no} A_0} - \dfrac{\omega_{no}}{GB_2} - \dfrac{\omega_{no}}{GB_3}\left(1 + \dfrac{R_6}{R_5}\right)\right]} \tag{12}$$

Since the argument of the arctangent of (12) is very large, this equation may be written

$$\theta_{H'B} \approx -\frac{\pi}{2} + \frac{1 + Q_o\left[\dfrac{\omega_2}{\omega_{no} A_0} - \dfrac{\omega_{no}}{GB_2} - \dfrac{\omega_{no}}{GB_3}\left(1 + \dfrac{R_6}{R_5}\right)\right]}{Q_o + \dfrac{\omega_{no}}{GB_3}\left(1 + \dfrac{R_6}{R_5}\right)} \tag{13}$$

Adding this to the phase shift provided by the inverting integrator [see (7) of Sec. 5.4] gives the total open-loop phase shift θ_T of the state-variable configuration as

$$\theta_T = \frac{\omega_1}{\omega_{no} A_0} - \frac{\omega_{no}}{GB_1} + \frac{1 + Q_o\left[\dfrac{\omega_2}{\omega_{no} A_0} - \dfrac{\omega_{no}}{GB_2} - \dfrac{\omega_{no}}{GB_3}\left(1 + \dfrac{R_6}{R_5}\right)\right]}{Q_o + \dfrac{\omega_{no}}{GB_3}\left(1 + \dfrac{R_6}{R_5}\right)} \tag{14}$$

Under ideal conditions, A_0 and all the quantities GB_i are very large; thus θ_T approaches $1/Q_o$. Therefore, for small deviations

$$Q \approx \frac{1}{\theta_T} = \cfrac{1}{\cfrac{\omega_1}{\omega_{no} A_0} - \cfrac{\omega_{no}}{GB_1} + \cfrac{1 + Q_o\left[\cfrac{\omega_2}{\omega_{no} A_0} - \cfrac{\omega_{no}}{GB_2} - \cfrac{\omega_{no}}{GB_3}\left(1 + \cfrac{R_6}{R_5}\right)\right]}{Q_o + \cfrac{\omega_{no}}{GB_3}\left(1 + \cfrac{R_6}{R_5}\right)}} \tag{15}$$

This result gives the actual value of Q in terms of the design value Q_o and the parameters of the realization.

Example 5.5-1 *Q Enhancement of a State-Variable Filter*   A state-variable filter is to be designed with $H_0 = 1$ and $Q_o = 10$. The 741-type operational amplifiers are to be used. It is desired to find the value of ω_n at which the realization becomes unstable. From Table 5.3-1 $R_6/R_5 = 1/10, \omega_1 = \sqrt{10}\,\omega_n$, and $\omega_2 = \sqrt{10}\,\omega_n$. Setting the denominator of (15) to zero and letting $GB_i = GB$ $(i = 1, 2, 3)$ give

$$\left(\frac{\sqrt{10}}{A_0} - \frac{\omega_n}{GB}\right)\left(Q_o + \frac{1.1\omega_n}{GB}\right) + 1 + Q_o\left(\frac{\sqrt{10}}{A_0} - \frac{2.1\omega_n}{GB}\right) = 0 \tag{16}$$

Since the $\sqrt{10}/A_0$ terms are negligible, this reduces to

$$\left(\frac{GB}{\omega_n}\right)^2 - 31\left(\frac{GB}{\omega_n}\right) - 1.1 = 0 \tag{17}$$

Solving for GB/ω_n gives approximately 31. Thus the maximum frequency of f_n for $GB = 1000$ kHz is 32.26 kHz. We also note this corresponds to a normalized GB, GB_n of 31.   □

We may apply the results of the analysis given above to the problem of compensating for the resulting Q enhancement. From (15) we may write

$$Q \approx \frac{1}{1/Q_o - 3.1\omega_n/GB} \tag{18}$$

where we have assumed that $GB_1 = GB_2 = GB_3 = GB$, $Q_o > \omega_n/GB$, $R_6/R_5 = 0.1$, and that the A_0 terms have little effect on the realization. Equation (18) can also be expressed as

$$f_n Q \approx \frac{1}{1/f_n Q_o - 3.1/GB} \tag{19}$$

where GB is now in terms of hertz. Similar results for the resonator will be obtained in Sec. 5.7 as

$$f_n Q \approx \frac{1}{1/f_n Q_o - 4/GB} \tag{20}$$

where GB again has the units of hertz. Equations (19) and (20) are useful in the design of the state-variable and resonator circuits when the $f_n Q_o$ product becomes large. This product has been plotted versus $f_n Q$ in Fig. 5.5-2 for both the state-variable and resonator realizations. The gain bandwidth of the operational amplifiers has been assumed to be 1 MHz. These curves may be used to predistort the value of Q used for a design. To do this, start with the value of $f_n Q_o$ and locate this value on the *horizontal* axis (*indicated as the $f_n Q$ axis*). Project this value upward until the proper curve (resonator or state-variable) is reached. Then project over to the vertical axis (*indicated as the $f_n Q_o$ axis*) and read the predistorted value of $f_n Q_o$. The Q_o obtained from dividing this product by f_n is the one to be used in the design procedure. If f_n is constant (which is only approximately true) then the actual Q will be close to the desired value of Q_o.

Example 5.5-2 *State-Variable and Resonator Filter Design Using a Predistorted Value of Q_o* It is desired to find the predistorted values of Q_o which are to be used to design a state-variable and a resonator realization with $Q = 10$ and $f_n = 10$ kHz. The gain bandwidth of the operational amplifiers is assumed to be 1 MHz. For both realizations, the specified value of $f_n Q_o$ is 100,000. For the state-variable circuit, Fig. 5.5-2 gives the predistorted value of $f_n Q_o$ as 75,760. Therefore, the predistorted value of Q_o is 7.57. For the resonator, the predistorted value of $f_n Q_o$ is found as 71,430.

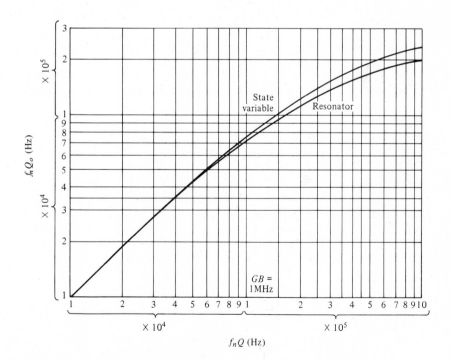

Figure 5.5-2 Curves for Q enhancement in the state-variable and resonator filters.

Thus the predistorted Q_o is 7.14. Note that less predistortion is necessary for the state-variable circuit. □

One of the problems in designing high-Q circuits is that of measuring the Q obtained. The conventional method requires the determination of the -3-dB bandwidth. This is difficult to determine in the high-Q case since it is very small. An alternate approach for measuring the Q of a second-order system is to apply a step input to the system and then to examine the resulting time-domain response. To see how this is done, consider a general transfer function given as

$$\frac{V_2(s)}{V_1(s)} = \frac{N(s)}{s^2 + (\omega_n/Q)s + \omega_n^2} = \frac{N(s)}{(s + \alpha)^2 + \beta^2} \tag{21}$$

where $N(s)$ is the numerator polynomial and where

$$\alpha = \frac{\omega_n}{2Q} \qquad \beta = \omega_n \sqrt{1 - \frac{1}{4Q^2}} \tag{22}$$

Note that $-\alpha$ and β are the real and imaginary components respectively of the complex poles. Solving for $V_2(s)$ and assuming a unit step input gives

$$V_2(s) = \frac{N(s)}{s[(s + \alpha)^2 + \beta^2]} \tag{23}$$

Taking the inverse Laplace transformation and defining $v_2(t) = \mathcal{L}^{-1}[V_2(s)]$ we find that the step response will have the form

$$v_2(t) = K_1 + K_2 e^{-\alpha t} \sin (\beta t + \phi) \tag{24}$$

where the explicit values of K_1, K_2, and ϕ depend on the polynomial $N(s)$. The general form of $v_2(t)$ will appear as shown in Fig. 5.5-3 where, since ϕ is arbitrary,

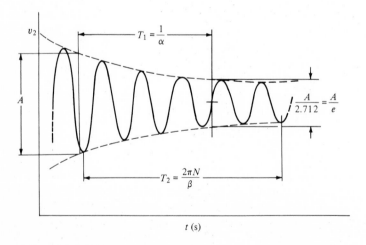

Figure 5.5-3 Finding the Q of a high-Q filter.

the waveform at $t = 0$ has not been shown. The quantity α may be found by measuring the time interval T_1 between any two points on the envelope which have amplitudes of A and $A/e = A/2.7183$ (e is the base of the natural logarithm) respectively as shown in Fig. 5.5-3. The quantity β may be found by measuring the time interval T_2 between N successive peaks or valleys of the response. Thus

$$\alpha = \frac{1}{T_1} \qquad \beta = \frac{2\pi N}{T_2} \tag{25}$$

From (22) we may write that

$$Q = \frac{\sqrt{\alpha^2 + \beta^2}}{2\alpha} \qquad \omega_n = \sqrt{\alpha^2 + \beta^2} \tag{26}$$

Since for a high-Q circuit the response shown in Fig. 5.5-3 decays very slowly, in practice this method of measuring ω_n and Q is very useful.

5.6 FREQUENCY-DEPENDENT SENSITIVITY

In this section we consider the frequency dependence of the filter realizations of this chapter, as evidenced by their pole sensitivity to GB. The frequency dependence of the infinite-gain filter realization may be analyzed in a manner similar to that done in Sec. 4.7 for finite-gain filters. In this case, σ_A of (20) of Sec. 4.7 equals $1/A_0 \simeq 0$. Substituting the values of the admittances Y_i from Sec. 5.1 for the low-pass infinite-gain realization into (17) of Sec. 4.7 ($q_i = c_i$, $i = 0, 1, 2$) gives

$$b_2 = C_5 C_6 \qquad b_1 = C_6(G_1 + G_2 + G_3) \qquad b_0 = G_2 G_3 \tag{1a}$$

$$q_2 = C_5 C_6 \qquad q_1 = C_6(G_1 + G_2 + G_3) + G_3 C_5 \qquad q_0 = G_1 G_3 + G_2 G_3 \tag{1b}$$

For this filter

$$\omega_n^2 = \frac{b_0}{b_2} = \frac{b_0}{q_2} \tag{2a}$$

$$\frac{\omega_n}{Q} = \frac{b_1}{b_2} = \frac{b_1}{q_2} \tag{2b}$$

$$\omega_n^2 H_0 = \frac{G_1 G_3}{b_2} = \frac{G_1 G_3}{q_2} \tag{2c}$$

From (1b) it is seen that $q_1 = b_1 + G_3 C_5$ and $q_0 = b_0 + G_1 G_3$. Substituting these expressions in (2) gives

$$\omega_n^2 = \frac{q_0}{q_2} - \frac{G_1 G_3}{q_2} \tag{3a}$$

$$\frac{\omega_n}{Q} = \frac{q_1}{q_2} - \frac{G_3 C_5}{q_2} \tag{3b}$$

Thus x and y of (10) of Sec. 4.7 become

$$x = -\frac{G_3 C_5 Q}{\omega_n} = -q_2 \frac{Q}{R_3 C_6 \omega_n} = -q_2 Q \sqrt{\frac{R_2 C_5}{R_3 C_6}} \tag{4a}$$

$$y = q_2 - \left(q_2 + \frac{G_1 G_3}{\omega_n^2}\right) = -H_0 q_2 \tag{4b}$$

Substituting these into (16) of Sec. 4.7 gives

$$|S_{GB}^{p_i}| = \frac{Q\omega_n}{GB\sqrt{4Q^2 - 1}} \left[\frac{R_2 C_5}{R_3 C_6} - \frac{|H_0|}{Q}\left(\frac{R_2 C_5}{R_3 C_6}\right)^{1/2} + H_0^2\right]^{1/2} \tag{5}$$

One design procedure used for the low-pass infinite-gain realization is to let $R_1 = R_2 = R_3 = R$ (equal R). For this choice (5) simplifies to

$$|S_{GB}^{p_i}| = \frac{Q\omega_n}{GB\sqrt{4Q^2 - 1}} \left[\frac{C_5}{C_6} - \frac{1}{Q}\left(\frac{C_5}{C_6}\right)^{1/2} + 1\right]^{1/2} \tag{6}$$

For this equal-R design, the Q is given as

$$Q = \tfrac{1}{3}\sqrt{C_5/C_6} \tag{7}$$

Combining this with (6) gives

$$|S_{GB}^{p_i}| = \frac{Q\omega_n}{GB\sqrt{4Q^2 - 1}} (9Q^2 - 2)^{1/2} \tag{8}$$

The third-order denominator of the transfer function for this filter is now found as

$$D(s) = q_2\left|\frac{s^3}{GB} + s^2\left[1 + \frac{\omega_n}{GBQ} + \frac{\omega_n}{GB}\left(\frac{R_2 C_5}{R_3 C_6}\right)^{1/2}\right] + s\left[\frac{\omega_n}{Q} + \frac{\omega_n^2 + \omega_n^2 H_0}{GB}\right] + \omega_n^2\right| \tag{9}$$

Rearranging this and assuming an equal-R design procedure we obtain

$$D(s) = \frac{q_2}{GB}\left[s^3 + s^2\left(GB + \frac{\omega_n}{Q} + 3Q\omega_n\right) + s\left(\frac{\omega_n GB}{Q} + 2\omega_n^2\right) + \omega_n^2 GB\right] \tag{10}$$

Normalizing by ω_n results in

$$D(s_n) = \frac{q_2 \omega_n^2}{GB_n}\left[s_n^3 + \left(GB_n + \frac{1}{Q} + 3Q\right)s_n^2 + \left(\frac{GB_n}{Q} + 2\right)s_n + GB_n\right] \tag{11}$$

The loci of the upper-half-plane complex pole for this denominator polynomial as a function of GB_n with Q as a parameter is shown in Fig. 5.6-1. Comparing this with Fig. 4.7-2 it is readily observed that the infinite-gain and the finite-gain realizations have a similar dependence upon GB. The bandpass and high-pass infinite-gain realizations may be analyzed in a manner similar to that given above. The results of such analyses are summarized in Table 5.6-1. The loci of the upper-half-plane complex pole as a function of GB_n with Q as a parameter are shown in Figs. 5.6-2 and 5.6-3.

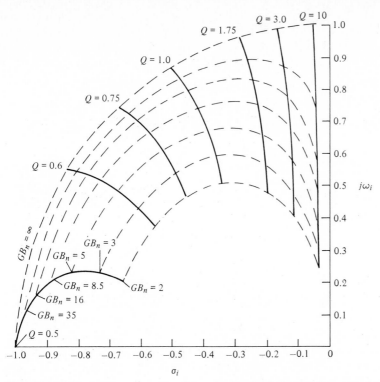

Figure 5.6-1 Effect of GB on the poles of the second-order low-pass infinite-gain single-amplifier filter of Sec. 5.1.

Table 5.6-1 Summary of the frequency-dependent sensitivities for infinite-gain realizations

Infinite-gain realizations	Constraints	$\left\lvert S_{\text{GB}}^{p_i} \right\rvert$ and $D(s_n)$
Low-pass (Fig. 5.1-2)	None	$\left\lvert S_{\text{GB}}^{p_i} \right\rvert = \dfrac{Q\omega_n}{\text{GB}\sqrt{4Q^2 - 1}} \left[\dfrac{R_2 C_5}{R_3 C_6} - \dfrac{\lvert H_0 \rvert}{Q} \left(\dfrac{R_2 C_5}{R_3 C_6} \right)^{1/2} + H_0^2 \right]^{1/2}$
	$R_1 = R_2 = R_3 = R$ (Fig. 5.6-1)	$D(s_n) = \dfrac{q_2 \omega_n^2}{\text{GB}_n} \left[s_n^3 + \left(\text{GB}_n + \dfrac{1}{Q} + 3Q \right) s_n^2 + \left(\dfrac{\text{GB}_n}{Q} + 2 \right) s_n + \text{GB}_n \right]$
Bandpass (Fig. 5.1-4)	None	$\left\lvert S_{\text{GB}}^{p_i} \right\rvert = \dfrac{\omega_n \lvert H_0 \rvert}{\text{GB}\sqrt{4Q^2 - 1}} \left(1 + \dfrac{R_1}{R_5} \right)$
	$C_2 = C_3 = C$ (Fig. 5.6-2)	$D(s_n) = \dfrac{q_2 \omega_n^2}{\text{GB}_n} \left[s_n^3 + \left(\text{GB}_n + \dfrac{1}{Q} + 2Q \right) s_n^2 + \left(\dfrac{\text{GB}_n}{Q} + 1 \right) s_n + \text{GB}_n \right]$
High-pass (Fig. 5.1-5)	None	$\left\lvert S_{\text{GB}}^{p_i} \right\rvert = \dfrac{Q\omega_n}{\text{GB}\sqrt{4Q^2 - 1}} \left[\dfrac{R_6 C_3}{R_5 C_2} - \dfrac{\lvert H_0 \rvert}{Q} \left(\dfrac{R_6 C_3}{R_5 C_2} \right)^{1/2} + H_0^2 \right]^{1/2}$
	$C_1 = C_2 = C_3 = C$ (Fig. 5.6-3)	$D(s_n) = \dfrac{q_2 \omega_n^2}{\text{GB}_n} \left[2s_n^3 + \left(\text{GB}_n + \dfrac{1}{Q} + 3Q \right) s_n^2 + \left(\dfrac{\text{GB}_n}{Q} + 1 \right) s_n + \text{GB}_n \right]$

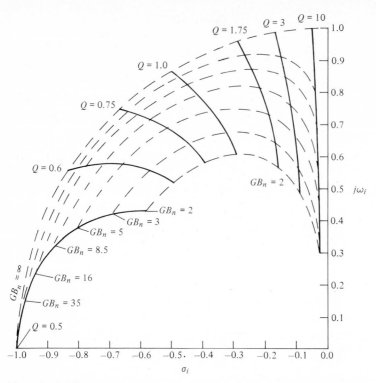

Figure 5.6-2 Effect of GB on the poles of the second-order bandpass infinite-gain single-amplifier filter of Sec. 5.1.

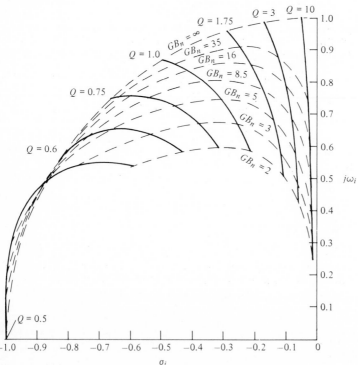

Figure 5.6-3 Effect of GB on the poles of the second-order high-pass infinite-gain single-amplifier filter of Sec. 5.1.

Example 5.6-1 *The Actual ω_n and Q of an Infinite-Gain Bandpass Filter*   It is desired to use the equal-C design procedure to realize an infinite-gain realization for a second-order bandpass filter with $Q = 8.53$, $f_n = 1000$ Hz, and $H_0 = 1$. The gain bandwidth of the amplifier is 1 MHz. The actual pole locations may be found using the realization of (13) of Sec. 5.1 and selecting $C_2 = C_3 = C = 0.01$ μF. Thus we obtain $R_1 = 135.8$ kΩ, $R_5 = 939$ Ω, and $R_6 = 271.6$ kΩ. The desired complex-pole locations can be found from p_1, $p_2 = \omega_n/2Q(-1 \pm j\sqrt{4Q^2 - 1})$. Thus, p_1, $p_2 = -368.8 \pm j6272$. The actual pole locations can be calculated from (11) of Sec. 4.7 using $GB_n = 1000$. Since in (10) of Sec. 4.7 $y = 0$ and $x = -2q_2 Q_2$ for the infinite-gain bandpass realization with $C_2 = C_3$, the real and imaginary parts of the pole locations may be found from

$$\text{Re} \left[\frac{\partial p_i}{\partial GB} \right] = -1/GB_n^2 = -10^{-6} \tag{12a}$$

$$\text{Im} \left[\frac{\partial p_i}{\partial GB} \right] = \frac{2Q^2 - 1}{GB_n^2 Q\sqrt{4Q^2 - 1}} = 0.99 \times 10^{-6} \simeq 10^{-6} \tag{12b}$$

The real and imaginary parts of the change in pole location can be found by multiplying by dGB to give

$$\text{Re} \, [dp_i] \approx \text{Re} \left[\frac{dp_i}{dGB} \right] dGB = 2\pi \tag{13a}$$

$$\text{Im} \, [dp_i] \approx \text{Im} \left[\frac{dp_i}{dGB} \right] dGB = -2\pi \tag{13b}$$

In this case dGB represents the change in GB from infinity to 1 MHz; thus $GB \simeq dGB < 0$. The actual pole locations are now found as $p_1, p_2 = -361 \pm j6266$. The actual f_n and Q designated as $f_n^{(\text{act})}$ and $Q^{(\text{act})}$ are calculated as $f_n^{(\text{act})} = 998.9$ Hz and $Q^{(\text{act})} = 8.67$. $\square$

The final category of RC-amplifier filters to be examined in this section is the multiple-amplifier type of realization. This includes the state-variable and resonator circuits introduced in Sec. 5.2. The state-variable realization will be considered first. Its circuit configuration is shown in Fig. 5.6-4. The voltage $V_{\text{in}3}(s)$ defined in the figure may be written as

$$V_{\text{in}3}(s) = V_1(s)\left(\frac{R_4}{R_3 + R_4}\right) + V_{\text{BP}}(s)\left(\frac{R_3}{R_3 + R_4}\right)$$

$$- V_{\text{LP}}(s)\left(\frac{R_6}{R_5 + R_6}\right) + V_{\text{HP}}(s)\left(\frac{R_5}{R_5 + R_6}\right) \tag{14}$$

Now let us define the following transfer functions

$$A_1(s) = \frac{V_{\text{BP}}(s)}{V_{\text{HP}}(s)} \qquad A_2(s) = \frac{V_{\text{LP}}(s)}{V_{\text{BP}}(s)} \qquad A_3(s) = \frac{V_{\text{HP}}(s)}{V_{\text{in}3}(s)} \tag{15}$$

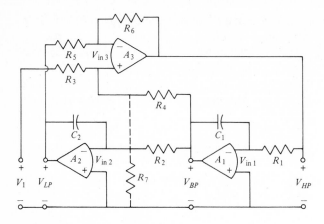

Figure 5.6-4 The state-variable filter.

Solving these equations for $V_{HP}(s)/V_1(s)$ we obtain

$$\frac{V_{HP}(s)}{V_1(s)} = \frac{A_3(s)\dfrac{R_4}{R_3 + R_4}}{1 + A_3(s)\dfrac{R_5}{R_5 + R_6} + A_1(s)A_2(s)\dfrac{R_3}{R_3 + R_4} + A_1(s)A_2(s)A_3(s)\dfrac{R_6}{R_5 + R_6}}$$

(16)

The denominator of this transfer function will be identical for the low-pass and bandpass cases and for the inverting or noninverting configurations. It can be rewritten as

$$D(s) = \frac{1}{A_1(s)A_2(s)A_3(s)} + \frac{1}{A_1(s)A_2(s)}\frac{R_5}{R_5 + R_6} + \frac{1}{A_3(s)}\frac{R_3}{R_3 + R_4} + \frac{R_6}{R_5 + R_6}$$

(17)

Using (3) of Sec. 5.4, $A_1(s)$ and $A_2(s)$ may be written in the form

$$A_1(s) \simeq \frac{-\omega_1\,GB_1}{s(s + \omega_1 + GB_1)} \qquad A_2(s) \simeq \frac{-\omega_2\,GB_2}{s(s + \omega_2 + GB_2)}$$

(18)

where $\omega_1 = 1/R_1 C_1$, $\omega_2 = 1/R_2 C_2$, and $\omega_a \ll GB$. Since $A_3(s)$ of (15) gives the open-loop gain of the operational amplifier A_3 of Fig. 5.6-4, it may be expressed as

$$A_3(s) \simeq GB_3/s$$

(19)

Combining the above results, $D(s)$ becomes the following fifth-order polynomial:

$$D(s) = \left(s^2 + \frac{\omega_n}{Q}s + \omega_n^2\right) + \frac{s}{GB}\left[s^2\left(3 + \frac{R_6}{R_5}\right) + s\left(\omega_1 + \omega_2 + \frac{\omega_n}{Q}\right) + \frac{\omega_2\omega_n}{Q}\right]$$

$$+ \left(\frac{s}{GB}\right)^2\left[s^2\left(3 + 2\frac{R_6}{R_5}\right) + s(\omega_1 + \omega_2)\left(2 + \frac{R_6}{R_5}\right) + \omega_1\omega_2\right]$$

$$+ \left(\frac{s}{GB}\right)^3\left[\frac{R_5 + R_6}{R_5}(s^2 + s\omega_1 + s\omega_2 + \omega_1\omega_2)\right] \tag{20}$$

where it can be shown that

$$\omega_n^2 = \omega_1\omega_2\frac{R_6}{R_5} \tag{21a}$$

$$\frac{\omega_n}{Q} = \left(\frac{1 + R_6/R_5}{1 + R_4/R_3 + R_4/R_7}\right)\omega_1 \tag{21b}$$

The resistor R_7 incorporated into the state-variable circuit, as shown in Fig. 5.6-4 (see also the universal active filter circuit of Fig. 5.2-1), only affects the quantity ω_n/Q given in (21b). If R_7 is not included, then, letting $R_7 = \infty$ in (21b) and neglecting the $(s/GB)^2$ and $(s/GB)^3$ terms of (20), for $D(s)$ we obtain the following simplified expression

$$D(s) \simeq \left(s^2 + \frac{\omega_n}{Q}s + \omega_n^2\right) + \frac{s}{GB}\left[\left(3 + \frac{R_6}{R_5}\right)s^2 + \left(\omega_1 + \omega_2 + \frac{\omega_n}{Q}\right)s + \frac{\omega_2\omega_n}{Q}\right] \tag{22}$$

Comparing this with (6) of Sec. 4.7 we may identify

$$k = 1 \qquad q_2 = 3 + \frac{R_6}{R_5} \qquad q_1 = \omega_1 + \omega_2 + \frac{\omega_n}{Q} \qquad q_0 = \frac{\omega_2\omega_n}{Q} \tag{23}$$

Substituting these values in (10) of Sec. 4.7 gives

$$x = 2 + \frac{R_6}{R_5} - \frac{Q}{\sqrt{R_6/R_5}}\left(\sqrt{\varepsilon} + \frac{1}{\sqrt{\varepsilon}}\right) \tag{24a}$$

$$y = 3 + \frac{R_6}{R_5} - \frac{1}{Q}\left(\frac{\sqrt{\varepsilon}}{\sqrt{R_6/R_5}}\right) \tag{24b}$$

where
$$\varepsilon = \omega_2/\omega_1 \tag{25}$$

Typical design values for the state-variable realization are $R_6/R_5 = \frac{1}{10}$ and $\varepsilon = 1$. Thus, for (24) we obtain

$$x = 2.1 - \sqrt{40}Q \qquad y = 3.1 - \frac{\sqrt{10}}{Q} \tag{26}$$

Assuming $Q > 1$ gives $x \simeq -6.32Q$ and $y \simeq 3.1$. Substituting these values into (16) of Sec. 4.7 yields

$$| S^{p_i}_{GB} | \simeq \frac{7 \omega_n Q}{GB \sqrt{4Q^2 - 1}} \tag{27}$$

Comparing this result to the ones obtained for the single amplifier realizations shows that the state-variable filter is similar to the positive-gain ones in regard to its dependence on the gain-bandwidth product. A plot of the loci for the upper-half-plane complex pole can be obtained from (22) by normalizing with respect to ω_n to get

$$D(s_n) \simeq \frac{\omega_n^2}{GB_n} \left[s_n^3 \left(3 + \frac{R_6}{R_5} \right) + s_n^2 \left(GB_n + \omega_{1n} + \omega_{2n} + \frac{1}{Q} \right) + s_n \left(\frac{GB_n}{Q} + \frac{\omega_{2n}}{Q} \right) + GB_n \right] \tag{28}$$

where $\omega_{1n} = \omega_1 / \omega_n$ and $\omega_{2n} = \omega_2 / \omega_n$. The loci obtained from this equation are shown in Fig. 5.6-5. In the figure, a portion of the $j\omega$ axis has been expanded for better detail. From these loci we may observe that the poles of the state-variable realization can cross the $j\omega$ axis for large values of Q. Since the state-variable filter

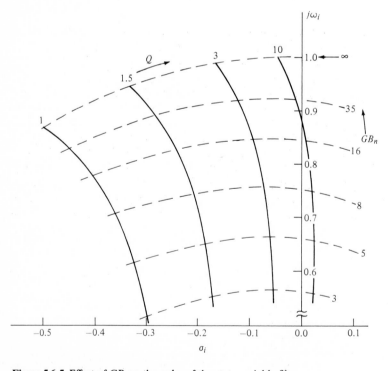

Figure 5.6-5 Effect of GB on the poles of the state-variable filter.

has the characteristic that its open-loop transfer function at ω_n is approximately $1/\underline{0}°$, small amounts of phase lag produced by the integrators and the summer will cause Q enhancement. This is especially true for large values of ω_n. This effect was considered in the previous section.

Now let us consider the resonator realization shown in Fig. 5.1-4. We solve for $V_{BP}(s)$ as

$$V_{BP}(s) = -A_{d1}(s)V_{in1}(s) = -A_{d1}(s)\left(\frac{\omega_1 R_1/R_4}{s + K\omega_1}\right)V_1(s)$$

$$- A_{d1}(s)\left(\frac{\omega_1 R_1/R_3}{s + K\omega_1}\right)V_x(s) - A_{d1}(s)\left(\frac{s + \omega_1}{s + K\omega_1}\right)V_{BP}(s) \qquad (29)$$

where

$$\omega_1 = \frac{1}{R_1 C_1} \qquad K = 1 + \frac{R_1}{R_4} + \frac{R_1}{R_3} \qquad (30)$$

and $A_{d1}(s)$ is the open-loop gain of operational amplifier A_1. From the figure we may define

$$A_3(s) = \frac{V_x(s)}{V_{LP}(s)} \qquad A_2(s) = \frac{V_{LP}(s)}{V_{BP}(s)} \qquad (31)$$

Substituting these expressions into (29) and solving for $V_{BP}(s)$ in terms of $V_1(s)$ give

$$\frac{V_{BP}(s)}{V_1(s)} = \frac{-A_{d1}(s)\dfrac{\omega_1(R_1/R_4)}{s + K\omega_1}}{1 + A_{d1}(s)\left(\dfrac{s + \omega_1}{s + K\omega_1}\right) + A_{d1}(s)A_2(s)A_3(s)\dfrac{\omega_1(R_1/R_3)}{s + K\omega_1}} \qquad (32)$$

The denominator of this transfer function can be written as

$$D(s) = \frac{s + K\omega_1}{A_{d1}(s)A_2(s)A_3(s)} + \frac{s + \omega_1}{A_2(s)A_3(s)} + \frac{\omega_1 R_1}{R_3} \qquad (33)$$

Substituting the relations

$$A_{d1}(s) \simeq \frac{-GB_1}{s} \qquad (34a)$$

$$A_2(s) \simeq \frac{-\omega_2 GB_2}{s(s + \omega_2 + GB_2)} \qquad (34b)$$

$$A_3(s) \simeq \frac{-GB_3/2}{s + GB_3/2} \qquad (34c)$$

in the above, we obtain

$$D(s) = \left(s^2 + \frac{\omega_n}{Q}s + \omega_n^2\right) + \frac{s}{GB}[4s^2 + (3\omega_1 + \omega_2 + \omega_1 K)s + \omega_1\omega_2]$$

$$+ \left(\frac{s}{GB}\right)^2[5s^2 + (2\omega_1 + 3\omega_2 + 3\omega_1 K)s + (2\omega_1\omega_2 + \omega_1\omega_2 K)]$$

$$+ \left(\frac{s}{GB}\right)^3[2s^2 + (2\omega_2 + 2\omega_1 K)s + 2\omega_1\omega_2] \tag{35}$$

where it can be shown that

$$\omega_n^2 = \omega_1\omega_2\frac{R_1}{R_3} \qquad \frac{\omega_n}{Q} = \omega_1 \qquad K = 1 + \frac{R_1}{R_4} + \frac{R_1}{R_3} \tag{36a}$$

$$\omega_1 = \frac{1}{R_1 C_1} \qquad \omega_2 = \frac{1}{R_2 C_2} \tag{36b}$$

Neglecting the $(s/GB)^2$ and $(s/GB)^3$ terms of (35) results in

$$D(s) \simeq \left(s^2 + \frac{\omega_n}{Q}s + \omega_n^2\right) + \frac{s}{GB}[4s^2 + (\omega_2 + 3\omega_1 + \omega_1 K)s + \omega_1\omega_2] \tag{37}$$

Proceeding as for the state-variable case

$$k = 1 \qquad q_2 = 4 \qquad q_1 = \omega_2 + \omega_1\left(4 + \frac{R_1}{R_3} + \frac{R_1}{R_4}\right) \qquad q_0 = \omega_1\omega_2 \tag{38}$$

Solving for x and y of (10) of Sec. 4.7 gives

$$x = -Q^2\frac{R_3}{R_1} - \frac{R_1}{R_3} - \frac{R_1}{R_4} = -\frac{\omega_2}{\omega_1} - \frac{Q}{\omega_2/\omega_1} - \frac{R_1}{R_4} \tag{39a}$$

$$y = 4 - \frac{R_3}{R_1} = 4 - \frac{\omega_1\omega_2}{\omega_n^2} \tag{39b}$$

Assuming $\omega_1 = \omega_2$ and $R_1/R_4 = H_0$ (bandpass) we get

$$x = -1 - Q^2 - H_0 \simeq -Q^2 \tag{40a}$$

$$y = 4 - \frac{1}{\omega_n^2} \simeq 4 \tag{40b}$$

for $Q_p > 1$. Thus the magnitude of the pole sensitivity becomes

$$|S_{GB}^{p_i}| \simeq \frac{\omega_n Q^2}{GB\sqrt{4Q^2 - 1}} \tag{41}$$

From the results obtained above we see that the performance of the resonator for $Q_p > 1$ is similar to that of the single-amplifier infinite-gain realization. A plot of the loci for the upper-half-plane complex pole can be obtained from (37) by normalizing with respect to ω_n to get

$$D(s_n) \simeq \frac{\omega_n^2}{GB_n}\left[4s_n^3 + s_n^2(GB_n + 3\omega_{1n} + \omega_{2n} + \omega_{1n}K) + s_n\left(\frac{GB_n}{Q} + \omega_{1n}\omega_{2n}\right) + GB_n\right]$$

(42)

where $\omega_{1n} = \omega_1/\omega_n$ and $\omega_{2n} = \omega_2/\omega_n$. The resulting loci as a function of GB_n with Q as a parameter are shown in Fig. 5.6-6. In obtaining the data for the figure, it has been assumed that $\omega_{1n} = 1/Q$, $\omega_{2n} = 1$, and $K = 1 + Q + R_1/R_4$. This corresponds to the design procedure given in Sec. 5.2. Note that unlike the infinite-gain realizations, the roots can cross the $j\omega$ axis for large values of GB_n. Comparing the state-variable filter with the resonator filter under the same design conditions, we conclude that the former is the more stable one.

In this section the effect of the frequency dependence of the amplifier on various types of filter realizations as indicated by $|S_{GB}^{p_i}|$ has been examined. If the factor of $(\omega_n/GB)/\sqrt{4Q^2 - 1}$ is removed from the sensitivities that have been calculated, and if $Q_p > 1$, then a comparison between the sensitivities of the various realizations can be made. The value of H_0 will be assumed to be approximately unity. For the positive-gain and the state-variable realizations, $|S_{GB}^{p_i}|$ is independent of Q. For the infinite-gain and the resonator realization, $|S_{GB}^{p_i}|$ is

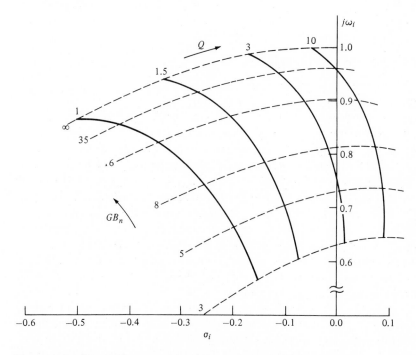

Figure 5.6-6 Effect of GB on the poles of the resonator filter.

proportional to Q. A similar examination of the properties of negative-gain realizations shows that $|S_{\text{GB}}^{p_i}|$ is proportional to Q^2. This provides another reason for the unpopularity of such realizations.

The information developed in this section is very important in the application of active filters. A straightforward use of the information is to incorporate the gain-bandwidth dependence into the design procedure. If ω_n and Q are given and if the GB of the amplifier is known, then the pole shift can be computed from the pole-loci plots or from (8) of Sec. 4.7. A compensated pole location can then be obtained by adding the pole shift to the desired pole location. Knowledge of the compensated pole locations will yield the compensated values of ω_n and Q which can then be used as the specifications for the normal design procedure. The success of this approach relies on the assumption that the pole shifts are linear in the vicinity of the desired pole location. For example, in Example 5.6-1, the compensated pole locations could have been calculated as p_1, $p_2 = -375.1 \pm j6278.3$ which corresponds with a compensated f_n and Q of 1001 and 8.36 respectively. Although such a procedure may not provide exact results, in general it will diminish the effect of the gain bandwidth of the amplifier upon the realization and make it easier to tune. The results contained in this section also permit the selection of a realization type which is appropriate for a given set of specifications. In addition, they may be used to define the required amplifier performance necessary for various specifications. Finally, these results provide the considerations and background necessary to develop realizations having improved frequency performance.

5.7 GAIN-BANDWIDTH-INDEPENDENT REALIZATIONS

In order to simultaneously achieve high-frequency performance and high Q in an active filter, it is important to minimize the sensitivity $|S_{\text{GB}}^{p_i}|$ of the filter with respect to the operational amplifier gain-bandwidth GB. In this section we show how this may be done by the use of compensation methods. Such methods apply primarily to filter configurations in which one input terminal of each operational amplifier is grounded, since otherwise the filter's frequency response is dependent not only on GB but also on many of the other operational amplifier parameters such as the input and output impedance levels, etc. One of the most versatile and useful filter configurations in which all the operational amplifiers have a grounded input terminal is the resonator introduced in Sec. 5.2. It is not only a useful and versatile second-order filter circuit, but it has also been shown to be a valuable building block for use in the realization of higher-order filter functions. Thus, its optimization is well worth considering.

The basic form of the resonator is shown in Fig. 5.7-1. This circuit can be analyzed by opening the closed feedback loop at the dotted line shown in the figure and investigating the open-loop transfer function $V_x'(s)/V_x(s)$. To do this we first consider the circuit shown in Fig. 5.7-2. This represents the first stage

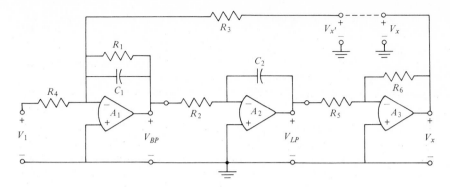

Figure 5.7-1 The resonator filter.

(amplifier A_1) of the resonator with $V_1(s)$ set to zero. The transfer function $T_1(s)$ for this circuit is

$$T_1(s) = \frac{V_{BP}(s)}{V'_x(s)} \simeq \frac{-\omega_1(R_1/R_3)GB_1}{s^2 + GB_1 s + \omega_1 GB_1 + \omega_{a1} K\omega_1} \tag{1}$$

where ω_{a1} and GB_1 are respectively the bandwidth ω_a and the gain-bandwidth GB of amplifier A_1. This relation has been derived by assuming that $\omega \gg \omega_{a1}$, and by letting

$$K = 1 + R_1/R_3 + R_1/R_4 \qquad \omega_1 = 1/R_1 C_1 \tag{2}$$

The phase of (1) can be written as

$$\arg[T_1(j\omega)] \approx \pi - \tan^{-1}\left(\frac{\omega GB_1}{\omega_1 GB_1 + \omega_{a1} K\omega_1 - \omega^2}\right) \tag{3}$$

Comparing the voltage transfer function of the resonator, as given in (20) of Sec. 5.2, to the standard second-order transfer function denominator [see (1) of Sec. 4.2] we obtain

$$\omega_1 = \frac{\omega_{no}}{Q_o} \tag{4}$$

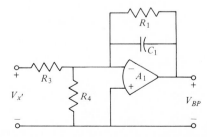

Figure 5.7-2 The A_1 stage of the resonator filter.

where ω_{no} and Q_o are the design values of ω_n and Q. Thus (3) can be written as

$$\arg\left[T_1(j\omega)\right] \approx \pi - \tan^{-1}\left(\frac{\omega GB_1}{\omega_{no} GB_1/Q_o + \omega_{a1} K\omega_{no}/Q_o - \omega^2}\right) \tag{5}$$

If $Q_o > 1$, then the argument of the arctangent becomes large and (5) simplifies to

$$\arg\left[T_1(j\omega)\right] \approx \frac{\pi}{2} + \frac{1}{Q_o}\frac{\omega_{no}}{\omega} + \frac{K}{Q_o A_{01}}\frac{\omega_{no}}{\omega} - \frac{\omega}{GB_1} \tag{6}$$

where A_{01} is the open-loop gain A_0 of amplifier A_1. Now let us consider the A_2 operational amplifier stage of Fig. 5.7-1. If we define $T_2(s)$ as the transfer function $V_{LP}(s)/V_{BP}(s)$, then from (7) of Sec. 5.4 we may write

$$\arg\left[T_2(j\omega)\right] = \frac{\pi}{2} + \frac{\omega_2}{\omega A_{02}} - \frac{\omega}{GB_2} \tag{7}$$

where

$$\omega_2 = \frac{1}{R_2 C_2} \tag{8}$$

Finally for the inverter A_3, which is the third amplifier in Fig. 5.7-1, defining $T_3(s) = V_x(s)/V_{LP}(s)$, from (9b) of Sec. 4.6 we obtain (assuming $R_5 = R_6$)

$$\arg\left[T_3(j\omega)\right] \approx \pi - \frac{2\omega}{GB_3} \tag{9}$$

Now let us define the overall open-loop transfer function as $T(s) = T_1(s)T_2(s)T_3(s)$. The total open-loop phase shift is thus given as

$$\arg\left[T(j\omega)\right] = \arg\left[T_1(j\omega)\right] + \arg\left[T_2(j\omega)\right] + \arg\left[T_3(j\omega)\right]$$

$$\approx \left(\frac{1}{Q_o} + \frac{K}{Q_o A_{01}} + \frac{\omega_2}{\omega_{no} A_{02}}\right)\left(\frac{\omega_{no}}{\omega}\right) - \left(\frac{\omega}{GB_1} + \frac{\omega}{GB_2} + \frac{2\omega}{GB_3}\right) \tag{10}$$

From the above discussion, we see that the A_3 inverter stage of the resonator has twice the phase lag of the inverting-integrator stages. This is one of the disadvantages of using an inverter in order to obtain a noninverting integrator. Now consider what happens as the quantities A_{0i} and $GB_i (i = 1, 2, 3)$, approach infinity. In this case (10) reduces to

$$\arg\left[T(j\omega)\right] \approx \frac{1}{Q}\frac{\omega_{no}}{\omega} \tag{11}$$

Thus we see that in this case Q and Q_o are the same. Therefore the actual Q may be expressed in terms of Q_o; the desired Q, by combining (10) and (11). The result is

$$Q \approx \frac{\omega_{no}/\omega}{\left(\dfrac{1}{Q_o} + \dfrac{K}{Q_o A_{01}} + \dfrac{\omega_2}{\omega_{no} A_{02}}\right)\dfrac{\omega_{no}}{\omega} - \left(\dfrac{\omega}{GB_1} + \dfrac{\omega}{GB_2} + \dfrac{2\omega}{GB_3}\right)} \tag{12}$$

If we use the design procedure (of Sec. 5.2) in which $R_2 = R_3 = R_4 = R_5 = R_6 = R$ ($H_0 = Q_o$ for bandpass), $C_1 = C_2 = C$, and $R_1 = Q_o R$, and assume that $\omega = \omega_{no}$ and that all amplifiers are matched, then (12) simplifies to

$$Q \approx \frac{1}{1/Q_o + 1/Q_o A_0 + 3/A_0 - 4\omega_{no}/\text{GB}} \tag{13}$$

Thus the Q actually obtained is higher than the design Q. This Q *enhancement* is caused by the loop phase lag, and it can actually make the network unstable if ω_{no}/GB becomes sufficiently large.

Example 5.7-1 *Q Enhancement of a Resonator Filter*   The resonator circuit of Fig. 5.7-1 is to be used to realize a second-order bandpass filter with $Q_o = 4$ and $f_{no} = 38,185$ Hz. The design procedure used is $R_2 = R_3 = R_4 = R_5 = R_6 = R$, $C_1 = C_2 = C$, and $R_1 = Q_o R$. The operational amplifiers are matched and have the characteristics $A_0 = 10^5$ and GB = 1 MHz. The value of Q actually obtained for this filter when $f = f_{no}$ may be found from (13) to have a value of 10.28. The Q enhancement that occurs is clearly evident.

□

Equation (13) can also be used to predict the frequency at which a resonator filter becomes unstable. To do this, we approximate the equation as

$$Q \approx \frac{1}{1/Q_o - 4\omega_{no}/\text{GB}} \tag{14}$$

From this we see that Q is infinite when $\omega_{no} = \text{GB}/4Q_o$. This result, however, is valid only for small changes in pole position, and as a consequence Q_o should be greater than 10 to give meaningful results. For example, using the figures in Example 5.7-1, the maximum value of ω_{no} predicted by (14) is 25 kHz. The filter, however, is stable at higher frequencies because the value of Q_o for this example is too small to justify the use of (14).

Now let us consider what happens if the inverting-integrator inverter configuration of stages A_2 and A_3 in Fig. 5.7-1 is replaced with the noninverting integrator of Fig. 5.4-10 (where R_2 and R_1 are replaced by R_5 and R_6). The result is the *modified resonator circuit* shown in Fig. 5.7-3. The loop phase shift can be found using (27) of Sec. 5.4 and (6). Thus we obtain (for $\omega_{no} < \text{GB}_2$)

$$\arg\left[T(j\omega)\right] \approx \left(\frac{1}{Q_o} + \frac{K}{Q_o A_{01}} + \frac{R_6/R_5}{A_{02}}\right)\frac{\omega_{no}}{\omega} + \left[\frac{\omega}{\text{GB}_3}\left(1 + \frac{R_5}{R_6}\right) - \frac{\omega}{\text{GB}_1} - \frac{\omega}{\text{GB}_3}\frac{R_6}{R_5}\right] \tag{15}$$

where we have assumed that the arguments of the arctangent functions are small. For the case where $R_5 = R_6$ and the gain bandwidths of the amplifiers are matched, this becomes

$$\arg\left[T(j\omega)\right] \approx \left(\frac{1}{Q_o} + \frac{K}{Q_o A_0} + \frac{1}{A_0}\right)\frac{\omega_{no}}{\omega} \tag{16}$$

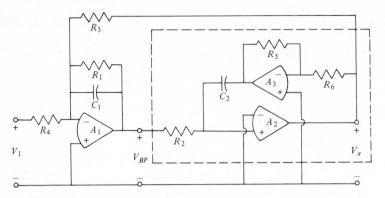

Figure 5.7-3 The modified resonator filter circuit.

Combining this with (11) we find that

$$Q \approx \frac{1}{1/Q_o + K/Q_o A_0 + 1/A_0} \tag{17}$$

This result shows that Q is independent of GB, and thus no Q enhancement occurs for the modified resonator circuit. The difference between the original resonator and the modified one is further illustrated in Fig. 5.7-4. The data plotted represent two realizations of a bandpass filter having $Q_o = 100$. The value of f_n was increased, starting with a value of 30 Hz. The original resonator became unstable at approximately 2000 Hz, whereas the Q of the modified resonator had not significantly changed at $f_n = 15$ kHz. These experimentally observed results are in good agreement with (14), which predicts instability at 2500 Hz (for GB = 1 MHz). A final advantage of the modified resonator circuit is obtained as a result of the temperature coefficients of the individual operational amplifier gain bandwidths being similar. Thus the excess unwanted lagging phase shift is cancelled with good temperature tracking.

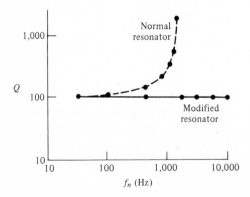

Figure 5.7-4 Comparison of the characteristics of the normal and the modified resonator filters.

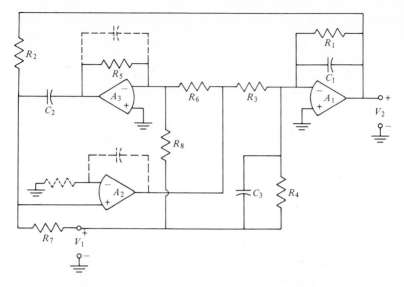

Figure 5.7-5 A high-frequency modification of the biquadratic resonator filter.

A general high-frequency second-order block can be developed by substituting the noninverting integrator of Fig. 5.4-10 in place of the inverter and inverting integrator in the biquadratic resonator realization given in Fig. 5.2-6. The resulting network is shown in Fig. 5.7-5.[1] The design techniques of Sec. 5.2 and Table 5.2-1 remain valid for this network. It provides the flexibility of a general design and yet has very good frequency capability. To see this, we note that the actual pole frequency ω_p can be expressed in terms of the desired pole frequency ω_{po} as

$$\omega_p \approx \omega_{po}\left(1 + \frac{2\omega_{po}}{GB_1} + \frac{\omega_{po}}{GB_3}\right)^{-1/2} \tag{18}$$

where the design assumes that $R_1 = R_2 = R_3 = R_5 = R_6 = R$, that R_4 and R_7 are much greater than R_5, and that $C_1 = C_2 = C$. The pole Q_p can similarly be expressed in terms of the desired Q_{po} as

$$Q_p \approx \frac{Q_{po}}{1 + Q_{po}\omega_p\left(\dfrac{1 + R_5/R_6 + R_5/R_8}{GB_3} - \dfrac{R_6/R_5}{GB_2} - \dfrac{1 + C_3/C_1}{GB_1}\right)} \tag{19}$$

Since the signs of the individual terms in the factor in parenthesis which multiplies $Q_{po}\omega_p$ in the denominator of (19) are both positive and negative, with a suitable

[1] D. Akerberg and K. Mossberg, "A Versatile Active RC Building Block with Inherent Compensation for the Finite Bandwidth of the Amplifier," *IEEE Trans. Circuits and Systems*, vol. CAS-21, no. 1, January 1974, pp. 75–78.

design procedure we may cancel the effects on Q of the gain bandwidths of the operational amplifiers. To see this, let us assume that $GB_1 = GB_2 = GB_3 = GB$, and define

$$k = \frac{R_6}{R_5} + \frac{C_3}{C_1} - \frac{R_5}{R_6} - \frac{R_5}{R_8} \tag{20}$$

Thus, our objective is to make $k = 0$. With reference to Table 5.2-1, we see that $C_3 = 0$ and $R_8 = \infty$ are required for the low-pass and bandpass cases. In this situation, $k = 0$ if

$$R_5 = R_6 \tag{21}$$

Similarly, for the high-pass and $j\omega$-axis zeros we require

$$\frac{R_6}{R_5} = \frac{C_3}{2C_1}\left[-1 \pm \sqrt{1 + \left(\frac{2C_1}{C_3}\right)^2}\right] \tag{22}$$

in order for k to be zero. It should be noted that use of (22) often results in values of R_6 and R_5 which give significant changes in the signal level from the junction of R_3 and R_6 to the output of A_3 of Fig. 5.7-5. This difference in signal levels will aggravate the effects of slew rate and to a certain extent undo the results achieved by inherent compensation of gain bandwidth. An alternate procedure to removing R_8 completely from the circuit (setting $R_8 = \infty$) is to detach it from V_1 and ground it. In this case, k is still given by (20); however, R_8 acts as if it had a value of infinity in Table 5.2-1. Now we may let $R_5 = R_6$ in the high-pass and $j\omega$-axis zero cases and still achieve $k = 0$ by letting

$$\frac{C_3}{C_2} = \frac{R_5}{R_8} \tag{23}$$

where R_8 is attached between the inverting terminal of A_3 and ground.

Example 5.7-2 *A Biquadratic GB-Independent Resonator Filter*   It is desired to use the configuration of Fig. 5.7-5 to realize a filter function having $Q_{po} = 50$, $f_{po} = 10$ kHz, $Q_{zo} = \infty$, and $f_{zo} = 7.5$ kHz. From (28) and (29) of Sec. 5.2 we obtain

$$\omega_{zo}^2 = \frac{R_6}{R_3 R_5 R_7 C_2 C_3} \tag{24a}$$

$$\omega_{po}^2 = \frac{R_6}{R_2 R_3 R_5 C_2 C_3} \tag{24b}$$

$$\frac{\omega_{po}}{Q_{po}} = \frac{1}{R_1 C_1} \tag{24c}$$

We may select $R_3 = R_5 = R_6 = 10$ kΩ and $C_1 = C_2 = C_3 = 3 \times 10^{-9}$ F.

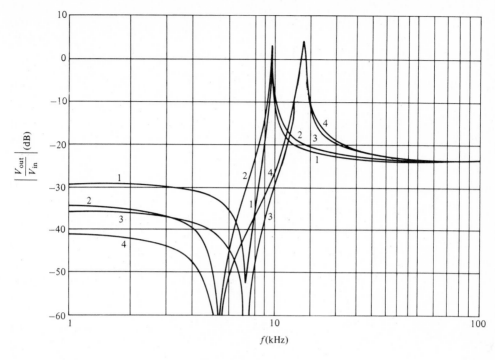

Figure 5.7-6 Frequency response of the filter of Example 5.7-2.

From (24) we get $R_2 = 2.81$ kΩ, $R_1 = 265.26$ kΩ, and $R_7 = 5$ kΩ. In order to make $k = 0$, we select $R_5 = R_8 = 10$ kΩ and ground the end of R_8 that is connected to V_1. Figure 5.7-6 shows the frequency response of this filter and illustrates the tuning process for it. Curve 1 corresponds to the nominal filter. In curve 2, R_7 has been changed to 9 kΩ to illustrate its effect on ω_z. In curve 3 (with R_7 restored to its original 5-kΩ value), R_2 has been changed to 1.3 kΩ to tune ω_p. Finally, curve 4 shows the effects of using $R_2 = 1.3$ kΩ and $R_7 = 9$ kΩ in order to tune both ω_p and ω_z.

In calculating the performance of this filter it should be noted that the original resonator filter circuit would become unstable when $f_p > \mathrm{GB}/4Q_{po} = 2.5$ kHz (assuming GB $= 500$ kHz). The improvement in the performance of the circuit by the modification may be seen in Fig. 5.7-7. In Fig. 5.7-7a, the percent change in Q_p is plotted as a function of temperature for both the original and the modified versions of the resonator. The percent change in the modified one is about 5 percent over 60°C change, whereas the original one would experience approximately 130 percent change over the same temperature range. Another characteristic of this filter is shown in Fig. 5.7-7b. Here we note that the percentage deviation of the actual Q from the desired one as a function of f_p is less than -2 percent for values of f_p from 1 to 30 kHz. $\square$

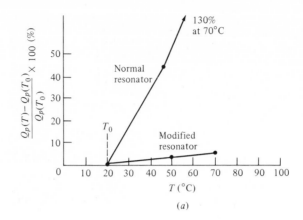

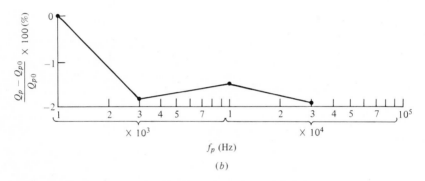

Figure 5.7-7 Performance characteristics for the filter of Example 5.7-2.

One of the disadvantages of the modified resonator circuit shown in Fig. 5.7-5 is that it requires stricter compensation of the amplifiers than the original resonator realization does. The reason for this can be seen by looking at the feedback loop consisting of amplifiers A_2 and A_3. Neither of these amplifiers has a capacitor in the feedback path around it. Thus there is no short circuit around it at high frequencies. Consequently the phase lag of the amplifiers at frequencies near GB can create sufficient phase shift to cause a high-frequency oscillation. Typically, the high-frequency oscillation is approximately equal to GB and is limited in amplitude by the slew rate of the amplifier. Frequently, the higher-frequency oscillation will not disturb the operation of the amplifier, and it may not even be observable at the filter output (due to the damping effect of the capacitor C_2). In practice, however, careful control of the amplifier's high-frequency response must be exercised in order to see that this high-frequency oscillation does not occur. The elements shown by the dashed lines in Fig. 5.7-5 are used to eliminate such a high-frequency oscillation if it is present.

A final consideration of the modified resonator filter is concerned with an unstable mode which may occur at high values of Q_p and f_p and which is caused by the phase shift due to the slew rate as discussed in Sec. 5.4. If the internal circuit signal levels become large enough, this will result in the circuit breaking into an unstable mode. Such an unstable mode can be prevented by paralleling the inputs of amplifiers A_2 and/or A_3 by two parallel silicon diodes of opposite polarity so that the input signal to the amplifiers is limited to $\pm V_d$, where V_d is the forward voltage drop of the diode.

The concept of inherent compensation of amplifiers may also be applied to the state-variable circuit. A method for achieving this is shown in Fig. 5.7-8a. If

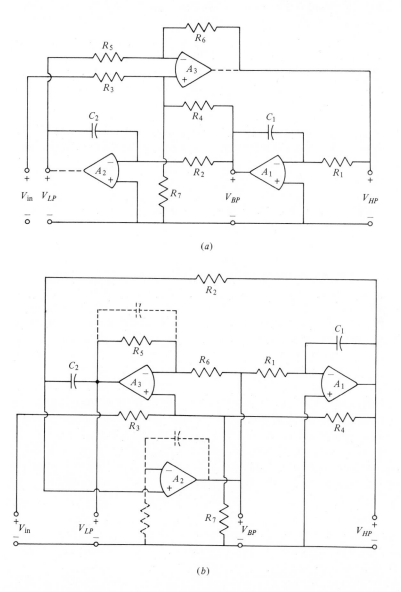

(a)

(b)

Figure 5.7-8 Modification of the state-variable filter to provide inherent compensation of the amplifiers.

the outputs of amplifiers A_2 and A_3 are interchanged, then the circuit is as shown in Fig. 5.7-8*b*. Here the polarity of amplifier A_2 has been inverted for reasons of stability. Relationships similar to those given for the modified resonator circuit can be developed to show that the operational amplifier gain bandwidths can be cancelled from the expression for Q_p. An example is given in the Problems. The modified state-variable filter requires the dotted components shown in Fig. 5.7-8*b* in order to prevent the high-frequency oscillation discussed above from occurring. In addition, pairs of silicon diodes may be required in parallel with the inputs of amplifiers A_2 and A_3 to prevent unstable modes due to high signal levels when Q_p and f_p are large.

So far in this section we have considered techniques for minimizing S_{GB}^Q. Frequently, it is also important to minimize $S_{GB}^{\omega_n}$ and/or $S_{GB}^{p_i}$. Of these, the minimization of $S_{GB}^{\omega_n}$ is often of less importance, because a large value of this sensitivity does not necessarily mean that the filter will be unstable, as is more certainly the case when S_{GB}^Q is large. Of course, minimizing $|S_{GB}^{p_i}|$ simultaneously minimizes S_{GB}^Q and $S_{GB}^{\omega_n}$. Techniques for achieving this on a first-order basis are possible (see the Problems) but are in need of further refinement before they are applicable on a general basis.

5.8 SUMMARY

In this chapter we have discussed several types of *RC*-amplifier filter configurations. In Sec. 5.1 we showed how a single operational amplifier may be used in multiple- or single-feedback configurations to realize most of the common filter functions. In Sec. 5.2 we introduced the well-known state-variable and resonator multiple-amplifier configurations. Both of these configurations are well suited to the realization of high-Q functions. In Sec. 5.3 a modification of the state-variable configuration called the *universal active filter* was discussed. In Sec. 5.4 a detailed treatment of the properties of the integrator used in the state-variable and resonator filters was given. In Sec. 5.5 it was shown how the nonideal properties of operational amplifiers led to Q enhancement in these same types of filters. In Sec. 5.6 the general effects of gain-bandwidth limitations of operational amplifiers on various filter realizations were discussed. Finally, in Sec. 5.7, a treatment of a gain-bandwidth-independent version of the resonator was given. A summary of the properties of the various *RC*-amplifier filter configurations given in this and the preceding chapter is given in Table 5.8-1. In this table a + sign indicates that the filter is exceptional in the indicated property, while a − sign indicates that it is average or below.

Table 5.8-1 Comparison of different RC-amplifier filter types

	Low number of passive elements	Low number of active elements	Low spread of element values	Simple design relations	Low sensitivity	Realizes high Q
Single-amplifier positive-gain (Secs. 4.2 and 4.3)	+	+	+	+	−	−
Single-amplifier infinite-gain multiple-feed-back (Sec. 5.1)	+	+	−	−	+	−
Single-amplifier infinite-gain single-feedback (Sec. 5.1)	−	+	+	+	+	−
Multiple-amplifier state-variable (Secs. 5.2 and 5.3)	−	−	−	−	+	+
Multiple-amplifier resonator (Sec. 5.2)	−	−	−	+	+	+

PROBLEMS

5-1 (*Sec. 5.1*)   Use the infinite-gain single-amplifier multiple-feedback filter configuration shown in Fig. 5.1-2 to realize a second-order low-pass Bessel function having a delay of 500 μs and a unity gain at direct current. (*Note:* Choose $C = .05 \times 10^{-6}$, $m = \frac{1}{4}$, and use the lower of the two possible values for R_2.)

5-2 (*Sec. 5.1*)   (*a*) For the low-pass infinite-gain multiple-feedback single-amplifier filter shown in Fig. 5.1-2, set $R_1 = R_2 = R_3 = R$, and find expressions for the quantities ω_n, $1/Q$, and H_0 of (1) of Sec. 4.2.

(*b*) In terms of these quantities, find expressions for R and C_6. Assume that C_5 is given some convenient value.

(*c*) Use the relations found above to design a filter meeting the specifications given in Example 5.1-1.

5-3 (*Sec. 5.1*)   (*a*) Design a bandpass filter with a 3-dB bandwidth from 725 to 800 Hz and with 0-dB gain at the resonant frequency. Use a low-pass to bandpass transformation on an appropriate two-pole Butterworth function to find the bandpass one. Use the single-amplifier infinite-gain multiple-feedback bandpass structure shown in Fig. 5.1-4 to realize the bandpass pole pairs.

(*b*) Repeat the problem assuming that the gain is to be 20 dB at the resonant frequency.

5-4 (*Sec. 5.1*)   Another single-amplifier infinite-gain multiple-feedback bandpass filter structure can be realized by letting $Y_1 = sC_1$ and $Y_3 = G_3$ in the circuit shown in Fig. 5.1-1. Determine the other element choices, find the transfer function, and develop a design procedure for this filter in terms of quantities ω_n, Q, and H_0 of (1) of Sec. 4.3.

5-5 (*Sec. 5.1*) Design a single-amplifier infinite-gain single-feedback bandpass filter to meet the specifications given in Example 5.1-2. Select an appropriate impedance normalization for the passive elements.

5-6 (*Sec. 5.1*) Sketch the magnitude of the frequency response of the filter circuit shown in Fig. P5-6. Label the values of the dc gain, break frequency, and other characteristics of this circuit.

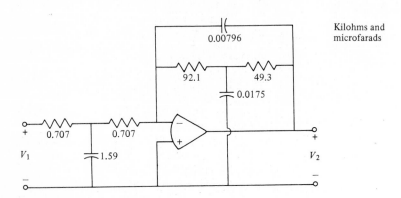

Figure P5-6

5-7 (*Sec. 5.1*) Repeat Prob. 5-6 for the circuit shown in Fig. P5-7.

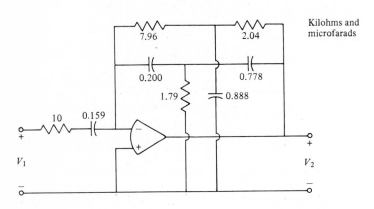

Figure P5-7

5-8 (*Sec. 5.2*) (*a*) Use the state-variable filter configuration given in Fig. 5.2-2 to realize a bandpass function with a Q of 100 and a resonant frequency f_n of 100 Hz. Let $C_1 = C_2 = 0.1 \ \mu$F, and $R_3 = R_5 = R_6 = 10 \ k\Omega$. Find the resulting value of H_0 in (1) of Sec. 4.3.

(*b*) If the same circuit is used to realize a low-pass function, what is the value of H_0 of (1) of Sec. 4.2?

(*c*) If the same circuit is used to realize a high-pass function, what is the value H_0 of (16) of Sec. 4.3?

5-9 (*Sec. 5.2*) (*a*) Use the resonator circuit shown in Fig. 5.2-4 to realize a bandpass second-order function with a Q of 100, a resonant frequency f_n of 100 Hz, and a value of $|H_0|$ in (1) of Sec. 4.3 of unity. Let $C_1 = C_2 = 1.0\ \mu F$, and $R_2 = R_3$.

(*b*) Repeat the design for a low-pass function with the same specifications.

5-10 (*Sec. 5.3*) (*a*) Use the design procedure given in Table 5.3-1 to design a bandpass filter in which $H_0 = 1$, $f_n = 1000$ Hz, and $Q = 50$ for the case where the passband gain has a phase of $\pm 180°$.

(*b*) Repeat the problem for the case where the phase shift is $0°$.

5-11 (*Sec. 5.3*) (*a*) For the filters designed in Prob. 5-10, if the magnitude of the voltage at the high-pass output is 1 V peak-to-peak, what are the magnitudes of the voltages at the low-pass and bandpass outputs when the excitation frequency $\omega = \omega_n/10$?

(*b*) Repeat the problem for an excitation frequency of $\omega = \omega_n$.

(*c*) Repeat the problem for an excitation frequency $\omega = 10\omega_n$.

5-12 (*Sec. 5.4*) What value of excess phase lag will be introduced at small signal levels at frequencies of $f = GB/100$, $GB/50$, $GB/10$, and $GB/5$ by the inverting integrator shown in Fig. 5.4-1 with $f_{RC} = 1$ kHz, $f_a = 10$ Hz, and $GB = 1$ MHz?

5-13 (*Sec. 5.4*) Use techniques similar to those given in Sec. 5.4 to find the excess phase shift of the damped integrator shown in Fig. P5-13. Assume $A_d(s)$ is given by (17) of App. B. Let $\omega_a \ll \omega \ll GB$.

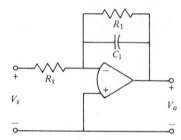

Figure P5-13

5-14 (*Sec. 5.4*) Find the pole and zero locations of the voltage transfer function for the integrator of Fig. 5.4-1 if $A_d(s)$ is given by (17) of App. B and R_o is not zero. Assume that $1/RC \ll GB$ and $\omega_a \approx 0$. Sketch the asymptotic magnitude and phase of the closed-loop frequency response if $1/R_o C < 4GB$.

5-15 (*Sec. 5.5*) Using the state-variable realization defined in Table 5.3-1, find the largest value of f_n in hertz possible if $Q_o = 10$ and $GB = 1$ MHz, and if the difference between Q and Q_o is to be no more than 10 percent.

5-16 (*Sec. 5.5*) Repeat Prob. 5-15 for the resonator filter whose circuit is given in Fig. 5.2-4.

5-17 (*Sec. 5.5*) Use the state-variable filter shown in Fig. 5.2-5 to realize a second-order all-pass network having $f_{zo} = f_{po} = 10$ kHz and $Q_p = -Q_z = 10$.

5-18 (*Sec. 5.5*) Repeat Prob. 5-17 if Q_z is equal to infinity.

5-19 (*Sec. 5.5*) If $\omega_n = 1000$ rad/s and $Q = 500$, what are the values of T_1 and T_2 of Fig. 5.5-3 (assume $N = 100$)?

5-20 (*Sec. 5.6*) (*a*) Find $|S_{GB}^{p_i}|$ of a state-variable filter realization in which $Q = 10$, $f_n = 10$ kHz, $\varepsilon = 1$, $R_6/R_5 = 0.1$, and $R_7 = \infty$. Assume the GB of the operational amplifiers is 1 MHz.

(*b*) Find the actual pole locations for this filter. What are the actual values of Q and f_n?

(*c*) If f_n is increased to 60 kHz, find the approximate pole locations from Fig. 5.6-5. What are the values of Q and f_n obtained in this case?

5-21 (*Sec. 5.6*) (*a*) Find $|S_{GB}^{p_i}|$ of a resonator filter realization in which $Q = 10$, $f = 10$ kHz, $R_2 = R_3 = R_4 = R_5 = R_6 = R$, $C_1 = C_2 = C$, and $R_1 = QR$. Assume the GB of the operational amplifiers is 1 MHz.

(*b*) Find the actual pole locations for this filter using (11). What are the actual values of Q and f_n?

5-22 (*Sec. 5.6*) If f_p of Prob. 5-21 is increased to 25 kHz, find the approximate pole locations from Fig. 5.6-6. What are the actual values of Q and f_n? Assume that $\omega_1 = \omega_n/Q$, $\omega_2 = \omega_n$, and $R_1/R_4 = H_0 = 1$.

5-23 (*Sec. 5.7*) The resonator circuit shown in Fig. 5.7-1 has operational amplifiers in which $A_0 = 10^5$ and GB $= 10^6$ Hz. If the circuit realizes a filter function in which $Q_o = 10$, find the frequency where it becomes unstable. Assume a design in which $R_2 = R_3 = R_4 = R_5 = R_6 = R$, $C_1 = C_2 = C$, and $R_1 = Q_{po}R$ is used.

5-24 (*Sec. 5.7*) At what value of Q_o will the filter described in Example 5.7-1 become unstable?

5-25 (*Sec. 5.7*) Use the circuit of Fig. 5.7-3 to design a bandpass realization with $Q_o = 5$, $f_{no} = 50$ kHz, and $H_0 = 1$. Select $R_3 = R_5 = R_6 = 10$ kΩ and $C_1 = C_2 = 3 \times 10^{-9}$ F.

5-26 (*Sec. 5.7*) Use the circuit of Fig. 5.7-5 to realize a low-pass filter having the same specifications as those given in Prob. 5-25.

5-27 (*Sec. 5.7*) Show that S_{GB}^Q for the second-order bandpass-modified infinite-gain filter shown in Fig. P5-27 is equal to zero if $s/GB \gg (s/GB)^2$. Both amplifiers are identical and have a frequency response given by $A_d(s) \approx -GB/s$. (*Hint:* Consider first just the passive network within the dotted line and write the nodal equation for it at the junction of the two C_3 capacitors and the inverting input of A_1. Next apply the amplifier constraints and solve for V_2/V_1.)

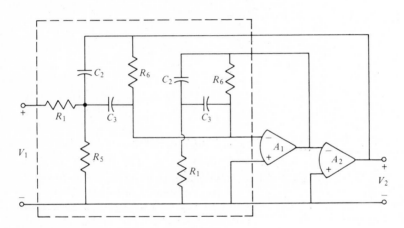

Figure P5-27

5-28 (*Sec. 5.7*) Use the techniques employed in Sec. 5.7 to find the Q of the state-variable realization of Fig. 5.7-8b and to show that Q is independent of the amplifier's gain bandwidth if the operational amplifiers are matched. (*Hint:* The material in Sec. 5.5 may be of help in this problem.)

PASSIVE NETWORK SIMULATION METHODS

In the first section of Chap. 4, we pointed out that there are two general methods for using active RC filters to realize network functions. The first such method was the subject of the last two chapters. It was the *cascade method* in which RC-amplifier filters were used to realize second-order functions, and a cascade of such realizations was used to obtain higher-order functions. In this chapter we discuss the second general method for using active RC filters to realize network functions. It is called the *direct method*, and has the characteristic that filter functions of any order are realized by a single circuit configuration. The first direct method that we shall consider starts with a passive prototype network and uses active RC circuits to simulate certain portions of the passive realization. Such a method is called a *passive network simulation method*. It has the advantage of low sensitivities, basically the same ones that characterize the prototype passive network. In addition, since tables of element values for passive network realizations are readily available, the synthesis procedures for such simulation methods become very straightforward. We shall encounter other advantages and disadvantages of these methods in connection with our discussion of individual filter types.

6.1 ACTIVE NETWORK ELEMENTS—INDUCTANCE SIMULATION

In this section we present a discussion of some of the various types of active elements used in passive network simulation methods. Such elements are best characterized in terms of their transmission (or $ABCD$) two-port parameters.

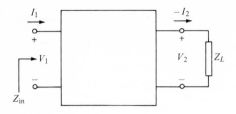

Figure 6.1-1 A two-port network terminated in an impedance.

They are defined by the equation

$$\begin{bmatrix} V_1(s) \\ I_1(s) \end{bmatrix} = \begin{bmatrix} A(s) & B(s) \\ C(s) & D(s) \end{bmatrix} \begin{bmatrix} V_2(s) \\ -I_2(s) \end{bmatrix} \tag{1}$$

where the voltage and current variables for the two-port are as shown in Fig. 6.1-1. From (1), the individual parameters are readily seen to be

$$A(s) = \frac{V_1(s)}{V_2(s)}\bigg|_{I_2(s)=0} \qquad B(s) = \frac{V_1(s)}{-I_2(s)}\bigg|_{V_2(s)=0} \tag{2a}$$

$$C(s) = \frac{I_1(s)}{V_2(s)}\bigg|_{I_2(s)=0} \qquad D(s) = \frac{I_1(s)}{-I_2(s)}\bigg|_{V_2(s)=0} \tag{2b}$$

As an example, the noninverting realization of the VCVS of Fig. 4.1-2a is readily shown to have $A(s) = R_1/(R_1 + R_2)$, and $B(s) = C(s) = D(s) = 0$. Obviously, from (2a), and from (1) of Sec. 4.1, we see that $A(s)$ is the reciprocal of the voltage transfer function of the device. The input impedance $Z_{in}(s)$ of a two-port network terminated in an impedance $Z_L(s)$, as shown in Fig. 6.1-1, is readily found by taking the ratio of the separate equations of (1) and using the relation $V_2(s) = -I_2(s)Z_L(s)$. Thus

$$Z_{in}(s) = \frac{V_1(s)}{I_1(s)} = \frac{A(s)V_2(s) - B(s)I_2(s)}{C(s)V_2(s) - D(s)I_2(s)} = \frac{A(s)Z_L(s) + B(s)}{C(s)Z_L(s) + D(s)} \tag{3}$$

We may now define several classes of active network elements in terms of the restrictions on their transmission parameters. There are two general classifications, namely the GIC (generalized immittance converter), which multiplies the load impedance by the quantity $A(s)/D(s)$, and the GIV (generalized immittance inverter), which multiplies the load impedance by $C(s)/B(s)$ and inverts it. These general classifications and several specific devices in each classification are summarized in Table 6.1-1. Further details may be found in the literature.[1,2,3]

---

[1] S. K. Mitra, *Analysis and Synthesis of Linear Active Networks*, John Wiley & Sons, Inc., New York, 1969, chap. 2.

[2] L. P. Huelsman, "A Fundamental Classification of Negative Immittance Converters," *IEEE Intern. Conv. Rec.*, vol. 13, p. 7, March 1965, pp. 113–118.

[3] S. K. Mitra, *Active Inductorless Filters*, IEEE Press, New York, 1971, pp. 215–221.

Table 6.1-1 Active two-port network elements

Device	Transmission parameters				Restrictions	Input impedance
	A	B	C	D		
GIC (generalized immittance converter)						
NIC (negative immittance converter)	Non-zero	0	0	Non-zero	$A < 0$ or $D < 0$	$Z_{\text{in}} = \dfrac{A}{D} Z_L$
PIC (positive immittance converter)					A and D same parity	
GIV (generalized immittance inverter)						
NIV (negative immittance inverter)	0	Non-zero	Non-zero	0	$B < 0$ or $C < 0$	$Z_{\text{in}} = \dfrac{B}{C}\dfrac{1}{Z_L}$
PIV (positive immittance inverter)					B and C same parity	
Gyrator					$B = \dfrac{1}{C}$	

The realization shown in Fig. 6.1-2 is probably the most used and practical one for both these elements. It was proposed in 1969,[1] and is available today as a hybrid integrated circuit.[2] The configuration shown in Fig. 6.1-2a, in which $Z_5(s)$ is considered as the load impedance, has the transmission parameters

$$A(s) = 1 \qquad B(s) = C(s) = 0 \qquad D(s) = \frac{Z_2(s)Z_4(s)}{Z_1(s)Z_3(s)} = \frac{1}{K_c(s)} \qquad (4)$$

Thus, this two-port device is a GIC. The quantity $K_c(s)$ is called the *GIC constant*. Since $A(s) = 1$, it is sometimes called a *current* generalized impedance converter.

[1] S. K. Mitra, *Analysis and Synthesis of Linear Active Networks*, John Wiley & Sons, Inc., New York, 1969, p. 494; also, A. Antoniou, "Realization of Gyrators Using Operational Amplifiers, and Their Use in RC-Active Network Synthesis," *Proc. IEEE*, vol. 116, no. 11, November 1969, pp. 1838–1850.

[2] ATF 431 Gyrator, Amperex Electronic Corp., Slatesville, R.I.; TCA 580 Gyrator, Signetics Corp., Sunnyvale, Calif.

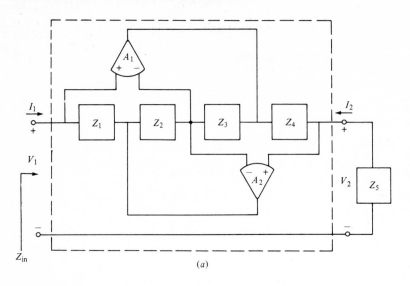

(a)

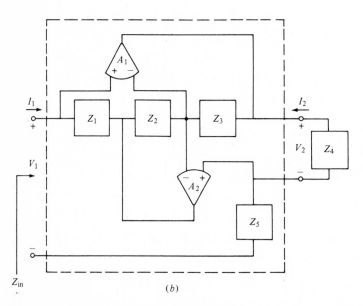

(b)

Figure 6.1-2 A GIC (*generalized immittance converter*) used to realize a GIN (*generalized immittance network*).

The configuration shown in Fig. 6.1-2b, in which $Z_4(s)$ is the load impedance, has the parameters

$$A(s) = D(s) = 0 \qquad B(s) = Z_5 \qquad C(s) = \frac{Z_2(s)}{Z_1(s)Z_3(s)} \tag{5}$$

Thus this two-port device is a realization of a GIV. If only the input impedance is considered, both circuits of Fig. 6.1-2 are identical. The input impedance is

$$Z_{in}(s) = \frac{Z_1(s)Z_3(s)Z_5(s)}{Z_2(s)Z_4(s)} \tag{6}$$

In this case the device is called a GIN (*generalized immittance network*), a two-terminal device in which one terminal is ground. This device is the starting point for the passive network simulation techniques covered in this and the following section.

One of the simplest approaches to the passive network simulation method is to take an *RLC* network which satisfies the specified filter requirements and to replace the inductors with active *RC* equivalents. The resulting elements are called *synthetic inductors*. The GINs of Fig. 6.1-2 are ideally suited for this purpose. If $Z_2(s)$ is a capacitor, and the remaining impedances are resistances, then the input impedance of the GIN and the equivalent value of inductance are

$$Z_{in}(s) = \frac{sR_1C_2R_3R_5}{R_4} \qquad L_{eq} = \frac{R_1C_2R_3R_5}{R_4} \tag{7}$$

Alternately, if $Z_4(s)$ is a capacitor we obtain

$$Z_{in}(s) = \frac{sR_1R_3C_4R_5}{R_2} \qquad L_{eq} = \frac{R_1R_3C_4R_5}{R_2} \tag{8}$$

A synthetic inductor realized by a GIN turns out to have excellent characteristics. If the amplifiers A_1 and A_2 of Fig. 6.1-2 are matched, then the effect of the nonideal parameters of the amplifiers on the synthetic inductance are minimized. In addition, since $Z_1(s)$ is always a resistance, the dc bias currents at the inputs of the amplifiers do not present difficulties (a further discussion of this problem is given in the following section). The only disadvantage of the synthetic inductance approach to network synthesis is that since the realization of the synthetic inductor is an active device, it must have one terminal grounded. Thus the *RLC* network must also be of a type which has all inductors grounded. This restricts the synthetic inductance approach primarily to high-pass realizations. The FDNR technique, to be discussed in Sec. 6.2, will be seen to apply to low-pass and bandpass realizations. Floating synthetic inductors can be made[1,2] but their performance and simplicity are not as good as those of the grounded synthetic inductors.

It is of interest to consider the signal levels at the amplifier outputs in the synthetic inductor. For this, consider Fig. 6.1-3 in which a GIN is driven by a

---

[1] T. N. Rao, "Readily Biased Wideband Gyrator Circuit for Floating or Earthed Inductors," *Electronic Letters*, vol. 5, July 1966, pp. 309–310.
[2] W. H. Holmes, "An Improved Version of a Floating Gyrator," *IEEE J. Solid-State Circuits*, vol. SC-4, no. 3, June 1969, pp. 162–163.

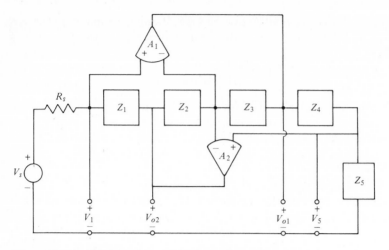

Figure 6.1-3 A GIN driven by a nonideal voltage source.

voltage source in series with a resistor. For $Z_2(s) = 1/sC_2$, and $Z_i(s) = R_i$ $(i \neq 2)$, we obtain

$$\frac{V_{o1}(s)}{V_1(s)} = 1 + \frac{R_4}{R_5} \tag{9a}$$

$$\frac{V_{o2}(s)}{V_1(s)} = 1 - \frac{R_4}{sR_3R_5C_2} = \frac{s - R_4/R_3R_5C_2}{s} \tag{9b}$$

For the alternate synthetic inductor case in which $Z_4(s) = 1/sC_4$ we have

$$\frac{V_{o1}(s)}{V_1(s)} = 1 + \frac{1}{sR_5C_4} = \frac{s + 1/R_5C_4}{s} \tag{10a}$$

$$\frac{V_{o2}(s)}{V_1(s)} = 1 - \frac{R_2}{sR_3R_5C_4} = \frac{s - R_2/R_3R_5C_4}{s} \tag{10b}$$

These transfer functions indicate that for a constant amplitude sinusoidal excitation applied at $V_1(s)$, the amplifier's output voltages do not increase as a function of frequency, and thus that saturation will not occur. However, since the synthetic inductor is usually driven at $V_s(s)$ rather than at $V_1(s)$, in practice it may be more meaningful to examine $V_{o1}(s)/V_s(s)$ or $V_{o2}(s)/V_s(s)$ for the case where R_s is a complex impedance. An example of such a situation is given in the problems.

The steps in applying the synthetic inductance technique are as follows:

1. Design an *RLC* high-pass normalized prototype network using the tables in App. A and the method of Sec. 2.4.
2. Design the synthetic inductors which will replace the inductors of the prototype. Use normalized resistors and capacitors.
3. Denormalize all the resistors and capacitors.

When constructing the realization, it is desirable to test the synthetic inductors before connecting them into the circuit. Since these elements simulate passive elements, such tests are readily made. For example, one can resonate the synthetic inductor with a known capacitor to measure the value of inductance. Alternately, one can drive the inductor using a voltage source with a known resistance R and obtain a value for L/R by examining the breakpoint of the magnitude response.

Example 6.1-1 *A Fifth-Order High-Pass Butterworth Synthetic-Inductance Filter* A high-pass filter is to be realized using the synthetic inductance approach. The cutoff frequency is to be 10^4 rad/s and the source and load resistors are 1000 Ω each. Using App. A and applying the low-pass to high-pass transformation of Sec. 2.4 we obtain the high-pass prototype network shown in Fig. 6.1-4a. The synthetic inductors are easily designed from (7) by letting $R_1 = R_3 = R_4 = 1.0 \ \Omega$, $C_2 = 1$ F, and $R_5 = 0.618 \ \Omega$. The resulting normalized realization is shown in Fig. 6.1-4b. In order to obtain the denormalized realization, all capacitors should be multiplied by 10^{-7} and all resistors by 10^3. The synthetic inductors should be tuned as discussed above. $\qquad \Box$

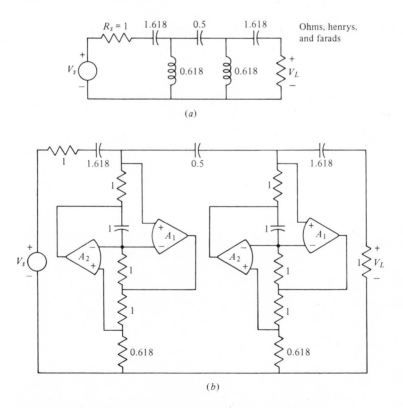

(a)

(b)

Figure 6.1-4 The fifth-order high-pass filter realized in Example 6.1-1.

The synthetic-inductance method described here is also applicable to high-pass realizations having finite $j\omega$-axis zeros. Such $j\omega$-axis zeros are created by the series connection of a capacitor and an inductor. Connecting one of the terminals of the inductor to ground permits the use of the synthetic inductance approach.

6.2 FREQUENCY-DEPENDENT NEGATIVE RESISTORS

One of the most useful devices for implementing passive network simulation methods is the FDNR (*frequency-dependent negative resistor*). This is a two-terminal (one-port) active network element with the symbol shown in Fig. 6.2-1, whose admittance is

$$Y(s) = \frac{I(s)}{V(s)} = s^2 D \tag{1}$$

where D is a positive-real constant. The units of D are farad-seconds (Fs). If in the relation of (1) we let $s = j\omega$, we find $Y(j\omega) = -\omega^2 D$, which is a negative function, is also dependent on frequency, and is real, i.e., resistive in nature. Thus, its characteristics are aptly described by the term *frequency-dependent negative resistor*. An FDNR can be realized using the GIN (*generalized immittance network*) of Fig. 6.1-2. For the general circuit, the input admittance is

$$Y_{in}(s) = \frac{Y_1(s)Y_3(s)Y_5(s)}{Y_2(s)Y_4(s)} \tag{2}$$

where $Y_i(s) = 1/Z_i(s)$. As one way to realize an FDNR we let $Y_1(s) = Y_3(s) = sC$ and $Y_2(s) = Y_4(s) = Y_5(s) = G$. Equation (2) then reduces to

$$Y_{in}(s) = \left(\frac{C^2}{G}\right)s^2 = Ds^2 \tag{3}$$

This is the input admittance of an FDNR in which $D = C^2/G$. The circuit is shown in Fig. 6.2-2.[1]

---

[1] Although other forms of the FDNR are possible, the one given here has several advantages which lead to good performance. It is available commercially as the ATF 431 Gyrator, Amperex Electronic Corp., Slatesville, R.I.

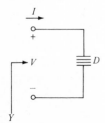

Figure 6.2-1 The FDNR (*frequency-dependent negative resistor*).

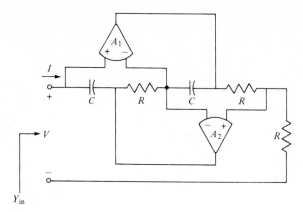

Figure 6.2-2 Using the GIN to realize an FDNR.

Let us now determine some properties of an FDNR. Since it has a real negative value for $s = j\omega$, it is a potentially unstable one-port network. To see this, consider the circuit of Fig. 6.1-3 in which a GIN is driven by a voltage source $V_s(s)$ having a source resistance of R_s. Since the input voltages across the operational amplifier inputs are zero, $V_1(s) = V_5(s)$. The voltage transfer function from V_s to V_1 or V_5 is

$$\frac{V_1(s)}{V_s(s)} = \frac{V_5(s)}{V_s(s)} = \frac{1}{1 + \dfrac{R_s Z_2(s) Z_4(s)}{Z_1(s) Z_3(s) Z_5(s)}} \tag{4}$$

There are two commonly used realizations for an FDNR. If $Z_1(s) = 1/sC_1$, $Z_2(s) = R_2$, $Z_3(s) = 1/sC_3$, $Z_4(s) = R_4$, and $Z_5(s) = R_5$, then (4) becomes (realization 1)

$$\frac{V_1(s)}{V_s(s)} = \frac{R_5/R_s R_2 R_4 C_1 C_3}{s^2 + R_5/R_s R_2 R_4 C_1 C_3} = \frac{\omega_0^2}{s^2 + \omega_0^2} \tag{5}$$

where

$$\omega_0 = \sqrt{\frac{R_5}{R_s R_2 R_4 C_1 C_3}} \tag{6}$$

The frequency ω_0 is the radian frequency at which the FDNR "resonates" with the source resistor R_s, that is, the frequency of potential instability. In practice, the instability predicted in the above may not occur, since the gains of A_1 and A_2, being finite, will cause the poles to shift slightly to the left of the $j\omega$ axis. The voltage transfer function of (4), however, will still have a high Q and thus be difficult to physically measure. As an alternate and more observable characteristic, consider the transfer function from $V_1(s)$ to $V_{o1}(s)$ and $V_{o2}(s)$. These are

$$\frac{V_{o1}(s)}{V_1(s)} = 1 + \frac{Z_4(s)}{Z_5(s)} \tag{7a}$$

$$\frac{V_{o2}(s)}{V_1(s)} = 1 - \frac{Z_2(s) Z_4(s)}{Z_3(s) Z_5(s)} \tag{7b}$$

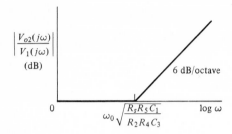

Figure 6.2-3 A Bode magnitude plot for (9).

Using the values for the impedances given above, (7) becomes

$$\frac{V_{o1}(s)}{V_1(s)} = 1 + \frac{R_4}{R_5} \tag{8a}$$

$$\frac{V_{o2}(s)}{V_1(s)} = -\frac{s - R_5/R_2 R_4 C_3}{R_5/R_2 R_4 C_3} \tag{8b}$$

If in (8b) we use the definition for ω_0 given in (6), we obtain

$$\frac{V_{o2}(s)}{V_1(s)} = -\frac{s - \omega_0\sqrt{R_5 R_s C_1/R_2 R_4 C_3}}{\omega_0\sqrt{R_5 R_s C_1/R_2 R_4 C_3}} \tag{9}$$

An asymptotic magnitude plot (Bode plot) for (9) is shown in Fig. 6.2-3. This plot and (9) can be used to experimentally verify the performance of an FDNR. As an example, if we choose $C_1 = C_3 = 0.01 \ \mu F$, and $R_2 = R_4 = R_5 = R_s = 10 \ k\Omega$, then D of (1) is 10^{-12} Fs, and ω_0 is 10^4 rad/s.

As an alternate FDNR realization, we may select the impedances of Fig. 6.1-2 as $Z_1(s) = 1/sC_1$, $Z_2(s) = R_2$, $Z_3(s) = R_3$, $Z_4(s) = R_4$, and $Z_5(s) = 1/sC_5$. In this case, ω_0 is defined as (realization 2)

$$\omega_0 = \sqrt{\frac{R_3}{R_s R_2 R_4 C_1 C_5}} \tag{10}$$

The voltage transfer functions from $V_1(s)$ to $V_{o1}(s)$ and $V_{o2}(s)$ now become

$$\frac{V_{o1}(s)}{V_1(s)} = 1 + sR_4 C_5 = \frac{s + \omega_0\sqrt{R_s R_2 C_1/R_3 R_4 C_5}}{\omega_0\sqrt{R_s R_2 C_1/R_3 R_4 C_5}} \tag{11a}$$

$$\frac{V_{o2}(s)}{V_1(s)} = -\left[\frac{s - \omega_0\sqrt{R_s R_3 C_1/R_2 R_4 C_5}}{\omega_0\sqrt{R_s R_3 C_1/R_2 R_4 C_5}}\right] \tag{11b}$$

Either of these expressions is useful for experimentally verifying the performance of realization 2. For example, if $C_1 = C_5 = 10^{-8}$ F and $R_2 = R_3 = R_4 = R_s = 10^4 \ \Omega$, we find $D = 10^{-12}$, $\omega_0 = 10^4$ rad/s, and the magnitudes of (11) are identical and are given by Fig. 6.2-4.

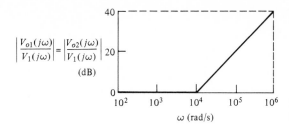

Figure 6.2-4 A Bode magnitude plot for (11).

To apply the FDNR to the synthesis of filters, it is necessary to first perform a transformation proposed by Bruton.[1] This transformation is called the *RLC* : *CRD transformation*. It has the form of the basic impedance normalization given in (4) of Sec. 1.4, but for which the impedance normalization constant z_n is a function of s. Specifically, $z_n = 1/s$. Thus we may write

$$Z(s) = z_n Z_n(s) = \frac{Z_n(s)}{s} \qquad Y(s) = \frac{1}{z_n} Y_n(s) = s Y_n(s) \tag{12}$$

As an example, consider the admittance of a general passive *RLC* network branch expressed as

$$Y_{RLC}(s) = G + \frac{1}{sL} + sC \tag{13}$$

Such a branch obviously consists of a paralleled resistor, inductor, and capacitor with values of $1/G$ ohms, L henrys, and C farads respectively. Applying the *RLC* : *CRD* transformation to this branch, we obtain

$$Y_{CRD}(s) = sG + s^2C + \frac{1}{L} \tag{14}$$

which consists of a paralleled capacitor, FDNR, and resistor, with values of G farads, C $(= D)$ farad-seconds, and L ohms respectively. The gain of a VCVS, since it is dimensionless, will be unaffected by such a transformation. Similarly, if the *RLC* : *CRD* transformation is applied to a two-port network, characterized by the transmission parameters defined in (2) of Sec. 6.1, then since the $A(s)$ and $D(s)$ parameters are dimensionless, they will not be affected by the transformation. The $B(s)$ and $C(s)$ parameters, however, since they have the dimensions of impedance and admittance, will be multiplied by $1/s$ and s respectively. These results are summarized in Table 6.2-1.

Since the FDNR is an active element, it is important from power supply considerations that one end of it always be grounded. This is equivalent to requiring that all the capacitors in the prototype network to which the *RLC* : *CRD*

---

[1] L. T. Bruton, "Network Transfer Functions Using the Concepts of Frequency-Dependent Negative Resistance," *IEEE Trans. Circuit Theory*, vol. CT-16, August 1969, pp. 406–408.

Table 6.2-1 Effect of the RLC: CRD transformation on various network elements

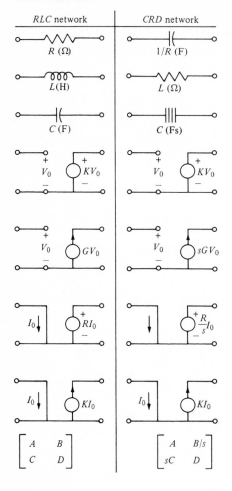

RLC network	CRD network
R (Ω)	$1/R$ (F)
L(H)	L (Ω)
C (F)	C (Fs)
V_0, KV_0	V_0, KV_0
V_0, GV_0	V_0, sGV_0
I_0, RI_0	$\dfrac{R}{s}I_0$
I_0, KI_0	I_0, KI_0
$\begin{bmatrix} A & B \\ C & D \end{bmatrix}$	$\begin{bmatrix} A & B/s \\ sC & D \end{bmatrix}$

transformation is to be applied also be grounded. Such a requirement is generally satisfied by low-pass filter realizations. Thus, the use of FDNRs is especially suitable to the realization of such filters. The design procedure may be outlined as follows:

1. Design an RLC low-pass normalized prototype network using the tables in App. A.
2. Transform the RLC network into a CRD normalized network using the relationships of Table 6.2-1.
3. Design the FDNRs using normalized resistor and capacitor values.

4. Provide paths for dc current from all amplifier input terminals to ground if they do not already exist. (Since the source and load resistors have been transformed to capacitors, it will be necessary to shunt one or both with resistors chosen so as to have negligible effect on the realization.)
5. Denormalize all the resistors and capacitors of the realization.

Example 6.2-1 *A Fifth-Order Low-Pass Butterworth FDNR Filter* It is desired to use FDNRs to realize a fifth-order double-terminated low-pass Butterworth filter having a cutoff frequency of 10^4 rad/s. The source and load resistors are required to have a value of 1000 Ω each. Using App. A we obtain the normalized low-pass prototype network shown in Fig. 6.2-5a. Applying

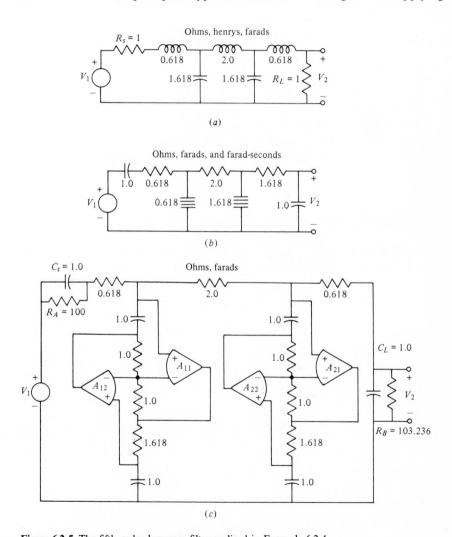

Figure 6.2-5 The fifth-order low-pass filter realized in Example 6.2-1.

the $RLC : CRD$ transformation produces the network of Fig. 6.2-5b. We now select FDNR realization 2, in which $Z_1(s) = 1/sC$, $Z_2(s) = Z_3(s) = R$, $Z_4(s) = R_4$, and $Z_5(s) = 1/sC$. From (1) and (2), $D = C^2 R_4$. Choosing C as 1.0 F provides a realization in which all capacitors are equal-valued. R can be any value since in (2) R_2 and R_3, which are equal to this value, cancel. Thus we may select $R = 1.0\ \Omega$. R_4 will be the element used to tune the FDNR. Since it is required that $D = 1.618$ Fs, the nominal value of R_4 is $1.618\ \Omega$. The normalized realization of the filter is shown in Fig. 6.2-5c. R_A and R_B have been added to permit dc current to flow from the positive terminals of A_{11} and A_{21} to ground. R_A and R_B are determined in the following manner. At dc we have

$$\frac{V_2(0)}{V_1(0)} = \frac{R_B}{R_A + R_B + 3.236} \tag{15}$$

The 3.236 corresponds to the total series resistance between R_A and R_B. If (15) is equal to $R_L/(R_s + R_L)$, then the circuit in Fig. 6.2-5c will have the same dc response as the prototype one in Fig. 6.2-5a. This provides one relationship for solving for R_A and R_B. The other relationship is found by letting R_A or $R_B \gg R_L$ or R_s. This is necessary so that at the normalized cutoff frequency of 1 rad/s, R_A and R_B do not load C_s and C_L respectively. Therefore by equating (15) to 0.5 and by selecting $R_A = 100\ \Omega$, we get $R_B = 103.24\ \Omega$, as shown in Fig. 6.2-5c. The final network is found by multiplying each capacitor by 10^{-7} and each resistor by 10^3. □

In physically constructing the filter described in the preceding example, each of the FDNRs should be tuned as shown in Fig. 6.2-6 by adjusting R_4 so that the break frequency of (11) occurs at 6180 rad/s. This value is found by substituting the ideal values of the passive components of the FDNR into (10) and (11). After each FDNR is tuned, the network of Fig. 6.2-5c may be constructed. After the filter circuit is assembled, very little large-scale tuning can be accomplished because of the complex way in which the individual components affect the overall circuit behavior. The filter characteristic, however, is usually quite sensitive to changes in the values of the R_4 resistors in the FDNRs and the values of C_s and C_L. Small changes in the filter response can thus be made by carefully adjusting these components.

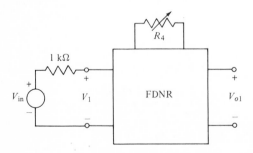

Figure 6.2-6 Tuning the FDNR.

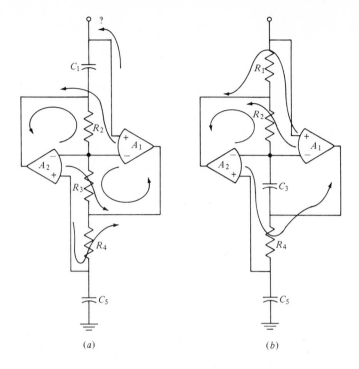

Figure 6.2-7 DC biasing currents in two types of FDNRs.

The need for the resistors R_A and R_B in the preceding example can be eliminated by using a different FDNR circuit. To see this, consider Fig. 6.2-7a, in which the possible paths for the dc current from each amplifier input for the FDNR used in the preceding example (realization 2) are shown. The biasing current problem occurs in the dc path to the plus terminal of the amplifier A_1. Thus, as shown, as long as $Z_1(s)$ is a capacitor, bias current paths as provided by R_A and R_B will be required. Now consider the FDNR circuit (realization 3) shown in Fig. 6.2-7b, in which $Z_3(s)$ and $Z_5(s)$ have been chosen as capacitors. In this case a dc path exists from each of the amplifier input terminals to ground. Therefore additional resistors are not required. Obviously, if no other considerations are present, this configuration is preferable to the others.[1] Another solution to the biasing problem described above is the use of operational amplifiers which have reduced bias current requirements.[2]

In the preceding paragraphs we have described the use of FDNRs to realize low-pass filter functions. Some additional considerations in connection with such

---

[1] Unfortunately the ATF-431 realization referred to previously does not permit the choice of $Z_3(s)$ as a capacitor.

[2] Such an amplifier is the LM155–157 series from National Semiconductor.

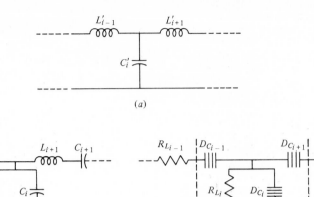

Figure 6.2-8 Using FDNRs to realize a bandpass filter.

realizations will be given in Chap. 7. We now consider a related application in the realization of bandpass functions. In Fig. 6.2-8a, an inner section of a passive low-pass structure is shown. If the low-pass to bandpass transformation of (8) of Sec. 2.4 is applied, the circuit of Fig. 6.2-8b results. If the $RLC : CRD$ transformation is applied, we obtain the circuit shown in Fig. 6.2-8c. This circuit contains two floating FDNRs, as well as a grounded FDNR and a resistor. To avoid having to directly realize the floating FDNRs (which would require floating power supplies), we instead use the circuit of Fig. 6.2-9, where a cascade of two GICs with a passive network is shown. The transmission parameters of the first GIC are

$$
T_{\mathrm{GIC1}} =
\begin{bmatrix}
1 & 0 \\
0 & \dfrac{Z_2(s)Z_4(s)}{Z_1(s)Z_3(s)}
\end{bmatrix}
\tag{16}
$$

The transmission parameters of the passive network will be designated as

$$
T_{\mathrm{passive}} =
\begin{vmatrix}
A(s) & B(s) \\
C(s) & D(s)
\end{vmatrix}
\tag{17}
$$

The transmission parameters of the second GIC are

$$
T_{\mathrm{GIC2}} =
\begin{bmatrix}
1 & 0 \\
0 & \dfrac{Z_1(s)Z_3(s)}{Z_2(s)Z_4(s)}
\end{bmatrix}
\tag{18}
$$

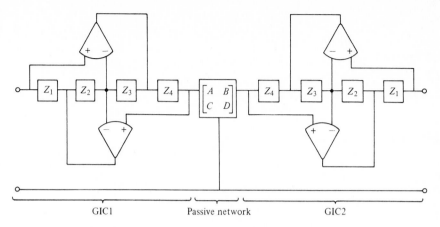

Figure 6.2-9 A cascade of two GICs with a passive network.

Cascading these three networks results in the overall transmission parameters

$$
T_{\mathrm{GIC1}} \times T_{\mathrm{passive}} \times T_{\mathrm{GIC2}} =
\begin{bmatrix}
A(s) & B(s)\dfrac{Z_1(s)Z_3(s)}{Z_2(s)Z_4(s)} \\[2ex]
C(s)\dfrac{Z_2(s)Z_4(s)}{Z_1(s)Z_3(s)} & D(s)
\end{bmatrix}
\tag{19}
$$

Thus the configuration of Fig. 6.2-9 is equivalent to a related passive network whose elements have been impedance-normalized by $Z_1(s)Z_3(s)/Z_2(s)Z_4(s)$. If $Z_1(s)$, $Z_2(s)$, $Z_3(s)$, and $Z_4(s)$ are selected so that

$$
\frac{Z_1(s)Z_3(s)}{Z_2(s)Z_4(s)} = \frac{1}{s^2}
\tag{20}
$$

then the elements of the passive network are subjected to a transformation which is the same as two $RLC : CRD$ transformations. Hence, Fig. 6.2-9 can be used to realize Fig. 6.2-8c by using a "pretransformed" passive network in which the FDNRs are replaced by resistors and the resistors are replaced by elements whose

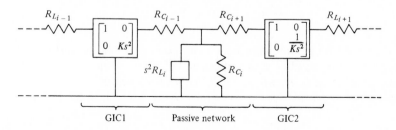

Figure 6.2-10 A realization of the filter section of Fig. 6.2-8c.

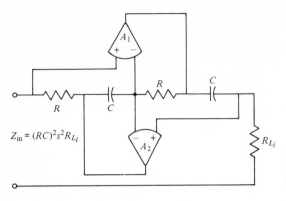

$Z_{in} = (RC)^2 s^2 R_{Li}$

Figure 6.2-11 A realization of an element with an impedance proportional to s^2.

impedance is proportional to s^2. As a result the structure of Fig. 6.2-8c can be realized through the use of two GICs and an element with the input impedance of $s^2 R_{Li}$ as shown in Fig. 6.2-10.[1] This latter element can be realized using the GIN of Fig. 6.2-11. The concepts illustrated in Figs. 6.2-8 through 6.2-11 can also be applied to use GICs and FDNRs to realize high-pass functions.

Example 6.2-2 *A Fourteenth-Order Bandpass Chebyshev FDNR Filter* The technique described above will be used to design a fourteenth-order bandpass Chebyshev filter with 0.5-dB passband ripple, $f_0 = 150$ Hz, a bandwidth of 150 Hz, and source and load resistors of 1000 Ω. From App. A we find that $L_1 = 1.7373$, $C_2 = 1.2582$, $L_3 = 2.6383$, $C_4 = 1.3443$, $L_5 = 2.6383$, $C_6 = 1.2582$, and $L_7 = 1.7373$, with units in henrys and farads. The network configuration is shown in Fig. 6.2-12a. Using the transformation from low-pass to bandpass of (9) of Sec. 2.4 results in the realization of Fig. 6.2-12b. The values are taken from the ones given above by using $C_i = 1/L_i$ (*i* odd) and $L_i = 1/C_i$ (*i* even). The next step is to make the *RLC* : *CRD* transformation resulting in Fig. 6.2-12c. As described above, the configuration of Fig. 6.2-10 can be used to realize the combination of D_{C3}, D_{C5}, R_{L4}, and D_{C4}. At the input end of the network, we may simultaneously realize both C_{RS} and D_{C1} by letting the GIC element $Z_5(s)$ be $R_s + 1/sC_1$. Therefore the input impedance of the GIC [with $Z_1(s) = 1/s$ and $Z_2(s) = Z_3(s) = Z_4(s) = 1$] becomes

$$Z_{in}(s) = \frac{1}{s^2 C_1} + \frac{R_s}{s} \tag{21}$$

which realizes the series combination of C_{RS} and D_{C1}. A similar procedure can be used to realize C_{RL} and D_{C7} at the output end of the network. The complete normalized realization is shown in Fig. 6.2-12d. The various complete GIC realizations are identified as *A* through *G* in both Fig. 6.2-12c and

---

[1] The units for such an element are usually taken as henry-seconds.

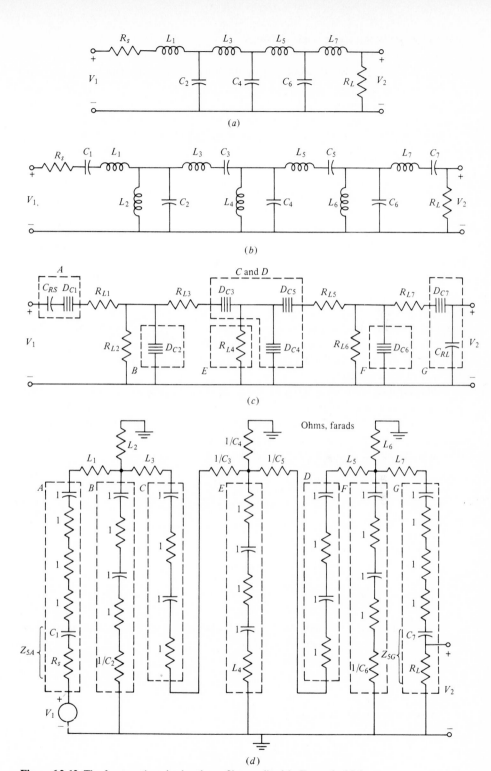

Figure 6.2-12 The fourteenth-order bandpass filter realized in Example 6.2-2.

Fig. 6.2-12d. The operational amplifiers have been omitted in Fig. 6.2-12d for purposes of clarity. The impedances $Z_1(s)$ through $Z_4(s)$ or $Z_5(s)$ are shown for each GIC in order from top to bottom. Note that in the GIC used for D_{C7} and C_{RL}, the load resistor has one side grounded. The remaining step is to denormalize the network using $z_n = 1000$ and $\Omega_n = 942.48$. ☐

6.3 LEAPFROG REALIZATION TECHNIQUES

In the preceding sections of this chapter we discussed two approaches to the direct method of filter synthesis, namely, the use of synthetic inductance and the use of FDNRs. Both of these techniques are examples of the simulation of the elements of a passive network through the use of active RC elements. In this section we shall present a discussion of a somewhat different simulation approach to the direct method of synthesis. It is known as the *leapfrog technique*,[1] and it uses negative feedback to simulate the voltage and current relations of a passive RLC ladder network. It results in a useful and stable realization, one advantage of which is the repeated use of nearly identical component circuits as building blocks of the overall realization.

The leapfrog technique is based on the use of an active RC circuit in which the voltages at various portions of the network are analogs of the shunt branch voltages and series branch currents of the prototype RLC passive network being simulated. To see this, consider the passive ladder network shown in Fig. 6.3-1. The branch relationships of this network can be written

$$I_1 = (V_1 - V_2)Y_1 \tag{1a}$$

$$V_2 = (I_1 - I_3)Z_2 \tag{1b}$$

$$I_3 = (V_2 - V_4)Y_3 \tag{1c}$$

$$V_4 = (I_3 - I_5)Z_4 \tag{1d}$$

$$I_5 = (V_4 - V_6)Y_5 \tag{1e}$$

$\cdots\cdots\cdots\cdots\cdots\cdots$

[1] F. E. J. Girling and E. F. Good, "Active Filters 12: The Leap-Frog or Active-Ladder Synthesis," *Wireless World*, vol. 76, July 1970, pp. 341–345.

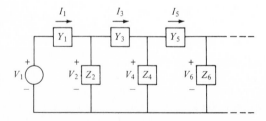

Figure 6.3-1 A passive ladder network.

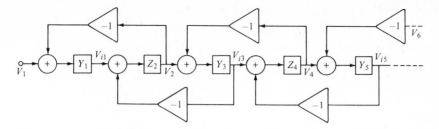

Figure 6.3-2 A block diagram of a simulation of the network of Fig. 6.3-1.

where, for convenience, we have deleted the (s) notation. A block diagram of a related network which simulates such a set of equations is shown in Fig. 6.3-2. This network consists of unity-gain inverting amplifiers represented by triangles, summers represented by circles, and voltage transfer functions with values Y_1, Z_2, Y_3, Z_4, Y_5, etc., represented by boxes.[1] The voltage at the output of each of the boxes can be written as

$$V_{i1} = (V_1 - V_2)Y_1 \tag{2a}$$

$$V_2 = (V_{i1} - V_{i3})Z_2 \tag{2b}$$

$$V_{i3} = (V_2 - V_4)Y_3 \tag{2c}$$

$$V_4 = (V_{i3} - V_{i5})Z_4 \tag{2d}$$

$$V_{i5} = (V_4 - V_6)Y_5 \tag{2e}$$

The equations of (2) are analogs of those of (1). Specifically, if we represent I_1, I_3, I_5, etc., of Fig. 6.3-1 by voltages V_{i1}, V_{i3}, V_{i5}, etc., of Fig. 6.3-2, then the equations are identical and the circuits will have the same performance.

A simpler realization than the one shown in Fig. 6.3-2 can be used to obtain an almost identical result. To see this, we may write (2) in the form

$$V_{i1} = (V_1 - V_2)Y_1 \tag{3a}$$

$$-V_2 = (V_{i1} - V_{i3})(-Z_2) \tag{3b}$$

$$-V_{i3} = (-V_2 + V_4)Y_3 \tag{3c}$$

$$V_4 = (-V_{i3} + V_{i5})(-Z_4) \tag{3d}$$

$$V_{i5} = (V_4 - V_6)Y_5 \tag{3e}$$

A block diagram of a network realization of these equations is shown in Fig. 6.3-3.

---

[1] Note that in this usage these quantities, being voltage transfer functions, are dimensionless, even though for convenience we have retained the Y and Z notation.

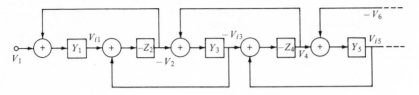

Figure 6.3-3 A simpler block diagram of a simulation of the network of Fig. 6.3-1.

Note that no unity-gain inverters are required; thus, the circuit realization is simplified. The variables V_2, V_{i3}, etc., in (3) have minus signs preceding them. This simply means that the transfer functions from V_1 to these variables will be inverting rather than noninverting. In addition, all the Z_i voltage transfer functions are now inverting. It will turn out that these inverting transfer functions are easier to realize than the noninverting ones.

An alternate realization may be developed by expressing (2) as

$$-V_{i1} = (V_1 - V_2)(-Y_1) \tag{4a}$$

$$-V_2 = (-V_{i1} + V_{i3})Z_2 \tag{4b}$$

$$V_{i3} = (-V_2 + V_4)(-Y_3) \tag{4c}$$

$$V_4 = (V_{i3} - V_{i5})Z_4 \tag{4d}$$

$$-V_{i5} = (V_4 - V_6)(-Y_5) \tag{4e}$$

These equations are realized by the configuration shown in Fig. 6.3-4. In this realization all the Y_i voltage transfer functions are inverting. Both Figs. 6.3-3 and 6.3-4 show valid realizations of the leapfrog structure.

Now let us consider the application of the leapfrog concept to the realization of a low-pass filter in which all the transmission zeros are at infinity. In this case, the series elements Y_i of Fig. 6.3-1 will be inductors and the shunt elements Z_i will be capacitors. The initial and final elements will include the source and load resistors respectively. If the realization is driven by a voltage source and is of odd order, the final element will be as shown in Fig. 6.3-5a (for $n = 5$). For the voltage-excitation even-order case, the final element will appear as in Fig. 6.3-5b (for

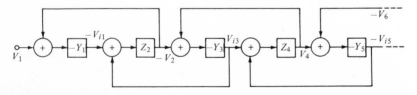

Figure 6.3-4 An alternate simplified block diagram of a simulation of the network of Fig. 6.3-1.

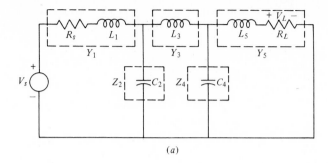

(a)

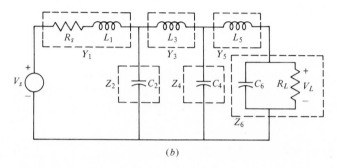

(b)

Figure 6.3-5 Passive low-pass filter configurations. (a) Fifth-order (odd). (b) Sixth-order (even).

$n = 6$). Other cases are covered in the Problems. For Fig. 6.3-5a, the voltage transfer functions of the boxes of Fig. 6.3-3 are found as

$$T_1(s) = Y_1(s) = \frac{1/L_1}{s + R_s/L_1} \tag{5a}$$

$$T_2(s) = -Z_2(s) = -\frac{1}{sC_2} \tag{5b}$$

$$T_3(s) = Y_3(s) = \frac{1}{sL_3} \tag{5c}$$

$$T_4(s) = -Z_4(s) = -\frac{1}{sC_4} \tag{5d}$$

$$T_5(s) = Y_5(s) = \frac{1/L_5}{s + R_L/L_5} \tag{5e}$$

Of these voltage transfer functions, $T_2(s)$ and $T_4(s)$ may be realized by conventional operational amplifier integrators, and $T_3(s)$ may be realized by such an

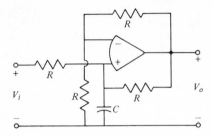

Figure 6.3-6 A noninverting integrator.

integrator in cascade with an inverter, or by a noninverting integrator, using the circuit of Fig. 6.3-6 for which

$$\frac{V_o(s)}{V_i(s)} = \frac{2}{sRC} \tag{6}$$

$T_1(s)$ and $T_5(s)$ may be realized by a damped inverting integrator. The term *damped* reflects the fact that the pole is displaced from the origin. A realization for such an integrator is shown in Fig. 6.3-7. Its voltage transfer function is

$$\frac{V_o(s)}{V_i(s)} = \frac{-1/RC_x}{s + 1/R_xC_x} \tag{7}$$

The resulting leapfrog realization of Fig. 6.3-5a, using inverters and using the form of Fig. 6.3-3, is given in Fig. 6.3-8a. All the resistor values are specified as functions of an arbitrary resistance value R, which may be chosen to achieve a desired impedance normalization. The values of the capacitors are given in terms of R and of the element values in Fig. 6.3-5a.

A similar realization of Fig. 6.3-5a using the configuration of Fig. 6.3-4 is possible. In this case, only two inverters are required, resulting in one less amplifier. The circuit is shown in Fig. 6.3-8b. For the network of Fig. 6.3-5b, the realization of Fig. 6.3-9a results if the configuration of Fig. 6.3-3 is used and that of Fig. 6.3-9b results if the configuration of Fig. 6.3-4 is used. It is of interest to note that in all the above circuits the impedance level of the leapfrog realizations are completely independent of the impedance level in the prototype network. For example, assuming ideal operational amplifiers, the leapfrog realizations will have an output impedance of zero and an input impedance of R.

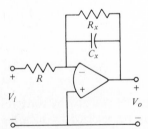

Figure 6.3-7 A damped inverting integrator.

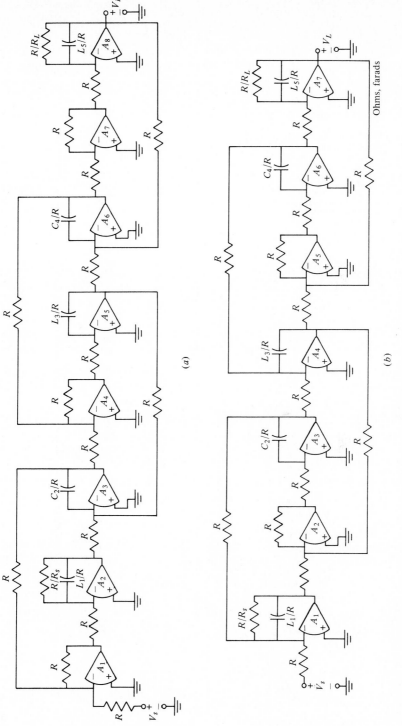

Figure 6.3-8 A realization of Fig. 6.3-5a. (a) Using the simulation of Fig. 6.3-3. (b) Using the simulation of Fig. 6.3-4.

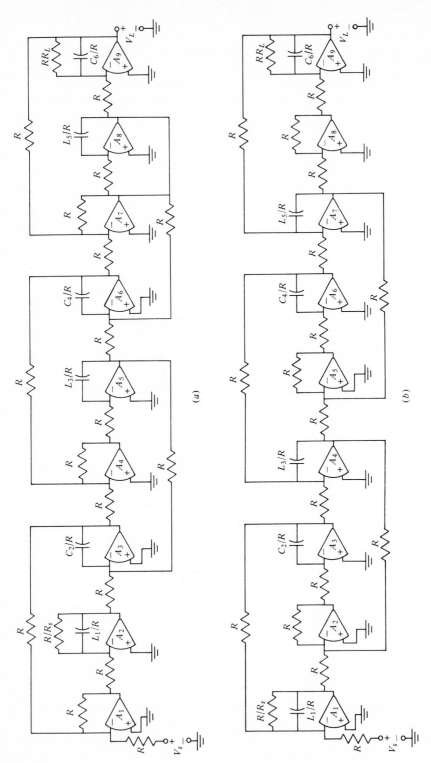

Figure 6.3-9 A realization of Fig. 6.3-5b. (*a*) Using the simulation of Fig. 6.3-3. (*b*) Using the simulation of Fig. 6.3-4.

298

The design procedure for leapfrog filters may be summarized as follows:

1. Design a normalized low-pass prototype using App. A.
2. Select the general configuration form of either Fig. 6.3-3 or Fig. 6.3-4.
3. Design the inner elements of the leapfrog filter using inverting and noninverting integrators and using a normalized value of R of unity.
4. Design the input and output elements of the leapfrog filter using (7) and Fig. 6.3-7.
5. Perform the necessary frequency and impedance denormalizations.

Example 6.3-1 *A Third-Order Low-Pass Butterworth Leapfrog Filter* It is desired to realize a third-order Butterworth low-pass filter with a cutoff frequency of 1 krad/s. The excitation voltage source is assumed to have zero internal resistance. From App. A we obtain the circuit shown in Fig. 6.3-10a. Using the leapfrog configuration of Fig. 6.3-4 and selecting $R = 1$, we obtain the filter circuit of Fig. 6.3-10b. Using a frequency denormalization of 10^3 and an impedance denormalization of 10^4, all the resistors become equal to 10 kΩ and all the capacitor values are multiplied by 10^{-7}. ☐

The leapfrog technique described above for low-pass filters is also applicable to bandpass filters with zeros at the origin and at infinity. In this case the low-pass to bandpass transformation is applied to each element of a prototype low-pass network. The general form of the resulting series and shunt branches is given in

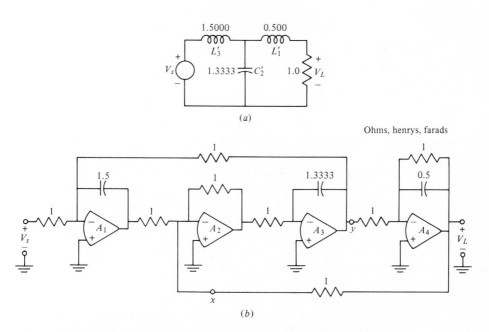

Figure 6.3-10 The third-order Butterworth leapfrog filter realized in Example 6.3-1.

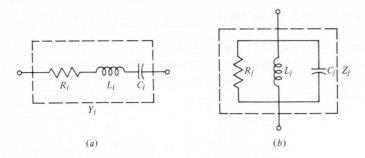

(a) (b)

Figure 6.3-11 Series and shunt branches of a bandpass network.

Fig. 6.3-11a and b respectively. Usually R_i in Fig. 6.3-11a will be zero and R_j in Fig. 6.3-11b will be infinity. The admittance of Fig. 6.3-11a is found as

$$Y_i(s) = \frac{(1/L_i)s}{s^2 + (R_i/L_i)s + 1/L_iC_i} \tag{8}$$

The impedance of Fig. 6.3-11b is found as

$$Z_j(s) = \frac{(1/C_j)s}{s^2 + (1/R_jC_j)s + 1/L_jC_j} \tag{9}$$

Obviously, for the bandpass case the voltage transfer functions required are second-order bandpass ones. Thus the integrators of the low-pass realization will be replaced by any of the second-order bandpass filter circuits developed in the preceding chapter. We see that the leapfrog technique provides another method of obtaining higher-order realizations from second-order ones. In practice the resulting realizations have excellent characteristics. There are seveal important requirements that such usage places on the second-order bandpass realization that is used. First, it must be capable of both inverting and noninverting operation, so that additional inverters are not required. Second, when $R_i = 0$ or $R_j = \infty$, the Q of the second-order realization must become infinite. The third requirement is that the realization be capable of summing two or more signals. In order to meet this last requirement, the single-feedback infinite-gain, resonator, or modified state-variable circuits are most appropriate. The infinite-gain realization will require an inverter, whereas the resonator and state-variable ones can realize both a noninverting and an inverting bandpass function. In order to get very high values of Q with the infinite-gain realization, one must also employ positive as well as negative feedback.[1] If the resonator is to be used, the circuit shown in Fig. 6.3-12 which

---

[1] P. R. Geffe, "Designers' Guide to: Active Bandpass Filters, Part 1," *Electronic Design News*, Feb. 5, 1974, pp. 68–73; "Part 2," *EDN*, Mar. 5, 1974, pp. 40–64; "Part 3," *EDN*, Apr. 5, 1974, pp. 46–52; "Part 4," *EDN*, May 5, 1974, pp. 63–71; "Part 5," *EDN*, June 5, 1974, pp. 64–72.

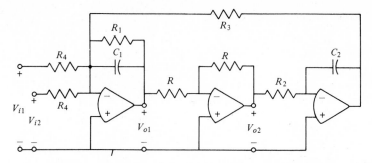

Figure 6.3-12 A modification of the resonator filter for use in bandpass leapfrog filter realizations. *Note:* V_{o1} is the inverting output and V_{o2} is the noninverting output.

provides a summing capability may be used. The transfer function is

$$V_{o2}(s) = -V_{o1}(s) = \left(\frac{s/R_4 C_1}{s^2 + s/R_1 C_1 + 1/R_2 R_3 C_1 C_2}\right)(V_{i1} + V_{i2}) \qquad (10)$$

Equating the transfer function of (10) to (8) and (9) produces the results given in Table 6.3-1. This table provides the relationships necessary to use the resonator circuit of Fig. 6.3-12 to realize a leapfrog filter.

Table 6.3-1 Design relationships for the implementation of Fig. 6.3-11 by the resonator of Fig. 6.3-12

Parameters* of Fig. 6.3-12	Fig. 6.3-11a series element $Y_i(s)$	Fig. 6.3-11b shunt element $Z_j(s)$
R_1	$\dfrac{R}{R_i}\sqrt{\dfrac{L_i}{C_i}}$	$RR_j\sqrt{\dfrac{C_j}{L_j}}$
R_4	$R\sqrt{\dfrac{L_i}{C_i}}$	$R\sqrt{\dfrac{C_j}{L_j}}$
If R is chosen as a convenient value:		
C	$\dfrac{\sqrt{L_i C_i}}{R}$	$\dfrac{\sqrt{L_j C_j}}{R}$
If C is chosen as a convenient value:		
R	$\dfrac{\sqrt{L_i C_i}}{C}$	$\dfrac{\sqrt{L_j C_j}}{C}$

* $R_2 = R_3 = R$ and $C_1 = C_2 = C$ with either R or C selected arbitrarily.

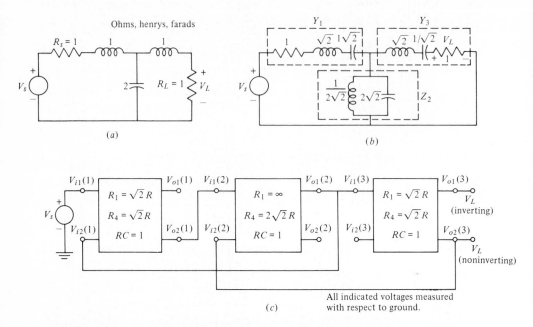

(a)

(b)

(c)   All indicated voltages measured with respect to ground.

Figure 6.3-13 The sixth-order bandpass leapfrog filter realized in Example 6.3-2.

Example 6.3-2 *A Sixth-Order Butterworth Bandpass Leapfrog Filter* It is desired to use the leapfrog technique to design a six-pole bandpass Butterworth filter with a center frequency of 1 kHz and a bandwidth of one octave. In order to have a stable first stage, we select the doubly terminated structure from App. A with $R = 1$. The low-pass prototype is shown in Fig. 6.3-13a. From Sec. 2.4 we find that the octave bandwidth for a center frequency of 1 rad/s is $1/\sqrt{2}$ rad/s. Applying the transformation from low-pass prototype to bandpass prototype with a bandwidth of $1/\sqrt{2}$ gives the circuit in Fig. 6.3-13b. If we use the circuit of Fig. 6.3-12 for the transfer function blocks of Fig. 6.3-13b, then the choice between Figs. 6.3-3 and 6.3-4 provides no simplification of circuitry. Selecting the form of Fig. 6.3-3, we obtain the realization of Fig. 6.3-13c. Each of the boxes represents the circuit of Fig. 6.3-12. If C is selected as 1 F then $R = 1$ Ω. The frequency denormalization is accomplished by dividing all capacitors by $2\pi \times 10^3$. The resistors may be scaled to an arbitrary level to give practical values of both resistors and capacitors. For example, impedance denormalizing by 10^4 gives $R = 10$ kΩ and $C = 0.159$ μF. R_1 and R_4 of each block may be calculated from the relationships given in Fig. 6.3-13c. □

The leapfrog method is probably one of the best methods available for realizing bandpass filters, since it makes use of the second-order bandpass blocks developed in Chap. 6, and yet it provides sensitivities very close to those of passive networks. In practice, in some cases, the individual second-order blocks can be detuned several percent before a noticeable change occurs in the frequency response. The leapfrog method can also be used to realize network functions with $j\omega$-axis zeros. The details may be found in the literature.[1]

6.4 THE PRIMARY RESONATOR BLOCK

In the preceding section we showed how active RC circuits, each realizing a second-order bandpass voltage transfer function, could be used as building blocks in a leapfrog structure to obtain a high-order bandpass filter. In this section we present a similar technique called the *PRB* (*primary resonator block*).[2,3,4] It also uses second-order bandpass realizations as building blocks to obtain a high-order bandpass filter. It has an advantage over the leapfrog technique, however, in that it does not require the second-order realizations to have infinite Q. It also has the disadvantage of having design equations somewhat more complicated than those of the leapfrog technique.

A block diagram of the general configuration for a PRB of order $2n$ is shown in Fig. 6.4-1. Each of the boxes represents a circuit realizing a second-order voltage transfer function, and having (ideally) zero output impedance. Feedback is provided from the outputs of the n cascaded second-order stages to the input of the operational amplifier by the resistors R_i $(i = 1, 2, \ldots, n)$. The overall voltage transfer function is

$$\frac{V_2(s)}{V_1(s)} = \frac{-a_0 T_1(s)T_2(s) \cdots T_n(s)}{1 + a_1 T_1(s) + a_2 T_1(s)T_2(s) + \cdots + a_n T_1(s)T_2(s) \cdots T_n(s)} \tag{1}$$

[1] G. Szentirmai, "Synthesis of Multiple-Feedback Active Filters," *Bell System Technical J.* vol. 52, no. 4, April 1973, pp. 527–555.

[2] G. Hurtig III, U.S. Patent 3,720,881, March 1973.

[3] ——, *Proc. Intern. Filter Symposium*, April 1972, p. 84.

[4] D. Johnson, J. Milburn, and F. Irons, "Higher-Order Multiple-Feedback Band-Pass Filters," *Proc. IEEE Region 3 Conf.*, April 1974.

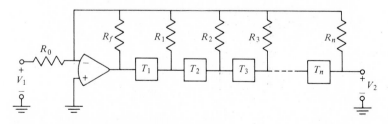

Figure 6.4-1 The PRB (*primary resonator block*).

where the $T_i(s)$ are the individual transfer functions of the second-order bandpass stages and where

$$a_i = \frac{R_f}{R_i} \qquad i = 0, 1, \ldots, n \tag{2}$$

The **PRB** technique may be used for bandpass realizations in a manner similar to that developed for the leapfrog technique. As a starting point, consider the inverting nth order low-pass, normalized (bandwidth of 1 rad/s) transfer function

$$N_{LP}(s) = \frac{-Hb_0}{b_n s^n + b_{n-1} s^{n-1} + \cdots + b_1 s + b_0} = \frac{-Hb_0}{P(s)} \tag{3}$$

The normalized low-pass to bandpass transformation of (9) of Sec. 2.4 may now be modified to the form

$$s = Q \frac{p^2 + 1}{p} \tag{4}$$

where the bandpass filter resulting from such a change of variable has a center frequency of 1 rad/s and a bandwidth of $1/Q$. Substituting (4) into (3) and redefining the p complex frequency variable as s, we obtain

$$N_{BP}(s) = N_{LP}\left(Q \frac{s^2 + 1}{s}\right) = \frac{V_2(s)}{V_1(s)} = \frac{-Hb_0 s^n/Q^n}{D_1(s)} \tag{5}$$

where
$$D_1(s) = \sum_{i=0}^{n} \frac{b_{n-i} s^i (s^2 + 1)^{n-i}}{Q^i} \tag{6}$$

Now let us assume that all the blocks in Fig. 6.4-1 are identical, and realize

$$T_i(s) = \frac{H_0 s/Q_p}{s^2 + \dfrac{1}{Q_p} s + 1} \qquad i = 1, 2, \ldots, n \tag{7}$$

For this case, (1) simplifies to

$$\frac{V_2(s)}{V_1(s)} = \frac{-a_0 H_0^n s^n/Q_p^n}{D_2(s)} \tag{8}$$

where

$$D_2(s) = \left(s^2 + \frac{1}{Q_p} s + 1\right)^n + \sum_{i=1}^{n} \frac{a_i H_0^i s^i [s^2 + (1/Q_p)s + 1]^{n-i}}{Q_p^i} \tag{9}$$

Applying the binomial theorem to $D_2(s)$ we obtain

$$D_2(s) = \sum_{i=0}^{n} \binom{n}{i} (s^2 + 1)^{n-i} (s/Q_p)^i$$

$$+ \sum_{i=0}^{n} \sum_{j=0}^{n-i} a_i H_0^i \binom{n-i}{j} (s^2 + 1)^{n-i-j} (s/Q_p)^{i+j} \tag{10}$$

The feedback circuit of Fig. 6.4-1 is obtained by equating (8) and (10) with (5) and (6), resulting in

$$\frac{a_0 H_0^n}{Q_p^n} = \frac{Hb_0}{Q^n} \tag{11}$$

and

$$\binom{n}{k}\frac{1}{Q_p^k} + \sum_{i=1}^{k} a_i H_0^i \binom{n-i}{k-i}\frac{1}{Q_p^k} = \frac{b_{n-k}}{Q^k} \qquad k = 1, 2, \ldots, n \tag{12}$$

One approach to solving (11) and (12) is to make the assignments

$$H_0 = 1/c \qquad Q_p = Q/c \tag{13}$$

where c is arbitrary, being chosen if possible so as to make the a_i nonnegative. Substituting (13) into (11) and (12) yields

$$a_0 = Hb_0 \tag{14}$$

and

$$a_1 = b_{n-1} - cn$$

$$a_k = b_{n-k} - \binom{n}{k}c^k - \sum_{i=1}^{k-1}\binom{n-i}{k-i}a_i c^{k-i} \qquad k = 2, 3, \ldots, n \tag{15}$$

where in (15) we have solved explicitly for a_k in terms of $a_1, a_2, \ldots, a_{k-1}$. Thus for a chosen value of c we may obtain successively $a_1, a_2, \ldots, a_n$.

We may also solve (15) explicitly for a_k in terms c, n, k, and the b_i. The pattern that emerges as each successive a_k is obtained is given by

$$a_k = (-1)^k \sum_{i=0}^{k} (-1)^i \binom{n-i}{k-i} c^{k-i} b_{n-i} \qquad k = 1, 2, \ldots, n \tag{16}$$

where we note that $b_n = 1$. Equation (16) may be readily verified by substitution into (15) and observing the vanishing of the various powers of c.

The range on c for nonnegative a_k depends on the low-pass prototype coefficients b_i. It is interesting to note from (16) that $a_n = P(-c)$, where $P(s)$ is the denominator polynomial defined in (3). Since $P(s)$ is strictly Hurwitz, all its zeros occur in the left half of the s plane. Also $P(0) > 0$. Therefore, denoting the real zero of $P(s)$ which is nearest the origin by $-\sigma$ ($\sigma > 0$), we note that if $0 < c < \sigma$ the result is $a_n > 0$. Of course, from the first equation of (15), $a_1 \geq 0$ if

$$0 < c < \frac{b_{n-1}}{n} \tag{17}$$

Other bounds on c are obtained. for the various cases, by considering the other equations of (15) or (16).

In applying the PRB technique, we note that if c is chosen so that one of the a_i is zero, then this eliminates one of the feedback resistors. For example, if $c = b_{n-1}/n$, then by (15), $a_1 = 0$ and thus R_1 is infinite (open-circuit). It is

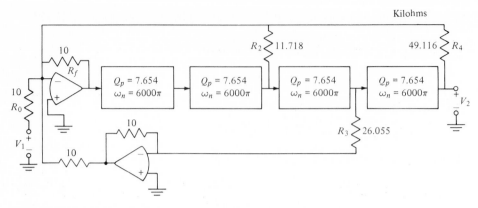

Figure 6.4-2 The eighth-order bandpass PRB filter realized in Example 6.4-1.

probably better, however, not to eliminate a resistor in this manner, since this eliminates some control over one of the states of the system. In any case, one would not want to make a_n zero, for then there would be no feedback from the output. Another consideration is that from (16) we see that the feedback resistors are independent of the gain and center frequency of the filter, and are determined solely by the b_i of the low-pass prototype filter. The gain H is set by the input resistor R_0, which by (2) and (14) is given by

$$R_0 = \frac{R_f}{Hb_0} \tag{18}$$

In Fig. 6.4-1, we have assumed that the component second-order bandpass realizations are noninverting. If inverting realizations are used, then additional inverters are required. An example is in Fig. 6.4-2. Finally, the constant c was found to be constrained as given in (17). This implies that c is normally less than unity, which by (13) causes Q_p to be large for a moderate value of Q. If we select $c = b_{n-1}/n$, then

$$Q = \frac{b_{n-1}Q_p}{n} \tag{19}$$

which gives the smallest value of Q_p for a specified Q.

Example 6.4-1 *A Maximally Flat Magnitude Eighth-Order Bandpass PRB Filter* It is desired to realize a maximally flat magnitude eighth-order bandpass filter with a center frequency of 3000 Hz, a -3-dB bandwidth of 600 Hz, and a -30-dB bandwidth of not more than 1500 Hz. The center-frequency gain is to be unity. We begin by letting Q_p be defined by (19). From the filter specifications we see that $Q = 5$ and $\omega_0 = 6000\pi$ rad/s. From Table 2.1-3a for $n = 4$, we obtain

$$\frac{V_2(s)}{V_1(s)} = \frac{1}{s^4 + 2.613126s^3 + 3.414214s^2 + 2.613126s + 1} \tag{20}$$

Thus from (3) we get $b_0 = 1$, $b_1 = b_3 = 2.613126$, and $b_2 = 3.414214$. If $c = b_{n-1}/n$, then $c = b_3/4 = 0.6532815$. Therefore

$$Q_p = \frac{Q}{c} = \frac{5}{0.6532815} = 7.654 \tag{21a}$$

$$H_0 = \frac{1}{c} = 1.5307 \tag{21b}$$

$$\omega_p = 6000\pi \text{ rad/s} \tag{21c}$$

where ω_p is the undamped natural frequency of (7). From (15) we get

$$a_1 = 0 \tag{22a}$$

$$a_2 = b_2 - \binom{4}{2}c^2 - \sum_{i=1}^{1}\binom{4-i}{2-i}a_i c^{2-i} = 0.8536 \tag{22b}$$

$$a_3 = b_1 - \binom{4}{3}c^3 - \sum_{i=1}^{2}\binom{4-i}{3-i}a_i c^{3-i} = 0.3838 \tag{22c}$$

$$a_4 = b_0 - c^4 - \sum_{i=1}^{3}a_i c^{4-i} = 0.2036 \tag{22d}$$

From (2) we can select R_f and solve for the various R_i. If $R_f = 10 \text{ k}\Omega$, then $R_1 = \infty$, $R_2 = 11.718 \text{ k}\Omega$, $R_3 = 26.055 \text{ k}\Omega$, and $R_4 = 49.116 \text{ k}\Omega$. The final design where the second-order bandpass filters are inverting and represented by blocks is given in Fig. 6.4-2. □

Many other filter configurations involving second-order building blocks and various combinations of feedback have appeared in the literature.[1,2,3,4,5] Although some excellent realizations have been developed, many of them are extremely difficult to design, requiring the use of a digital computer in the process. The procedures given in this and the preceding section represent good compromises between filter performance and design simplicity.

[1] G. Hurtig III, "Voltage Tunable Multiple-Bandpass Active Filters," *Proc. Intern. Symp. Circuits and Systems*, April 1974, pp. 569–572.

[2] G. Szentirmai, "On Multiple-Feedback Active Filter Structures," *Proc. 7th Asilomar Conf. Circuits, Systems, Computers*, November 1973, pp. 368–377.

[3] J. Tow, "Design and Evaluation of Shifted Companion Form (Follow the Leader Feedback) Active Filters," *Proc. Intern. Symp. Circuits and Systems*, April 1974, pp. 650–660.

[4] K. Laker and M. Ghausi, "A Low Sensitivity Multiloop Feedback Active-RC Filter," *Proc. Intern. Symp. Circuit Theory*, April 1973, pp. 126–129.

[5] J. Tow and Y. Kuo, "Coupled-Biquad Active Filters," *Proc. Intern. Symp. Circuit Theory*, April 1972, pp. 164–168.

6.5 PARALLEL-CASCADE METHOD

In this section we present our last direct method for realizing high-order network functions. It is capable of realizing all types of filter approximations with or without finite $j\omega$-axis zeros. It is also capable of realizing complex-conjugate zeros anywhere in the complex-frequency plane or simple zeros anywhere on the real axis. This method uses the GIC described by (4) of Sec. 6.1 and shown in Fig. 6.1-2a. The GIC constant $D(s)$ of (4) in Sec. 6.1 is given as

$$K_c(s) = \frac{1}{D(s)} = s^2 K_1 \qquad K_1 = R_1 R_3 C_2 C_4 \tag{1}$$

This method was proposed by Antoniou[1] and results in a very versatile realization scheme.

The parallel-cascade method starts from the voltage transfer function of a two-port network with y parameters $y_{ij}(s)$. This transfer function will be designated as $T(s)$ and expressed as

$$T(s) = \frac{V_2(s)}{V_1(s)} = \frac{-y_{21}(s)}{y_{22}(s)} = \frac{a_0 + a_1 s + a_2 s^2 + \cdots + a_n s^n}{b_0 + b_1 s + b_2 s^2 + \cdots + b_n s^n} = \frac{\sum_{i=0}^{n} a_i s^i}{\sum_{i=0}^{n} b_i s^i} \tag{2}$$

where $b_i > 0$ and $b_i \geq |a_i|$. The numerator coefficients may be positive, negative, or zero. If their magnitudes violate the latter inequality, the entire numerator may be multiplied by a sufficiently small constant. From (2) we see that $-y_{21}(s)$ and $y_{22}(s)$ can be expressed as

$$-y_{21}(s) = \sum_{i=0}^{n} a_i s^i \qquad y_{22}(s) = \sum_{i=0}^{n} b_i s^i \tag{3}$$

Now consider splitting the original network into two parallel subnetworks designated as N_1 and N_2 in Fig. 6.5-1. If we designate $y_{ij}^{(1)}(s)$ and $y_{ij}^{(2)}(s)$ as the

---

[1] A. Antoniou, "Novel RC-Active Network Synthesis Using Generalized Immittance Converters," *IEEE Trans. Circuit Theory*, vol. CT-17, May 1970, pp. 212–217.

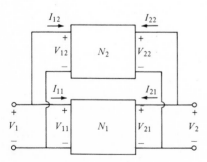

Figure 6.5-1 Two parallel-connected subnetworks.

admittance parameters of networks 1 and 2 respectively, then $T(s)$ may be expressed as

$$T(s) = \frac{-y_{21}^{(1)}(s) - y_{21}^{(2)}(s)}{y_{22}^{(1)}(s) + y_{22}^{(2)}(s)} = \frac{\sum\limits_{i=0}^{n} a_i s^i}{\sum\limits_{i=0}^{n} b_i s^i} \tag{4}$$

For our purposes, it is convenient to decompose the numerator and denominator of (4) in the following manner:

$$-y_{21}^{(1)}(s) = a_0 + a_1 s \tag{5a}$$

$$y_{22}^{(1)}(s) = b_0 + b_1 s \tag{5b}$$

$$-y_{21}^{(2)}(s) = \sum_{i=2}^{n} a_i s^i \tag{5c}$$

$$y_{22}^{(2)}(s) = \sum_{i=2}^{n} b_i s^i \tag{5d}$$

Thus the open-circuit voltage transfer functions of networks 1 and 2 are

$$T_1(s) = \frac{-y_{21}^{(1)}(s)}{y_{22}^{(1)}(s)} = \frac{a_0 + a_1 s}{b_0 + b_1 s} \tag{6a}$$

and

$$T_2(s) = \frac{-y_{21}^{(2)}(s)}{y_{22}^{(2)}(s)} = \frac{\sum\limits_{i=2}^{n} a_i s^i}{\sum\limits_{i=2}^{n} b_i s^i} \tag{6b}$$

Let us consider $T_1(s)$ first. It can be realized by the circuit of Fig. 6.5-2a (with $i = 1$) whose transfer function is

$$\frac{V_{21}(s)}{V_{11}(s)} = \frac{sC_{11} + G_{11}}{s(C_{11} + C_{21}) + (G_{11} + G_{21})} \tag{7}$$

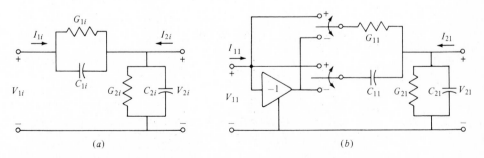

(a) $\qquad\qquad\qquad\qquad\qquad\qquad\qquad (b)$

Figure 6.5-2 Subnetwork N_1 of Fig. 6.5-1.

Equating the ratio of (5a) and (5b) to (7) results in the following design equations for the circuit elements:

$$G_{11} = a_0 \qquad C_{11} = a_1 \qquad G_{21} = b_0 - a_0 \qquad C_{21} = b_1 - a_1 \qquad (8)$$

If either or both the coefficients a_0 and a_1 are negative, then an inverting unity-gain amplifier can be placed in series with the network as shown in Fig. 6.5-2b. If a_0 is negative, then G_{11} is connected to the minus position. Similarly, if a_1 is negative, C_{11} is connected to the minus position.

Now consider the realization of $T_2(s)$. If we can cause $-y_{21}^{(2)}(s)$ and $y_{22}^{(2)}(s)$ to become similar to the original $-y_{21}(s)$ and $y_{22}(s)$ (but of lower order) then another type 1 network may be removed, and the procedure repeated. The circuit of Fig. 6.5-3 provides the means for accomplishing this objective. To see this, we use (4) of Sec. 6.1 to obtain

$$V_{22}(s) = V_{23}(s) \tag{9a}$$

$$I_{22}(s) = \frac{Z_1(s)Z_3(s)}{Z_2(s)Z_4(s)} I_{23}(s) \tag{9b}$$

If $Z_1(s) = R_1$, $Z_2(s) = 1/sC_2$, $Z_3(s) = R_3$, and $Z_4(s) = 1/sC_4$, then (9b) becomes

$$I_{22}(s) = s^2 R_1 C_2 R_3 C_4 I_{23}(s) = s^2 K_1 I_{23}(s) \tag{10}$$

where $K_1 = R_1 C_2 R_3 C_4$. The admittance parameters of network 3 can now be written as

$$I_{13}(s) = y_{11}^{(3)}(s)V_{13}(s) + y_{12}^{(3)}(s)V_{23}(s) \tag{11a}$$

$$I_{23}(s) = y_{21}^{(3)}(s)V_{13}(s) + y_{22}^{(3)}(s)V_{23}(s) \tag{11b}$$

From Figs. 6.5-1 and 6.5-3, however, we see that $V_{13}(s) = V_{12}(s)$ and $I_{13}(s) = I_{12}(s)$. Furthermore (9a) and (10) relate $I_{23}(s)$ and $V_{23}(s)$ in terms of $I_{22}(s)$ and $V_{22}(s)$. Thus (11) can be used to develop the admittance parameters of network 2.

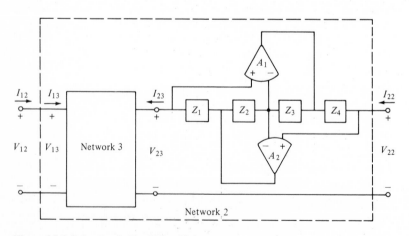

Figure 6.5-3 Subnetwork N_2 of Fig. 6.5-1.

These may be written as

$$I_{12}(s) = y_{11}^{(3)}(s)V_{12}(s) + y_{12}^{(3)}(s)V_{22}(s) = y_{11}^{(2)}(s)V_{12}(s) + y_{12}^{(2)}(s)V_{22}(s) \tag{12a}$$

$$I_{22}(s) = s^2 K_1 y_{21}^{(3)}(s)V_{12}(s) + s^2 K_1 y_{22}^{(3)}(s)V_{22}(s) = y_{21}^{(2)}(s)V_{12}(s) + y_{22}^{(2)}(s)V_{22}(s) \tag{12b}$$

We see that from (12)

$$y_{21}^{(3)}(s) = \frac{y_{21}^{(2)}(s)}{s^2 K_1} \tag{13a}$$

$$y_{22}^{(3)}(s) = \frac{y_{22}^{(2)}(s)}{s^2 K_1} \tag{13b}$$

Substitution of (5c) and (5d) into (13) results in

$$y_{21}^{(3)}(s) = \frac{-1}{s^2 K_1} \sum_{i=2}^{n} a_i s^i = \frac{-1}{K_1} \sum_{i=2}^{n} a_i s^{i-2} \tag{14a}$$

$$y_{22}^{(3)}(s) = \frac{1}{s^2 K_1} \sum_{i=2}^{n} b_i s^i = \frac{1}{K_1} \sum_{i=2}^{n} b_i s^{i-2} \tag{14b}$$

The open-circuit voltage transfer function of network 3 using the results of (14) is now given as

$$T_3(s) = \frac{-y_{21}^{(3)}(s)}{y_{22}^{(3)}(s)} = \frac{\displaystyle\sum_{i=2}^{n} a_i s^{i-2}}{\displaystyle\sum_{i=2}^{n} b_i s^{i-2}}$$

$$= \frac{a_2 + a_3 s + a_4 s^2 + \cdots + a_n s^{n-2}}{b_2 + b_3 s + b_4 s^2 + \cdots + b_n s^{n-2}} \tag{15}$$

This is of the same form as (2) but of lower order, so that the entire process may now be repeated. At this point the network realization appears as shown in Fig. 6.5-4. The general form of the parallel-cascade method after additional cycles is shown in Fig. 6.5-5. Note that the value of the CGIC constant K_i is not a factor in the design process and may be selected arbitrarily. Typically it should be chosen to minimize the value of r_{1i}, c_{2i}, r_{3i}, and c_{4i} and to optimize the voltage-handling capability. Normally $r_{1i} = r_{3i} = r_i$ and $c_{2i} = c_{4i} = c_i$. To illustrate what is optimum, consider the CGIC of Fig. 6.5-6 with the magnitudes of the voltages across the elements indicated as shown. Note that $V_{o2} - V_{o1}$ is equal to $I(r_3 + 1/j\omega c_2)$ where I is the current which flows through c_2 and r_3. The maximum value of $V_{o2} - V_{o1}$ will be constrained by the amplifier output voltage limitation. If either r_3 or c_2 is chosen abnormally large or small respectively, then the value of $V_{o2} - V_{o1}$ will quickly approach the limits of the amplifiers. A good rule of thumb is to let $r_3 = 1/\omega c_2$ where $\omega \simeq \omega_c$, the cutoff frequency, and select r_3 in the 1–10 kΩ range.

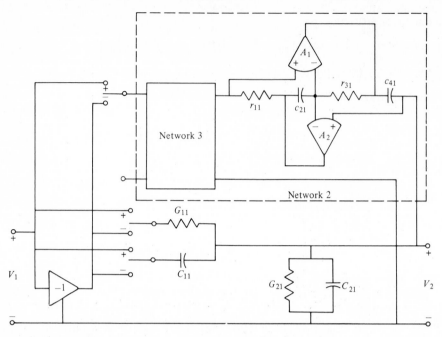

Figure 6.5-4 The parallel-connected subnetworks of Figs. 6.5-2 and 6.5-3.

The first-order passive network elements shown in Fig. 6.5-2a can be expressed on a general basis as

$$G_{1i} = a_{2i-2} \qquad G_{2i} = b_{2i-2} - a_{2i-2}$$
$$C_{1i} = a_{2i-1} \qquad C_{2i} = b_{2i-1} - a_{2i-1} \tag{16}$$

where i is the iteration number. Each iteration in the design procedure produces such a passive network as well as a CGIC except for the first one ($i = 1$), which produces only a passive network. If n is even, then we see that on the $1 + n/2$ iteration

$$G_{1,\,1+n/2} = a_n \qquad G_{2,\,1+n/2} = b_n - a_n \tag{17a}$$
$$C_{1,\,1+n/2} = 0 \qquad C_{2,\,1+n/2} = 0 \tag{17b}$$

since from (3) there are no $n + 1$ coefficients. If n is odd, then on the $(n + 1)/2$ iteration we have

$$G_{1,\,(n+1)/2} = a_{n-1} \qquad G_{2,\,(n+1)/2} = b_{n-1} - a_{n-1} \tag{18a}$$
and
$$C_{1,\,(n+1)/2} = a_n \qquad C_{2,\,(n+1)/2} = b_n - a_n \tag{18b}$$

Therefore, for n even, the last passive network has no capacitors, while for n odd the last passive network has both resistors and capacitors. For every subnetwork with negative numerator coefficients there corresponds one negative unity-gain amplifier. As a final step in the realization, these can all be replaced by a single amplifier, as shown in Fig. 6.5-5.

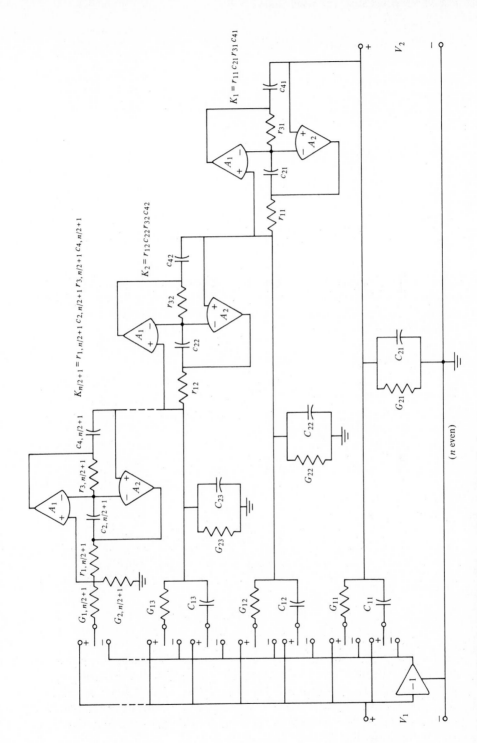

Figure 6.5-5 The general form of the parallel-cascade method.

313

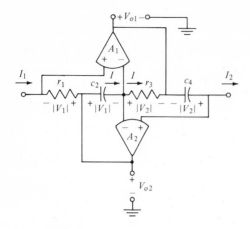

Figure 6.5-6 Voltage levels in the GIC used in the parallel-cascade method.

Example 6.5-1 *A Third-Order Low-Pass Elliptic Parallel-Cascade Filter* It is desired to use the parallel-cascade method to obtain an elliptic low-pass filter realization with a cutoff frequency of $2\pi \times 10^3$ rad/s, and with the (normalized) specifications given in Example 2.3-1. The normalized voltage transfer function is

$$T(s) = \frac{V_2(s)}{V_1(s)} = \frac{0.105891(s^2 + 5.153209)}{(s + 0.539958)(s^2 + 0.434067s + 1.010594)} \tag{19}$$

Rearranging this to identify the coefficients gives

$$T(s) = \frac{0.545678 + 0.105891s^2}{0.545678 + 1.244972s + 0.974025s^2 + s^3}$$

$$= \frac{a_0 + a_2 s^2}{b_0 + b_1 s + b_2 s^2 + b_3 s^3} \tag{20}$$

From (16) we get:
First iteration

$$G_{11} = a_0 = 0.545678 \qquad G_{21} = b_0 - a_0 = 0$$

$$C_{11} = a_1 = 0 \qquad C_{21} = b_1 - a_1 = 1.244972$$

Second iteration

$$G_{12} = a_2 = 0.105891 \qquad G_{22} = b_2 - a_2 = 0.868134$$

$$C_{12} = a_3 = 0 \qquad C_{22} = b_3 - a_3 = 1.0$$

The resulting realization is shown in Fig. 6.5-7. Denormalizing the frequency and using a 1000-Ω impedance denormalization gives the elements of the realizations as $R_{11} = 1/G_{11} = 1.833$ kΩ, $C_{21} = 1.981 \times 10^{-7}$ F, $R_{12} = 1/G_{12} = 9.444$ kΩ, $C_{22} = 1.592 \times 10^{-7}$ F, and $R_{22} = 1/G_{22} = 1.152$ kΩ. The CGIC impedances are selected as $r_{11} = r_{31} = 1$ kΩ and $c_{21} = c_{41} = 1.59 \times 10^{-7}$ F. The GIC has $K_1 = 2.528 \times 10^{-8}$. □

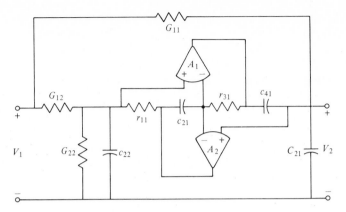

Figure 6.5-7 The third-order low-pass parallel-cascade filter realized in Example 6.5-1.

There are several interesting properties of the parallel-cascade technique. As one of these, it is suitable for bandpass realizations which have $j\omega$-axis zeros. As another, it has the ability to synthesize a low- and a high-pass realization with the same circuit for certain cases. For example, for a low-pass filter with no finite $j\omega$-axis zeros, in Fig. 6.5-2 we have $G_{1i} = 0$ for $i \neq 1$ and $C_{1i} = 0$ for all i. The output is taken across G_{1n} or C_{1n}, depending on whether n is even or odd respectively. For realizations where the denominator coefficients are symmetric, that is, $b_i = b_{n-i}$ as found in Butterworth filters, we may obtain the equivalent high-pass response from the low-pass realization by inserting the input signal in series with G_{2n} (or C_{2n}) and taking the output across G_{11}, which is grounded. A second-order realization capable of realizing low-pass, bandpass, and high-pass transfer functions using the same structure is given in the literature.[1]

In summary the parallel-cascade method has the desirable features of complete generality, and a simple design procedure. In addition it typically requires relatively few precision capacitors. For example, an eighth-order Butterworth parallel-cascade filter needs twelve capacitors. Of these, however, eight are used to determine arbitrary-valued GIC constants, thus only four need be precisely determined. A leapfrog realization, on the other hand, would require eight precisely determined capacitors to realize the same filter. The sensitivities of the parallel-cascade method are comparable to those of the other synthesis methods presented in this chapter. They will be investigated in more detail in a later section of this chapter.

6.6 SENSITIVITY

Methods for simulating passive networks by the use of active RC subnetworks have been introduced in this chapter. Here we present a discussion of various

[1] S. K. Mitra, "Transfer Matrix Realization Using RC: GIC Networks," *Circuit Theory and Applications*, vol. 3, 1975, pp. 81–85.

aspects of the sensitivity problem as it relates to these methods. Our treatment will be considerably different from the one given in the previous chapters, since in passive network simulation methods the order of the network function realized is usually much greater than 2. As a result the most appropriate sensitivity treatment is to consider $N(s)$, the filter's network function. In general this will have the form

$$N(s) = \frac{A(s)}{B(s)} = \frac{a_0 + a_1 s + a_2 s^2 + a_3 s^3 + \cdots + a_m s^m}{b_0 + b_1 s + b_2 s^2 + b_3 s^3 + \cdots + b_n s^n} \tag{1}$$

where $n \geq m$. For such a function, we can find the magnitude and phase relations which show the dependence of the frequency response upon some network parameter x by finding the function sensitivity $S_x^{N(s)}$ defined in Sec. 3.2. To accomplish this, as a first step we note that in passive network simulation methods, some of the passive elements are replaced by active RC equivalents. Let such a replaced impedance be designated as Γs^j where j is a positive or negative integer. For example, if $j = 0$, then the impedance is a resistor and $\Gamma = R$. Similarly, if $j = -1$ then $\Gamma = 1/C$, if $j = 1$ then $\Gamma = L$, etc. As a next step, the coefficient sensitivities $S_\Gamma^{a_i}$ and $S_\Gamma^{b_i}$ defined in Sec. 3.3 are calculated. Following this, using (7) of Sec. 3.3, the function sensitivity may be expressed in terms of the coefficient sensitivities. Finally, since Γ is a function of some parameter x, S_x^Γ, the sensitivity of Γ with respect to x may be calculated. Combining the above steps we see that the transfer function sensitivity for the function $N(s)$ defined in (1) is given as

$$S_x^N(s) = \frac{\sum\limits_{i=0}^{m} (S_\Gamma^{a_i} S_x^\Gamma) a_i s^i}{A(s)} - \frac{\sum\limits_{i=0}^{n} (S_\Gamma^{b_i} S_x^\Gamma) b_i s^i}{B(s)} \tag{2}$$

Since S_x^Γ is independent of the summations, this can be rewritten in the form

$$S_x^N(s) = S_x^\Gamma \left[\frac{\sum\limits_{i=0}^{m} S_\Gamma^{a_i} a_i s^i}{A(s)} - \frac{\sum\limits_{i=0}^{n} S_\Gamma^{b_i} b_i s^i}{B(s)} \right] \tag{3}$$

Now let us consider the calculation of the sensitivities $S_\Gamma^{a_i}$ and $S_\Gamma^{b_i}$ which are used in the above expression. Such a calculation is facilitated by assuming that the network has the general ladder structure shown in Fig. 6.6-1. In this structure, the series elements are shown as impedances and the shunt elements as admittances. Assuming, for simplicity, that the ladder is terminated with the element Y_6, its

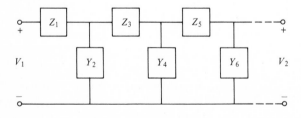

Figure 6.6-1 A general ladder structure.

voltage transfer function can be expressed in terms of the polynomials $A(s)$ and $B(s)$ defined in (1), specifically

$$A(s) = 1 \tag{4a}$$

$$B(s) = 1 + Z_1 Y_2 + Z_1 Y_4 + Z_1 Y_6 + Z_3 Y_4 + Z_3 Y_6 + Z_5 Y_6$$
$$+ Z_1 Y_2 Z_3 Y_4 + Z_1 Y_2 Z_3 Y_6 + Z_1 Y_2 Z_5 Y_6 + Z_1 Y_4 Z_5 Y_6$$
$$+ Z_3 Y_4 Z_5 Y_6 + Z_1 Y_2 Z_3 Y_4 Z_5 Y_6 \tag{4b}$$

This expression is easily used for less complicated structures by setting the appropriate Y_i and Z_i terms to zero. Similar expressions may be found for ladders with additional elements. The procedure for finding $S_x^{N(s)}$ may now be outlined as follows:

1. Starting with a prototype passive network realization having the form shown in Fig. 6.6-1, make any necessary transformations such as the $RLC : CRD$ transformation of Table 6.3-1.
2. Substitute the values of the impedances and admittances into (4) and find the expression for the rational function having the form of (1).
3. Calculate $S_\Gamma^{a_i}$ and $S_\Gamma^{b_i}$ for any desired impedance Γs^j.
4. Calculate S_x^Γ where x is some parameter of the RC active subnetwork used to realize the impedance Γs^j.

Example 6.6-1 *Function Sensitivity of an FDNR Filter Realization* As an example of the procedure for finding function sensitivity, consider the low-pass third-order network shown in Fig. 6.6-2a. Applying the $RLC : CRD$ transformation we obtain the result shown in Fig. 6.6-2b. Comparing this with Fig. 6.6-1, we define $Z_1 = R_1$, $Y_2 = s^2 D_2$, $Z_3 = R_3$, $Y_4 = sC_4$, and $Z_5 = Y_6 = 0$. Thus, from (4), $N(s)$ can be written as

$$N(s) = \frac{1}{1 + Z_1 Y_2 + Z_1 Y_4 + Z_3 Y_4 + Z_1 Y_2 Z_3 Y_4}$$

$$= \frac{1}{(R_1 D_2 R_3 C_4)s^3 + (R_1 D_2)s^2 + (R_1 C_4 + R_3 C_4)s + 1} \tag{5}$$

Ohms, henrys,
farads, and farad-seconds

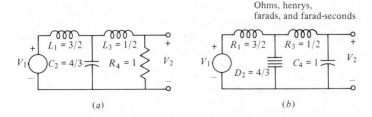

(a)　　　　　　　　(b)

Figure 6.6-2 The third-order low-pass network of Example 6.6-1.

Let us now investigate the sensitivity of this function to the FDNR. Thus we consider Γ as D_2. From (5) we find

$$S_{D_2}^{a_0} = S_{D_2}^{b_0} = S_{D_2}^{b_1} = 0 \qquad S_{D_2}^{b_2} = S_{D_2}^{b_3} = 1 \tag{6}$$

Using the circuit of Fig. 6.1-2 with $Y_1 = sC_{12}$, $Y_2 = G_{22}$, $Y_3 = sC_{32}$, $Y_4 = G_{42}$, and $Y_5 = G_{52}$ ($Y_i = 1/Z_i$) to realize the FDNR, D_2 may be expressed as

$$D_2 = \frac{C_{12} C_{32} G_{52}}{G_{22} G_{42}} \tag{7}$$

The sensitivities of D_2 with respect to the FDNR elements are now found as

$$S_{C_{12}}^{D_2} = S_{C_{32}}^{D_2} = S_{G_{52}}^{D_2} = 1 \qquad S_{G_{22}}^{D_2} = S_{G_{42}}^{D_2} = -1 \tag{8}$$

Using the above results in (3), the sensitivity of $N(s)$ with respect to x is found to be

$$S_x^{N(s)} = -\frac{b_3 s^3 + b_2 s^2}{b_3 s^3 + b_2 s^2 + b_1 s + b_0} \tag{9}$$

where x may represent any of the quantities C_{12}, C_{32}, or G_{52}. The sensitivity to G_{22} or G_{42} is simply the negative of the expression given in (9). Evaluating the coefficients b_i for the network element values shown in Fig. 6.6-2, we obtain

$$S_x^{N(s)} = -\frac{s^3 + 2s^2}{s^3 + 2s^2 + 2s + 1} \tag{10}$$

For $s = j\omega$, this may be written

$$S_x^{N(j\omega)} = \frac{2\omega^2 - 2\omega^4 - \omega^6}{1 + \omega^6} - j\frac{3\omega^3}{1 + \omega^6} \tag{11}$$

Thus (see property 14 of Table 3.1-1)

$$S_x^{|N(j\omega)|} = \frac{2\omega^2 - 2\omega^4 - \omega^6}{1 + \omega^6} \tag{12}$$

A plot of (12) is shown in Fig. 6.6-3. From this it is readily seen that the sensitivity of $|N(j\omega)|$ with respect to x is equal to or less than $\frac{1}{2}$ in the passband from 0 to 1 rad/s. The value of the sensitivity outside the passband, although larger, is usually of less concern.

The multiparameter sensitivity (see Sec. 3.2) of the network function to all the elements of the FDNR can be found as

$$\frac{d|N(j\omega)|}{|N(j\omega)|} = \text{Re} \sum_{i=1}^{5} S_{x_i}^{N(j\omega)} \frac{dx_i}{x_i} \tag{13}$$

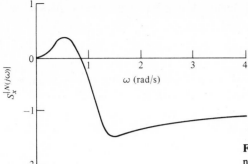

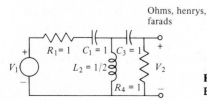

Figure 6.6-3 Magnitude sensitivity for the network of Fig. 6.6-2.

where $x_1 = C_{12}$, $x_2 = G_{22}$, $x_3 = C_{32}$, $x_4 = G_{42}$, and $x_5 = G_{52}$. Thus

$$\frac{d|N(j\omega)|}{|N(j\omega)|} = \frac{2\omega^2 - 2\omega^4 - \omega^6}{1 + \omega^6}\left(\frac{dC_{12}}{C_{12}} - \frac{dG_{22}}{G_{22}} + \frac{dC_{32}}{C_{32}} - \frac{dG_{42}}{G_{42}} + \frac{dG_{52}}{G_{52}}\right) \quad (14)$$

Obviously, by proper choice of the variation of the individual passive elements of the FDNR realization, this sensitivity can be minimized. ☐

Now let us consider the synthetic-inductor method of passive network simulation. The function sensitivity for this method may be found by a technique similar to that described above.

Example 6.6-2 *Function Sensitivity of a Synthetic-Inductor Filter Realization* A third-order doubly terminated high-pass filter is shown in Fig. 6.6-4. It is desired to find the sensitivity of the magnitude of the voltage transfer function with respect to each of the passive elements of the synthetic inductor used to provide a realization. The GIN of Fig. 6.1-2 is used to obtain the synthetic inductor by choosing $Z_1 = R_{12}$, $Z_2 = 1/sC_{22}$, $Z_3 = R_{32}$, $Z_4 = R_{42}$, and $Z_5 = R_{52}$. Equating the elements of Figs. 6.6-1 and 6.6-4 and letting $N(s) = V_2(s)/V_1(s)$ we obtain

$$N(s) = \frac{(C_1 L_2 C_3 R_4)s^3}{(C_1 L_2 C_3 R_4 + R_1 C_1 L_2 C_3)s^3 + (R_1 C_1 C_3 R_4 + L_2 C_3 + C_1 L_2)s^2 + (C_3 R_4 + R_1 C_1)s + 1} \quad (15)$$

Ohms, henrys, farads

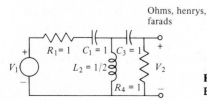

Figure 6.6-4 The third-order high-pass filter of Example 6.6-2.

Calculation of the coefficient sensitivities with respect to L_2 gives

$$S_{L_2}^{a_3} = 1 \qquad S_{L_2}^{b_2} = \frac{L_2(C_1 + C_3)}{R_1 C_1 C_3 R_4 + L_2(C_1 + C_3)} = \frac{1}{2} \qquad S_{L_2}^{b_3} = 1 \qquad (16)$$

The other coefficient sensitivities with respect to L_2 are zero. The value for L_2 given from the synthetic-inductor realization is

$$L_2 = \frac{R_{12} C_{22} R_{32} R_{52}}{R_{42}} \qquad (17)$$

From (3) the transfer function sensitivity is

$$S_x^{N(s)} = S_x^{L_2}\left(\frac{1}{a_3} - \frac{0.5b_2 s^2 + b_3 s^3}{b_0 + b_1 s + b_2 s^2 + b_3 s^3}\right)$$

$$= S_x^{L_2}\left(\frac{s^3 + 3s^2 + 4s + 1}{s^3 + 2s^2 + 2s + 1}\right) \qquad (18)$$

where x may be any of the quantities $R_{12}, C_{22}, R_{32}, R_{42}$, and R_{52}. Substituting $s = j\omega$ we obtain

$$S_x^{N(j\omega)} = S_x^{L_2}\left(\frac{2 + \omega^2 + \omega^6}{1 + \omega^6} - j\frac{\omega^3 + \omega^5}{1 + \omega^6}\right) \qquad (19)$$

Thus, for $x = R_{12}, C_{22}, R_{32}$, and R_{52},

$$S_x^{|N(j\omega)|} = \frac{2 + \omega^2 + \omega^6}{1 + \omega^6} \qquad (20)$$

while for $x = R_{42}$, we obtain the negative of the above expression. $\qquad \square$

Sensitivity analyses of the type illustrated above can be readily used to predict the influence of an element change upon the filter performance. As an illustration of this, consider Example 6.6-2. If the capacitor C_{22} used in the synthetic inductor is a Mylar one, with a temperature coefficient of $+600$ ppm/°C, then at $\omega = 1$ rad/s, $S_{C_{22}}^{|N(j1)|} = 2$. Therefore $d|N(j\omega)|/|N(j\omega)|$ at $\omega = 1$ rad/s is equal to 1200 ppm/°C. This amounts to a 1.2 percent change in $|N(j1)|$ produced by C_{22}, if the temperature changes by 10°C. Similar sensitivity analyses can be made for the other passive elements of the active RC subnetworks, as well as for the other passive elements of the filter.

Another passive network simulation method introduced in this chapter was the leapfrog technique. This was used to synthesize low-pass and bandpass filter functions which did not have finite $j\omega$-axis zeros. The technique required that the immittances of a ladder network be realized as voltage transfer functions. The sensitivity analysis follows the general procedure outlined above, namely, it involves finding S_x^Γ where Γ is a passive element and x is any of the parameters of the active RC subnetwork which realizes the voltage transfer function which

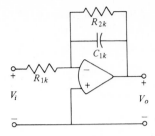

Figure 6.6-5 Circuit used to simulate the functions of (21).

synthesizes the passive element. For the low-pass case, the general form of the Z_j and Y_i elements shown in Fig. 6.3-1 is

$$Y_i = \frac{1/L_i}{s + R_i/L_i} \qquad Z_j = \frac{1/C_j}{s + 1/R_jC_j} \qquad (21)$$

where R_i may be zero and R_j may be infinity. Note that these Z and Y designations are reversed from the ones shown in Fig. 6.6-1. The circuit of Fig. 6.6-5 is used to simulate these functions. Its voltage transfer function is

$$T_k(s) = \frac{V_o(s)}{V_i(s)} = \frac{-1/R_{1k}C_{1k}}{s + 1/R_{2k}C_{1k}} \qquad (22)$$

The necessary choices of the elements are indicated in Table 6.6-1.

In the leapfrog structure, it may happen that a series or shunt immittance has the form $U_i + jV_i$. When such an immittance is simulated, then it is possible that both U_i and V_i are functions of the same parameter x of the subnetwork performing the simulation. In this case, calculation of $S_x^{N(s)}$ may be made directly, using

$$S_x^{N(s)} = \frac{\sum\limits_{i=0}^{m} S_x^{a_i} a_i s^i}{A(s)} - \frac{\sum\limits_{i=0}^{n} S_x^{b_i} b_i s^i}{B(s)} \qquad (23)$$

rather than finding the intermediate $S_{\Gamma}^{a_i}$ and $S_{\Gamma}^{b_i}$ sensitivities as given in (3).

Table 6.6-1 Necessary choices of elements for leapfrog realization

Passive element	Γ as a function of the elements of Fig. 6.6-5	
Y_i	$R_i = \dfrac{R_{1i}}{R_{2i}}$	$L_i = R_{1i}C_{1i}$
Z_j	$R_j = \dfrac{R_{2j}}{R_{1j}}$	$C_j = R_{1j}C_{1j}$

Example 6.6-3 *Function Sensitivity of a Leapfrog Filter Realization* It is desired to find the function sensitivity with respect to R_{11}, R_{21}, and C_{11} of Fig. 6.6-5 if the leapfrog technique is used to realize the low-pass filter shown in Fig. 6.6-6. Its voltage transfer function is

$$N(s) = \frac{V_2(s)}{V_1(s)} = \frac{R_4}{(L_1 C_2 L_3)s^3 + (R_1 C_2 L_3 + L_1 C_2 R_4)s^2}$$
$$+ (L_1 + L_3 + R_1 C_2 R_4)s + (R_1 + R_4)$$

$$= \frac{1}{2s^3 + 4s^2 + 4s + 2} \tag{24}$$

From Table 6.6-1, we find $R_1 = R_{11}/R_{21}$ and $L_1 = R_{11} C_{11}$ for Fig. 6.6-5 with $k = 1$. Thus, $N(s)$ becomes

$$N(s) = \frac{R_4 R_{21}}{(R_{11} C_{11} R_{21} C_2 L_3)s^3 + (R_{11} C_2 L_3 + R_{11} R_{21} C_{11} C_2 R_4)s^2}$$
$$+ (R_{11} R_{21} C_{11} + R_{21} L_3 + R_{11} C_2 R_4)s + (R_{11} + R_{21} R_4)$$
$$\tag{25}$$

The nonzero coefficient sensitivities with respect to R_{11}, R_{21}, and C_{11}, assuming unity values for these components, are

$$S_{R_{11}}^{b_0} = \frac{R_{11}}{b_0} = \frac{1}{2} \qquad S_{R_{11}}^{b_1} = \frac{R_{11}(R_{21} C_{11} + C_2 R_4)}{b_1} = \frac{3}{4}$$

$$S_{R_{11}}^{b_2} = S_{R_{11}}^{b_3} = 1 \qquad S_{R_{21}}^{a_0} = 1$$

$$S_{R_{21}}^{b_0} = \frac{R_{21} R_4}{b_0} = \frac{1}{2} \qquad S_{R_{21}}^{b_1} = \frac{R_{21} L_3}{b_1} = \frac{1}{4} \tag{26}$$

$$S_{C_{11}}^{b_2} = \frac{R_{11} R_{21} C_{11} C_2 R_4}{b_1} = \frac{1}{2} \qquad S_{C_{11}}^{b_3} = 1$$

Using these results in (23) we obtain

$$S_{R_{11}}^{N(s)} = -\frac{1}{2} \frac{1 + 3s + 4s^2 + 2s^3}{1 + 2s + 2s^2 + s^3} \tag{27a}$$

$$S_{R_{21}}^{N(s)} = \frac{1}{2} \frac{1 + 3s + 4s^2 + s^3}{1 + 2s + 2s^2 + s^3} \tag{27b}$$

$$S_{C_{11}}^{N(s)} = -\frac{1}{2} \frac{s + 2s^2 + s^3}{1 + 2s + 2s^2 + s^3} \tag{27c}$$

Ohms, henrys, farads

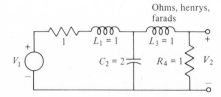

Figure 6.6-6 The third-order low-pass filter of Example 6.6-3.

Substituting $s = j\omega$ and solving for the real part we find

$$S_{R_{11}}^{|N(j\omega)|} = -\frac{1}{2}\frac{1 + \omega^4 + 2\omega^6}{1 + \omega^6} \tag{28a}$$

$$S_{R_{21}}^{|N(j\omega)|} = \frac{1}{2}\frac{1 + 2\omega^2 - \omega^4 + \omega^6}{1 + \omega^6} \tag{28b}$$

$$S_{C_{11}}^{|N(j\omega)|} = -\frac{1}{2}\frac{4\omega^2 - 3\omega^4 + \omega^6}{1 + \omega^6} \tag{28c}$$

□

The sensitivity analysis required when the leapfrog technique is applied to bandpass functions is similar to that described above. In such a case the immittances will have the form shown in Fig. 6.3-11. If the order of the bandpass realization is 4 or greater, the calculations become quite lengthy. In this case experimental or computer-generated sensitivity information may well be more useful to the user.

Another realization method presented in this chapter was the PRB one. A sensitivity analysis of such a realization is readily made by determining the $S_x^{\omega_n}$ and S_x^Q sensitivities of each of the second-order blocks, and relating these to the overall network function. From (1) of Sec. 6.4, if all the second-order blocks have identical transfer functions $T_i(s)$, the overall network function is

$$N(s) = \frac{V_2(s)}{V_1(s)} = \frac{A(s)}{B(s)} = \frac{-a_0 T_i(s)^n}{1 + \sum\limits_{j=1}^{n} a_j T_i(s)^j} \tag{29}$$

Using Table 3.1-1 this may be written as

$$S_x^{N(s)} = S_x^{A(s)} - S_x^{B(s)} = S_{T_i(s)}^{A(s)} S_x^{T_i(s)} - S_{T_i(s)}^{B(s)} S_x^{T_i(s)} \tag{30}$$

where

$$S_{T_i(s)}^{A(s)} = n \qquad S_{T_i(s)}^{B(s)} = \frac{\sum\limits_{j=1}^{n} j a_j T_i(s)^j}{1 + \sum\limits_{j=1}^{n} a_j T_i(s)^j} \tag{31}$$

Thus (30) becomes

$$S_x^{N(s)} = S_x^{T_i(s)} \left[\frac{n + \sum\limits_{j=1}^{n} (n - j) a_j T_i(s)^j}{1 + \sum\limits_{j=1}^{n} a_j T_i(s)^j} \right] = S_x^{T_i(s)}[M(s)] \tag{32}$$

The sensitivity of $|N(j\omega)|$ with respect to x is

$$S_x^{|N(j\omega)|} = \text{Re}\left[S_x^{T_i(j\omega)} M(j\omega) \right] \tag{33}$$

Since both $S_x^{T_i(j\omega)}$ and $M(j\omega)$ are complex quantities, (33) may be written as

$$S_x^{|N(j\omega)|} = \text{Re } [S_x^{T_i(j\omega)}] \text{ Re } [M(j\omega)] - \text{Im } [S_x^{T_i(j\omega)}] \text{ Im } [M(j\omega)] \qquad (34)$$

Assuming both ω_n and Q_p are functions of x, then

$$S_x^{|T_i(j\omega)|} = S_{1/Q_p}^{|T_i(j\omega)|} S_x^{1/Q_p} + S_{\omega_n}^{|T_i(j\omega)|} S_x^{\omega_n} \qquad (35)$$

where

$$T_i(s) = \frac{H_0(\omega_n/Q_p)s}{s^2 + \dfrac{\omega_n}{Q_p}s + \omega_n^2} \qquad (36)$$

After considerable algebra we find

$$S_{1/Q_p}^{|T_i(j\omega)|} = \frac{b_i^2}{1 + b_i^2} \qquad (37)$$

and

$$S_{\omega_n}^{|T_i(j\omega)|} = -\frac{b_i Q_p}{1 + b_i^2}\left(\frac{\omega}{\omega_n} + \frac{\omega_n}{\omega}\right) \qquad (38)$$

where

$$b_i = Q_p\left(\frac{\omega_n}{\omega} - \frac{\omega}{\omega_n}\right) \qquad (39)$$

Since $\text{Re } [S_x^{T_i(j\omega)}] = S_x^{|T_i(j\omega)|}$, from the above equations we obtain

$$\text{Re } [S_x^{T_i(j\omega)}] = \frac{b_i^2}{1 + b_i^2} S_x^{1/Q_p} - \frac{b_i Q_p}{1 + b_i^2}\left(\frac{\omega}{\omega_n} + \frac{\omega_n}{\omega}\right) S_x^{\omega_n} \qquad (40)$$

Similarly it may be shown that

$$\text{Im } [S_x^{T_i(j\omega)}] = \text{arg } [T_i(j\omega)] S_x^{\text{arg } [T_i(j\omega)]}$$

$$= \text{arg } [T_i(j\omega)]\{S_{1/Q_p}^{\text{arg } [T_i(j\omega)]} S_x^{1/Q_p} + S_{\omega_n}^{\text{arg } [T_i(j\omega)]} S_x^{\omega_n}\} \qquad (41)$$

Evaluating $S_{1/Q_p}^{\text{arg } [T_i(j\omega)]}$ and $S_{\omega_n}^{\text{arg } [T_i(j\omega)]}$ gives

$$S_{1/Q_p}^{\text{arg } [T_i(j\omega)]} = \frac{-b_i}{\text{arg } [T_i(j\omega)](1 + b_i^2)} \qquad (42)$$

and

$$S_{\omega_n}^{\text{arg } [T_i(j\omega)]} = \frac{Q_p\left(\dfrac{\omega}{\omega_n} + \dfrac{\omega_n}{\omega}\right)}{(1 + b_i^2) \text{ arg } [T_i(j\omega)]} \qquad (43)$$

Substituting (42) and (43) into (41) gives

$$\text{Im } [S_x^{T_i(j\omega)}] = -\frac{b_i^2}{1 + b_i^2} S_x^{1/Q_p} + \frac{Q_p\left(\dfrac{\omega}{\omega_n} + \dfrac{\omega_n}{\omega}\right)}{1 + b_i^2} S_x^{\omega_n} \qquad (44)$$

The desired expression is obtained by substituting (40) and (44) into (34) to get

$$S_x^{|N(j\omega)|} = \{\text{Re } [M(j\omega)] + \text{Im } [M(j\omega)]\}\left(\frac{b_i^2}{1 + b_i^2}\right)S_x^{1/Q_p}$$

$$- Q_p\left(\frac{\omega_n}{\omega} + \frac{\omega}{\omega_n}\right)\left|\left(\frac{b_i}{1 + b_i^2}\right)\text{Re } [M(j\omega)] + \left(\frac{1}{1 + b_i^2}\right)\text{Im } [M(j\omega)]\right|S_x^{\omega_n}$$

$$(45)$$

Although the general evaluation of (45) is complex, it can be shown that $S_x^{|N(j\omega)|}$ has multiple zeros of sensitivity in the passband which causes the PRB to exhibit sensitivity characteristics similar to the leapfrog techniques and other simulated passive network approaches.

Example 6.6-4 *Function Sensitivity of a PRB Filter Realization* It is desired to find the function $M(s)$ defined in (32) for a second-order maximally flat magnitude bandpass filter with $Q = 5\sqrt{2}$ and a center frequency of 1 rad/s. The low-pass transfer function has $b_0 = 1$ and $b_1 = \sqrt{2}$. The gain of the realization will be $H_0 = \sqrt{2}$ and the second-order blocks will have $Q_p = 10$ and $\omega_n = 1$. Following the development in Sec. 6.4 we find $a_0 = 1$, $a_1 = 0$, and $a_2 = \frac{1}{2}$. The resulting PRB structure will have only one feedback path. It is shown in Fig. 6.6-7. Evaluation of $M(s)$ gives

$$M(s) = \frac{2 + a_1 T_i(s)}{1 + a_1 T_i(s) + a_2 T_i(s)^2} = \frac{2(s^2 + s/10 + 1)^2}{(s^2 + s/10 + 1)^2 + (s/10)^2} \qquad (46)$$

It should be noted that $M(s)$ has zeros at the poles of $T_i(s)$. Since the passband of the overall filter represents an operating frequency on the $j\omega$ axis which is close to the complex-frequency plane zero locations, the passband sensitivity will be minimized. The actual situation is complicated somewhat by the b_i terms. □

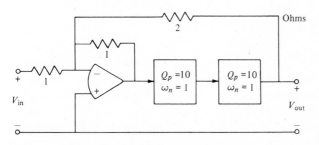

Figure 6.6-7 The PRB filter of Example 6.6-4.

The last synthesis method presented in this chapter is the parallel-cascade one. In this method the design procedure is directly based on the coefficients of $N(s)$. Thus the sensitivity analysis is relatively straightforward. Equation (16) of Sec. 6.5 gives the network elements in terms of the coefficients. Solving for the coefficients in terms of the network elements we obtain

$$a_{2i-2} = G_{1i} \qquad b_{2i-2} = G_{2i} + G_{1i}$$

$$a_{2i-1} = C_{1i} \qquad b_{2i-1} = C_{2i} + C_{1i} \tag{47}$$

Thus

$$S_{G_{1i}}^{a_{2i-2}} = S_{C_{1i}}^{a_{2i-1}} = 1$$

$$S_{G_{2i}}^{b_{2i-2}} = \frac{G_{2i}}{b_{2i-2}} \qquad S_{G_{1i}}^{b_{2i-2}} = \frac{G_{1i}}{b_{2i-2}} \tag{48}$$

$$S_{C_{2i}}^{b_{2i-1}} = \frac{C_{2i}}{b_{2i-1}} \qquad S_{C_{1i}}^{b_{2i-2}} = \frac{C_{1i}}{b_{2i-1}}$$

These coefficient sensitivities can be substituted into the expression

$$S_x^{N(s)} = \frac{\sum\limits_{i=0}^{m} S_x^{a_i} a_i s^i}{A(s)} - \frac{\sum\limits_{i=0}^{n} S_x^{b_i} b_i s^i}{B(s)} \tag{49}$$

to find the sensitivity of $N(s)$ with respect to any network parameter x.

Since each coefficient of a network function realized by the parallel-cascade method depends only on one or two of the passive elements, the sensitivity of this realization is not quite as good as the sensitivities of the previous methods discussed in this section. One advantage of the parallel-cascade approach, however, is that the active network elements, i.e., the GICs, do not have any influence on the realization. This, of course, only holds true while these elements behave ideally.

6.7 ACTIVE NETWORK ELEMENTS

In this section we consider how the nonideal properties of operational amplifiers affect the overall characteristics of the active network elements introduced in Secs. 6.1 and 6.2. Among these were the GIN (*generalized immittance network*), synthetic inductor, FDNR (*frequency-dependent negative resistor*), and GIC (*generalized immittance converter*). First, let us consider the GIN, originally introduced in Sec. 6.1. This consists of five admittances $Y_i(s)$ and two operational amplifiers connected as shown in Fig. 6.7-1. A circuit suitable for calculation of the input admittance $Y_{in}(s) = I_1(s)/V_1(s)$ when the operational amplifier gains are

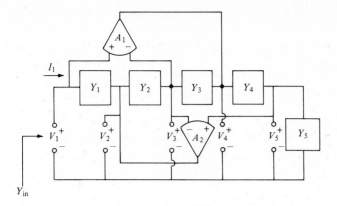

Figure 6.7-1 A GIN (*generalized immittance network*).

not infinite is given in Fig. 6.7-2. For this we obtain

$$Y_{in}(s) = \frac{I_1(s)}{V_1(s)}$$

$$= Y_1 \left[\frac{A_{d1}A_{d2}Y_3Y_5 + A_{d1}Y_3(Y_4+Y_5) + A_{d2}Y_2(Y_4+Y_5) + (Y_2+Y_3)(Y_4+Y_5)}{A_{d1}A_{d2}Y_2Y_4 + A_{d1}Y_3(Y_4+Y_5) + A_{d2}Y_2(Y_4+Y_5) + (Y_2+Y_3)(Y_4+Y_5)} \right]$$

$$(1)$$

where for convenience we have deleted the (s) notation in the terms of the right member. It is interesting to note that the numerator and denominator are identical except for the term which contains $A_{d1}A_{d2}$. As this product approaches infinity, it is easily seen that $Y_{in}(s)$ approaches the value given in (6) of Sec. 6.1, namely, $Y_1(s)Y_3(s)Y_5(s)/Y_2(s)Y_4(s)$. The GIN may be used to realize either a synthetic inductor or an FDNR. The synthetic inductor will be considered first. If we select $Y_2(s) = sC_2$ and the other admittances $Y_i = G_i$ $(i = 1, 3, 4, 5)$, then (case 1)

$$Z_{in}(s) = \frac{1}{Y_{in}(s)} = \frac{C_2 G_4}{G_1 G_3 G_5} s = sL_{eq} \qquad (2)$$

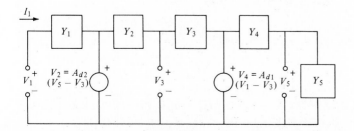

Figure 6.7-2 A GIN with noninfinite operational amplifier gains.

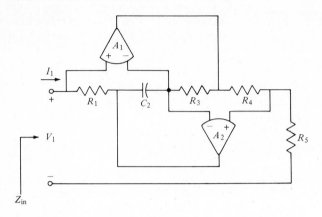

Figure 6.7-3 A GIN realization of a synthetic inductor.

Alternately, we could select $Y_4 = sC_4$ and $Y_i = G_i$ $(i = 1, 2, 3, 5)$, for which (case 2)

$$Z_{in}(s) = \frac{1}{Y_{in}(s)} = \frac{G_2 C_4}{G_1 G_3 G_5} s = sL_{eq} \tag{3}$$

Choosing case 1, we obtain the circuit of Fig. 6.7-3. Assuming that $A_{d1}(s)$ and $A_{d2}(s)$ are represented by a one-pole model [see (17) of App. B], (1) can be written as

$$Y_{in}(s) = G_1 \left[\frac{\begin{array}{c} GB_1 GB_2 G_3 G_5 + GB_1(s + \omega_{a2})G_3(G_4 + G_5) + GB_2(s + \omega_{a1})sC_2(G_4 + G_5) \\ + (s + \omega_{a1})(s + \omega_{a2})(sC_2 + G_3)(G_4 + G_5) \end{array}}{\begin{array}{c} sGB_1 GB_2 C_2 G_4 + GB_1(s + \omega_{a2})G_3(G_4 + G_5) + GB_2(s + \omega_{a1})sC_2(G_4 + G_5) \\ + (s + \omega_{a1})(s + \omega_{a2})(sC_2 + G_3)(G_4 + G_5) \end{array}} \right] \tag{4}$$

Neglecting terms which do not contain GB_1 or GB_2 gives

$$Y_{in}(s) \approx G_1$$

$$\left\{ \frac{[GB_2 C_2(G_4 + G_5)]s^2 + [GB_1 G_3(G_4 + G_5) + \omega_{a1}GB_2 C_2(G_4 + G_5)]s + (GB_1 GB_2 G_3 G_5)}{[GB_2 C_2(G_4 + G_5)]s^2 + [GB_1 GB_2 C_2 G_4]s + [\omega_{a2}GB_1 G_3(G_4 + G_5)]} \right\} \tag{5}$$

Rearranging (5) results in

$$Y_{in}(s) \simeq \frac{G_1 G_3 G_5}{sC_2 G_4}$$

$$\times \frac{[C_2(G_4 + G_5)/GB_1 G_3 G_5]s^2 + [(G_4 + G_5)/GB_2 G_5 + \omega_{a1}C_2(G_4 + G_5)/GB_1 G_3 G_5]s + 1}{[(G_4 + G_5)/GB_1 G_4]s + 1 + [\omega_{a2}G_3(G_4 + G_5)/GB_2 C_2 G_4]/s} \tag{6}$$

Letting $s = j\omega$, assuming that $[C_2(G_4 + G_5)/GB_1 G_3 G_5]\omega^2 \ll 1$, and inverting the result we obtain

$$Z_{in}(j\omega) \approx \frac{j\omega C_2 G_4}{G_1 G_3 G_5}$$

$$\times \frac{1 - j/\omega[\omega_{a2} G_3(G_4 + G_5)/GB_2 C_2 G_4] + j\omega[(G_4 + G_5)/GB_1 G_4]}{1 + j\omega[(G_4 + G_5)/GB_2 G_5 + \omega_{a1} C_2(G_4 + G_5)/GB_1 G_3 G_5]} \quad (7)$$

To analyze this expression we first define a function $Q(\omega)$ by the relation

$$Q(\omega) \triangleq \frac{\text{Im } [Z(j\omega)]}{\text{Re } [Z(j\omega)]} \quad (8)$$

where $Z(j\omega)$ is the driving-point impedance of a nonideal inductor modeled as a series connection of a resistor and an ideal conductor. The phase of $Z(j\omega)$ may be written as

$$\arg [Z(j\omega)] = \tan^{-1} \frac{\text{Im } [Z(j\omega)]}{\text{Re } [Z(j\omega)]} \quad (9)$$

If Q is large, then $\text{Im } [Z(j\omega)] \gg \text{Re } [Z(j\omega)]$, so that the $\tan^{-1}$ function of (9) may be simplified to

$$\arg [Z(j\omega)] \simeq \frac{\pi}{2} - \frac{\text{Re } [Z(j\omega)]}{\text{Im } [Z(j\omega)]} \quad (10)$$

Using (8) in this we obtain

$$\arg [Z(j\omega)] \approx \frac{\pi}{2} - \frac{1}{Q(\omega)} \quad (11)$$

Solving for $Q(\omega)$ and applying the result to the synthetic inductor case by using (7) to define $\arg [Z(j\omega)]$ we find

$$Q(\omega) \simeq \frac{1}{\omega\left[\dfrac{G_4 + G_5}{GB_2 G_5} + \dfrac{\omega_{a1} C_2(G_4 + G_5)}{GB_1 G_3 G_5} - \dfrac{G_4 + G_5}{GB_1 G_4}\right] + \dfrac{1}{\omega} \dfrac{\omega_{a2} G_3(G_4 + G_5)}{GB_2 C_2 G_4}} \quad (12)$$

where $Q(\omega) > 1$. This may be rewritten as

$$Q(\omega) \approx \frac{1}{\omega(G_4 + G_5)\left(\dfrac{1}{GB_2 G_5} - \dfrac{1}{GB_1 G_4}\right) + \omega\dfrac{C_2(G_4 + G_5)}{A_{01} G_3 G_5} + \dfrac{1}{\omega} \dfrac{G_3(G_4 + G_5)}{A_{02} C_2 G_4}} \quad (13)$$

If $GB_2 G_5 = GB_1 G_4$, then $Q(\omega)$ becomes

$$Q_0(\omega) \approx \frac{1}{\omega[C_2(G_4 + G_5)/A_{01} G_3 G_5] + (1/\omega)[G_3(G_4 + G_5)/A_{02} C_2 G_4]} \quad (14)$$

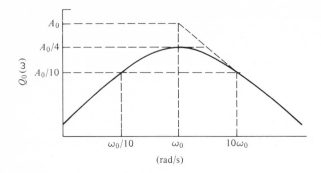

Figure 6.7-4 Behavior of Q in a synthetic inductor realized with matched operational amplifiers.

where $Q_0(\omega)$ is defined as $Q(\omega)$ for $Z_{in}(j\omega)$ of (7) with the restrictions that the operational amplifiers are matched and that $G_4 = G_5$. Thus we see that, on a first-order basis, the effects of the finite-gain bandwidth of the operational amplifiers can be cancelled in the synthetic inductor. Hence $Q_0(\omega)$ represents the limit of the Q of a synthetic inductor for matched amplifiers. If we assume that $A_{01} = A_{02} = A_0$, $G_1 = G_3 = G_4 = G_5 = G$, and $Y_2(s) = sC$, (14) simplifies to

$$Q_0(\omega) = \frac{1}{(2/A_0)(\omega/\omega_0 + \omega_0/\omega)} \tag{15}$$

where $\omega_0 = 1/RC$. Thus we see that $Q_0(\omega)$ has a maximum value of $A_0/4$ at $\omega = \omega_0$ which is the largest Q this circuit is capable of producing. For example, if $A_0 = 10^5$, then $Q_0(\omega_0) = 25,000$. This value only occurs at $\omega = \omega_0$, and it will decrease for any other frequency as shown by Fig. 6.7-4.

Now let us consider the behavior of $Q(\omega)$ in (13) for the case where the amplifiers are not matched, i.e., where their gain-bandwidth values are different. If we plot $Q(\omega_0)$, the value of $Q(\omega)$ at ω_0, versus various values of ω_0, we find that three possibilities exist, depending upon the relative values of GB_1 and GB_2. These are illustrated in Fig. 6.7-5. Assuming that $G_4 = G_5$, if $GB_1 = GB_2$, (13)

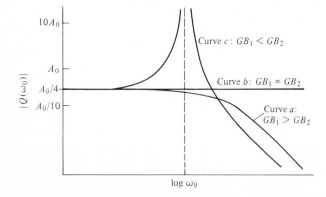

Figure 6.7-5 Behavior of Q in a synthetic inductor realized with unmatched operational amplifiers.

reduces to the form of (14) and the value of $Q(\omega_0)$ for each different ω_0 is simply $A_0/4$. This case corresponds to curve b in Fig. 6.7-5. On the other hand, if $GB_1 > GB_2$, then $1/GB_2 - 1/GB_1$ will always be positive, and a response similar to that shown as a curve a in Fig. 6.7-5 results. Finally, when $GB_1 < GB_2$, then $1/GB_2 - 1/GB_1$ is negative. This causes $Q(\omega_0)$ to be infinite at some value of ω_0. Thus, at even higher values of ω_0, the circuit of Fig. 6.7-3 will be unstable unless the resistance of the driving source is sufficiently large. Curve c in Fig. 6.7-5 is a typical plot of $|Q(\omega_0)|$ for this case. In the following example we show that for curve c it is possible for the maximum value of $Q(\omega)$ to exceed that of $Q(\omega_0)$.

Example 6.7-1 *The $Q(\omega)$ of a Synthetic Inductor*   It is desired to plot $|Q(\omega)|$ for a GIN realization of a synthetic inductor using amplifiers with $A_{01} = A_{02} = 10^5$, $GB_1 = 1$ MHz, $GB_2 = 1.1$ MHz, and $\omega_0 = 10^4$ rad/s. From (13) we get

$$Q(\omega) = \frac{A_0}{2(\omega/\omega_0 + \omega_0/\omega) - 0.01818\omega}$$

Solving for the frequency at which $Q(\omega) = \infty$ gives $\omega = 1054$ rad/s. Thus the plot of $|Q(\omega)|$ has the form shown in Fig. 6.7-6.   □

$Q(\omega)$ has been used to characterize the performance of the synthetic inductor. Ideally, the b curve of Fig. 6.7-5 is the most desirable one. Achieving it is not unrealistic since the gain bandwidths of monolithic operational amplifiers built on the same chip can be expected to match and track with temperature quite closely. Thus excellent synthetic characteristics are readily achieved in practice by letting $R_4 = R_5$ and using matched operational amplifiers. The second synthetic inductor case, in which $Y_4 = sC_4$, produces similar results.

Now let us consider the second GIN-realized active network element, the FDNR. It can be analyzed in a similar manner. If the elements of the GIN are

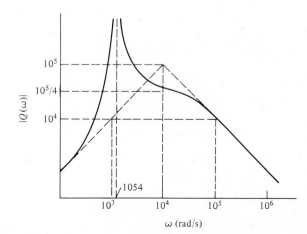

Figure 6.7-6 Plot of Q for the synthetic inductor realized in Example 6.7-1.

selected as $Y_1(s) = sC_1$, $Y_2(s) = G_2$, $Y_3(s) = sC_3$, $Y_4(s) = G_4$, and $Y_5(s) = G_5$, then (1) may be written as

$$Y_{in}(s) = sC_1 \left[\frac{\begin{matrix} GB_1 GB_2 sC_3 G_5 + GB_1(s + \omega_{a2})sC_3(G_4 + G_5) \\ + GB_2(s + \omega_{a1})G_2(G_4 + G_5) + (s + \omega_{a1})(s + \omega_{a2})(G_4 + G_5) \end{matrix}}{\begin{matrix} GB_1 GB_2 G_2 G_4 + GB_1(s + \omega_{a2})sC_3(G_4 + G_5) \\ + GB_2(s + \omega_{a1})G_2(G_4 + G_5) + (s + \omega_{a1})(s + \omega_{a2})(G_4 + G_5) \end{matrix}} \right]$$

(16)

Neglecting terms which do not contain GB_1 and GB_2 we obtain

$$Y_{in}(s) \approx sC_1 \frac{[GB_1 C_3(G_4 + G_5)]s^2 + (GB_1 GB_2 C_3 G_5)s + (G_4 + G_5)GB_2 G_2 \omega_{a1}}{[GB_1 C_3(G_4 + G_5)]s^2 + [(G_4 + G_5)(GB_1 \omega_{a2} C_3 + GB_2 G_2)]s + GB_1 GB_2 G_2 G_4}$$

(17)

Factoring out the ideal FDNR admittance and dividing the numerator and denominator of the above by the product $GB_1 GB_2$ give

$$Y_{in}(s) = \frac{s^2 C_1 C_3 G_5}{G_2 G_4} \frac{\dfrac{G_4 + G_5}{G_5 GB_2} s + 1 + \dfrac{(G_4 + G_5)G_2 \omega_{a1}}{C_3 G_5 GB_1} \dfrac{1}{s}}{\dfrac{C_3(G_4 + G_5)}{G_2 G_4 GB_2} s^2 + \dfrac{G_4 + G_5}{G_2 G_4}\left(\dfrac{\omega_{a2} C_3}{GB_2} + \dfrac{G_2}{GB_1}\right)s + 1}$$

(18)

Inverting, letting $s = j\omega$, and assuming that

$$\frac{\omega^2 C_3(G_4 + G_5)}{GB_2 G_2 G_4} < 1$$

(19)

we obtain the FDNR input impedance

$$Z_{in}(j\omega) \approx \frac{-G_2 G_4}{\omega^2 C_1 C_3 G_5} \frac{1 + j\omega\left[\dfrac{G_4 + G_5}{G_4 GB_1} + \dfrac{(G_4 + G_5)C_3 \omega_{a2}}{G_2 G_4 GB_2}\right]}{1 + j\omega\left[\dfrac{G_4 + G_5}{G_5 GB_2}\right] - j\dfrac{1}{\omega}\left[\dfrac{(G_4 + G_5)G_2 \omega_{a1}}{C_3 G_5 GB_1}\right]}$$

(20)

Since the FDNR ideally has a phase shift of π rad, Re $[Z_{in}(j\omega)]$ will be negative and much larger than Im $[Z_{in}(j\omega)]$. Thus the phase shift of (20) may be approximated as

$$\arg[Z_{in}(j\omega)] \approx \pi + \frac{\text{Im }[Z_{in}(j\omega)]}{\text{Re }[Z_{in}(j\omega)]}$$

(21)

A quality factor for the FDNR may now be defined as

$$Q_{\text{FDNR}}(\omega) = \frac{\text{Re }[Z_{in}(j\omega)]}{\text{Im }[Z_{in}(j\omega)]}$$

(22)

This factor is a measure of how close the FDNR approaches the ideal characteristics of a phase shift of 180° and a magnitude of $|G_2 G_4/\omega^2 C_1 C_3 G_5|$. From (21) and (22) we obtain

$$Q_{FDNR}(\omega) \approx \frac{1}{\arg\left[Z_{in}(j\omega)\right] - \pi} \tag{23}$$

From (20) we find

$$\arg\left[Z_{in}(j\omega)\right] = \pi + \tan^{-1}\left\{\omega\left[\frac{G_4 + G_5}{G_4 GB_1} + \frac{(G_4 + G_5)C_3\omega_{a2}}{G_2 G_4 GB_2}\right]\right\}$$

$$- \tan^{-1}\left\{\omega\left[\frac{G_4 + G_5}{G_5 GB_2}\right] - \frac{1}{\omega}\left[\frac{(G_4 + G_5)G_2\omega_{a1}}{GB_1 C_3 G_5}\right]\right\} \tag{24}$$

If $\omega < GB$, then using this result in (23) we obtain

$$Q_{FDNR}(\omega)$$

$$= \frac{1}{\omega(G_4 + G_5)\left[\dfrac{1}{G_4 GB_1} - \dfrac{1}{G_5 GB_2}\right] + (G_4 + G_5)\left[\dfrac{\omega C_3}{G_2 G_4 A_{02}} + \dfrac{G_2}{A_{01}\omega C_3 G_5}\right]} \tag{25}$$

Comparing (25) for the FDNR to (13) for the synthetic inductor, we note that, except for subscripts, the two Q factors are identical. Consequently, if $GB_1 G_4 = GB_2 G_5$, then the high-frequency response of $Q_{FDNR}(\omega)$ will be independent of the gain bandwidth of the operational amplifiers. For matched amplifiers, the low-frequency value of $Q_{FDNR}(\omega)$ is seen to have a maximum value of $A_0/4$ if $A_{01} = A_{02} = A$, $G_2 = G_4 = G_5 = G$, and $C_1 = C_3 = C$. This maximum value of $Q_{FDNR}(\omega)$ occurs at $\omega_0 = G/C$. Other choices for the capacitive reactances than $Z_1(s)$ and $Z_3(s)$ will give similar performance characteristics. Some examples of this may be found in the Problems. Summarizing the above we see that the independence from amplifier gain-bandwidth property of the GIN is carried over to its application in the realization of synthetic inductors and FDNRs.

The remaining active network element introduced in Sec. 6.1 was the GIC. It was used in the FDNR realization technique and also in the parallel-cascade method. One approach to the analysis of the GIC is to calculate the GIC constant $K_c(s)$ defined in (1) of Sec. 6.5. Figure 6.7-7 shows a suitable circuit for such a

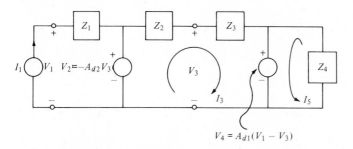

Figure 6.7-7 A GIC with noninfinite operational amplifier gains.

calculation. In this circuit, the output voltage $V_2(s)$ defined in Fig. 6.1-2a has been grounded. By mesh analysis we find that

$$K_c(s) = -\frac{I_1(s)}{I_5(s)} = \frac{Z_1 Z_3 A_{d1} A_{d2} + (Z_1 Z_2 + Z_1 Z_3)A_{d1}}{Z_2 Z_4 (A_{d1} A_{d2} + A_{d1} + 1) + Z_3 Z_4 (1 + A_{d2})} \qquad (26)$$

If, as in (1) of Sec. 6.5, we define

$$K_c(s) = (s^2 R_1 R_3 C_2 C_4)E(s) \qquad (27)$$

where $E(s)$ represents a multiplicative term which accounts for the nonideal behavior of the GIC, then using an approach similar to the one used for the synthetic inductor and the FDNR it may be shown that

$$\arg \left[E(j\omega) \right] \approx \pi - \left(\frac{\omega R_3 C_2}{A_{01}} + \frac{1}{A_{02}\omega R_3 C_2} \right) \qquad (28)$$

If $A_{01} = A_{02} = A_0$ and $\omega_0 = 1/R_3 C_2$, (28) becomes

$$\arg \left[E(j\omega) \right] \approx \pi - \frac{1}{A_0} \left(\frac{\omega}{\omega_0} + \frac{\omega_0}{\omega} \right) \qquad (29)$$

In deriving this result, advantage has been taken of the fact that a simplification occurs due to the subtraction of terms involving the gain bandwidths of the two operational amplifiers. In practice, of course, these results are modified by the specific characteristics of the load impedance $Z_5(s)$ in Fig. 6.1-2a. In general, however, we can expect the GIC to offer performance similar to that of the synthetic inductor or the FDNR. Thus, all these circuits are useful in the construction of high-performance active filter realizations. It is also possible to consider the large-signal analysis of these circuits. Such an analysis is tedious due to the complications created by the presence of both positive and negative feedback in the circuits. As such it is beyond the scope of our treatment here.

Since the effects of the amplifier gain bandwidth can be eliminated in most of the applications involving the GIC or GIN, it is important to examine what the mechanism is that actually limits the frequency capabilities of these networks. We shall illustrate this by using the synthetic inductor as an example. To begin, let us consider the GIN of Fig. 6.7-1 in which $Y_4(s) = sC_4$ and $Y_i(s) = 1/R_i$ ($i = 1, 2, 3, 5$), as shown in Fig. 6.7-8. The operational amplifier model to be used in this

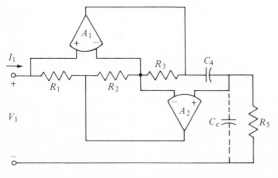

Figure 6.7-8 A GIN realization of a synthetic inductor.

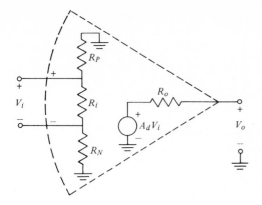

Figure 6.7-9 An operational amplifier model for use in Fig. 6.7-8.

analysis is shown in Fig. 6.7-9. It includes input, output, and common-mode impedances. Substituting this model into Fig. 6.7-8, and simplifying by assuming that $R_1 = R_2 = R_3 = R_5 = R$ and $C_4 = C$, we obtain the result shown in Fig. 6.7-10. This circuit has been analyzed[1,2] to show that Fig. 6.7-11 is an equivalent model of Fig. 6.7-10 when

$$G_{p1} = \frac{1}{R_{P1}} + \frac{1}{RA_0}\left(2 + \frac{R_o}{R}\right) + \frac{R}{R_i(2R + 3R_o + GBR^2C)} \tag{30a}$$

$$G_{p2} = -\left(\frac{1}{R_{N1}} + \frac{1}{R_{N2}} + \frac{2R_o}{R(2R + 3R_o + GBR^2C)}\right) \tag{30b}$$

$$G_{p3} \simeq \frac{2\omega^2}{GB^2R}\left[1 + \frac{4R_o}{R} + \frac{1}{2 + GBRC}\left(\frac{R}{R_i} - \frac{R_o}{R}\right)\right] \tag{30c}$$

$$C_p \simeq \frac{1}{GBR}\left(2 + \frac{R_o}{R}\right) \tag{30d}$$

$$r_{p4} = \frac{2R}{A_0} + \frac{3R_o}{A_0} + \frac{R^2}{R_C} \tag{30e}$$

$$r_{p5} = \frac{2\omega^2}{GB^2}(R + 6R_o) \tag{30f}$$

$$L_o = R^2C \tag{30g}$$

$$L_p = \frac{2R}{GB} + \frac{3R_o}{GB} \tag{30h}$$

[1] A. Antoniou and K. S. Naidu, "Modeling of a Gyrator Circuit," *IEEE Trans. Circuit Theory*, vol. CT-20, no. 5, September 1973, pp. 533–540.

[2] A. Antoniou and K. S. Naidu, "A Compensation Technique for a Gyrator and Its Use in the Design of a Channel Bank Filter," *IEEE Trans. Circuits and Systems*, vol. CAS-22, no. 4, April 1975, pp. 316–323.

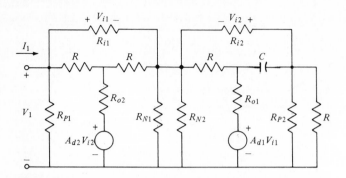

Figure 6.7-10 The circuit of Fig. 6.7-8 using the model of Fig. 6.7-9.

where $R_{i1} = R_{i2} = R_i$, $R_{o1} = R_{o2} = R_o$, $A_{01} = A_{02} = A_0$, $GB_1 = GB_2 = GB$, and R_C is the parallel resistance of the capacitor C. If the operational amplifier has ideal impedance levels and $R_C = \infty$, then

$$G_{p1} = \frac{2}{RA_0} \tag{31a}$$

$$G_{p2} = 0 \tag{31b}$$

$$G_{p3} \simeq \frac{2}{R}\left(\frac{\omega}{GB}\right)^2 \tag{31c}$$

$$C_p \simeq \frac{2}{RGB} \tag{31d}$$

$$r_{p4} = \frac{2R}{A_0} \tag{31e}$$

$$r_{p5} = 2R\left(\frac{\omega}{GB}\right)^2 \tag{31f}$$

$$L_o = R^2C \tag{31g}$$

$$L_p = \frac{2R}{GB} \tag{31h}$$

It is convenient to separate the model of Fig. 6.7-11 into a low-frequency and a high-frequency equivalent. At low frequencies we find that $r_{p4} \gg r_{p5}$, G_{p1} or

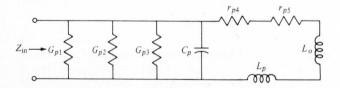

Figure 6.7-11 An equivalent model for the circuit of Fig. 6.7-10.

$|G_{p2}| \gg G_{p3}$, and $\omega C_p \ll 1/(\omega L_p + \omega L_o)$. Thus the input admittance can be written

$$Y_{in}^{(lf)}(j\omega) = G_{p1} + G_{p2} + \frac{1}{r_{p4} + j\omega(L_o + L_p)} = G_L + \frac{1}{j\omega L_L} \tag{32}$$

where

$$G_L = G_{p1} + G_{p2} + \frac{r_{p4}}{r_{p4}^2 + \omega^2(L_o + L_p)^2} \tag{33}$$

and

$$L_L = \frac{r_{p4}^2 + \omega^2(L_o + L_p)^2}{\omega^2(L_o + L_p)} \simeq L_o + L_p \tag{34}$$

and where it has been assumed that $r_{p4} \ll \omega(L_o + L_p)$. The low-frequency Q of the synthetic inductance is thus given as

$$Q_L = \frac{-\text{Im}\,[Y_{in}^{(lf)}(j\omega)]}{\text{Re}\,[Y_{in}^{(lf)}(j\omega)]} = \frac{1}{\omega L_L G_L} = \frac{\omega(L_o + L_p)}{(G_{p1} + G_{p2})[r_{p4}^2 + \omega^2(L_o + L_p)^2] + r_{p4}} \tag{35}$$

Since $G_{p1} > 0$ and $G_{p2} < 0$, Q_L may be positive or negative depending upon the relative magnitudes of the parameters of the operational amplifier parameters. In either case, the magnitude of Q_L is large since G_{p1}, G_{p2}, and r_{p4} are very small quantities. Q_L can be made positive by increasing G_{p1}, which is equivalent to putting a shunt resistor across the input of the synthetic inductor. At high frequencies where $r_{p5} \gg r_{p4}$ and $G_{p3} \gg G_{p1}$ or $|G_{p2}|$, then the input admittance of the synthetic inductor can be written as

$$Y_{in}^{(hf)}(j\omega) = G_H + \frac{1}{j\omega L_H} = G_{p3} + j\omega C_p + \frac{1}{r_{p5} + j\omega(L_o + L_p)} \tag{36}$$

where

$$G_H = G_{p3} + \frac{r_{p5}}{r_{p5}^2 + \omega^2(L_o + L_p)^2} \tag{37}$$

and

$$L_H = \frac{(L_o + L_p)^2 + r_{p5}^2/\omega^2}{(L_o + L_p) - C_p[r_{p5}^2 + \omega^2(L_o + L_p)^2]} \tag{38}$$

The high-frequency Q of the synthetic inductance is given as

$$Q_H = \frac{-\text{Im}\,[Y_{in}^{(hf)}(j\omega)]}{\text{Re}\,[Y_{in}^{(hf)}(j\omega)]} \simeq \frac{\omega(L_o + L_p)}{G_{p3}[r_{p5}^2 + \omega^2(L_o + L_p)^2] + r_{p5}} \tag{39}$$

where it has been assumed that $r_{p5} < \omega(L_o + L_p)$ and $C_p\omega^2(L_o + L_p)^2 < 1$.

The expressions for Q_L and Q_H given above are unfortunately too complex to permit generalization. They do, however, show that the synthetic-inductor characteristics are a function of amplifier impedance levels as well as of GB and A_0. An illustration of these effects is shown in Fig. 6.7-12. In this figure the inductance deviation in percent from L_o is given as a function of frequency for various values of L_o with R constant and for various values of R with L_o constant. This data is taken for a 741 operational amplifier whose parameters are $A_0 = 2 \times 10^5$, GB = 1.2 MHz, $R_o = 75\,\Omega$, $R_i = 2 \times 10^6\,\Omega$, and $R_P = R_N = 500\,\text{M}\Omega$. From the

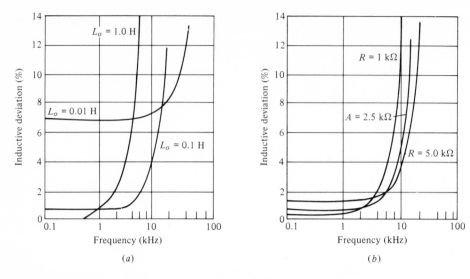

Figure 6.7-12 Effect of amplifier parasitics on synthetic-inductor characteristics.

figure it is apparent that serious inductance deviations occur for frequencies above 10 kHz. A method of compensating for the effects shown in Fig. 6.7-12 can be achieved by adding the capacitor C_c shown as a dotted element in Fig. 6.7-8. The inductance deviation for various values of C_c is shown in Fig. 6.7-13. In this plot the previously cited 741 operational amplifier parameter values have been used, together with $L_o = 0.088$ H and $R = 2$ kΩ. Thus, by this technique the useful frequency range of the active element can be extended to at least 20 kHz.

It should be noted that the results obtained in this section regarding the characteristics of GIN and GIC realizations must be modified somewhat when nonideal impedance levels of the operational amplifier are considered. This is

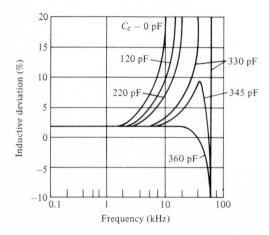

Figure 6.7-13 Use of compensation to improve synthetic-inductor characteristics.

especially true of the GB cancelling property. The reason the impedance levels affect GIN and GIC realizations is that in these configurations neither of the input terminals of the operational amplifiers is at signal ground. Note that if one of the terminals were at signal ground, then the impedance level at that point would be that of a virtual ground and thus the nonideal input and common mode impedances would have little or no effect on the realization. Obviously active filter realizations, such as the leapfrog structure in which the operational amplifiers have one terminal grounded, have an advantage in their ability to achieve high-frequency performance.

6.8 FREQUENCY-DEPENDENT SENSITIVITY

In Secs. 4.7 and 5.6 we introduced the concept of a frequency-dependent sensitivity. This sensitivity was used to relate the pole position of an RC-amplifier filter to the gain-bandwidth product of the operational amplifiers used in the realization. In this section we extend the concept to filters realized by the passive network simulation methods discussed in this chapter. In such filters, since the functions realized are usually of fairly high order, it is more appropriate to consider the sensitivity of the network function rather than that of the pole positions. Such a conclusion is in agreement with a similar one reached concerning the sensitivities discussed in Sec. 6.6.

To begin our treatment of frequency-dependent sensitivity for passive network simulation filters, we consider first a prototype network in which a series resistor of resistance R_i is present for each inductor L_i, and a shunt resistor of conductance G_j is present for each capacitor C_j. In addition we shall require that for all values of i and j

$$\frac{L_i}{R_i} = Q_L \qquad \frac{C_i}{G_i} = Q_C \qquad (1)$$

Such a network is said to have *semiuniform dissipation*.[1] The sensitivity performance of such a network when it has the form of a resistively terminated ladder has been investigated by Blostein.[2] His investigation was made in terms of the attenuation function $\alpha(\omega)$ defined in (5) of Sec. 3.2. He showed that the change in attenuation $\Delta\alpha(\omega)$ could be expressed as

$$\Delta\alpha(\omega) = 4.34 \left[\left(\frac{1}{Q_L} + \frac{1}{Q_C} \right) \omega\tau(\omega) + \frac{1}{2} \left(\frac{1}{Q_L} - \frac{1}{Q_C} \right) \text{Im} \left(\rho_1 + \rho_2 \right) \right] \qquad \text{dB} \qquad (2)$$

where $\tau(\omega)$ is the delay function and ρ_1 and ρ_2 are terms based on the power reflected at the network terminations. The multiplicative constant (4.34) is introduced to change the units of $\alpha(\omega)$ from nepers to decibels. In most filters, the

---

[1] If $Q_L = Q_C$ the network is said to have *uniform dissipation*.

[2] M. L. Blostein, "Sensitivity Analysis of Parasitic Effects in Resistance-Terminated LC Two-Ports," *IEEE Trans. Circuit Theory*, vol. CT-14, March 1967, pp. 21–25.

first term on the right side of (2) dominates for values of ω in the passband and in the transition band between the passband and the stopband. It is in these regions that $\tau(\omega)$ is usually large. Thus for these frequencies, (2) can be simplified as

$$\Delta\alpha \approx 4.34\left(\frac{1}{Q_L} + \frac{1}{Q_C}\right)\omega\tau(\omega) \tag{3}$$

This equation, although derived for a restricted class of filters, is of considerable importance in evaluating passive network simulation methods. The only difficulty in applying it is in obtaining an expression for the delay function $\tau(\omega)$, as this must be computed using the definition given in (30) of Sec. 2.5. It can also be obtained from the literature for standard filter functions.[1] For an approximate analysis $\tau(\omega)$ can be equated to a constant over the passband range. For example, for an nth-order frequency-normalized (1 rad/s bandwidth) low-pass Butterworth filter, a useful approximation is $\tau(\omega) \approx n$ sec.

Now let us apply (3) to various passive network simulation methods. First we consider the ones which use the GIC or the GIN. Here we may apply the results of (39) of Sec. 6.7 in which an expression for the high-frequency Q factor Q_H for a simulated inductance is given. Let typical parameters for a 741 operational amplifier be $A_0 = 2 \times 10^5$, $GB = 1.2$ MHz, $R_o = 75 \Omega$, $R_i = 2$ MΩ, and $R_P = R_N = 500$ MΩ.[2] In the GIN, assuming the resistance in parallel with the capacitor C is $R_C = 100$ MΩ, and also that $R = 2$ kΩ and $C = 22$ nF, then we find $Q_H \approx 2292$ at 10 kHz. At 20 kHz, $Q_H \approx 313$. Thus, the Q of the simulated inductance is seen to decrease rapidly as frequency increases, thereby limiting the frequency performance of the realization. The influence of the Q of the synthetic inductance upon the transfer function can be examined through the use of (3). As an example of such use, consider Fig. 6.8-1, in which the attenuation and delay functions of a low-pass fifth-order 1-dB-ripple Chebyshev function are shown.

[1] A. I. Zverev, *Handbook of Filter Synthesis*, John Wiley & Sons, Inc., New York, 1967.

[2] R_P and R_N are the common-mode input resistances at the positive and negative operational amplifier inputs as defined in Fig. 6.7-9.

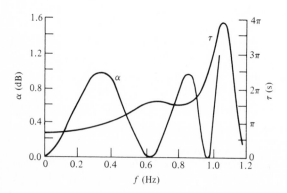

Figure 6.8-1 Attenuation (α) and delay (τ) functions of a low-pass fifth-order 1-dB-ripple Chebyshev function.

Assume that a realization of this filter function is to be used as a prototype circuit for the realization (after a low-pass to high-pass transformation) of a high-pass filter with a cutoff frequency of 10 kHz. At the cutoff frequency the value of $\omega\tau(\omega)$ is approximately 3.3π. If the capacitors in the realization are polystyrene, then the data of Table 3.8-2 give $Q_C = 5000$. Using Q_H as Q_L in (3) gives $\Delta\alpha = 0.029$ dB. This implies that the deviation from the attenuation $\alpha(\omega)$ will be 0.032 dB at 10 kHz. If this realization is used to design a high-pass filter with a cutoff frequency of 20 kHz then Q_L becomes 313, and the deviation in attenuation calculated from (3) is $\Delta\alpha = 0.153$ dB.

Considerations similar to those given above apply to passive network simulation methods which use FDNRs or GICs. For FDNRs realized by using a GIN, the operational amplifier model given in Fig. 6.7-9 is useful. When such a model is inserted in a circuit, however, the calculation of Q becomes quite complex, frequently requiring the use of a computer-aided circuit-analysis program.[1] An operational-amplifier model suitable for use in such programs is given in Fig. 6.8-2. The situation is comparable when GICs used as one-port networks are required. Other GIC applications, such as in the parallel-cascade method, cannot be treated in this manner. In such cases the frequency response of the realization is usually limited by the nonideal input impedance levels of the operational amplifiers. This is the result of the fact that in such circuits the amplifier input terminals are ungrounded. To date, however, no comprehensive analysis of the frequency performance of this technique has appeared in the literature.

The last passive network simulation filter to be considered here is the leapfrog technique. This can also be analyzed using (3). To see how this is done, recall that the method requires that driving-point immittance functions be realized as voltage transfer functions of active networks. If a typical voltage transfer function is designated as $A(s)$ and the driving-point function (assumed to be an

---

[1] L. W. Nagle and D. O. Pederson, "Simulation Program with Integrated Circuit Emphasis (SPICE)," *Electronics Research Lab. Rep.* ERL-M382, University of California, Berkeley, April 1973.

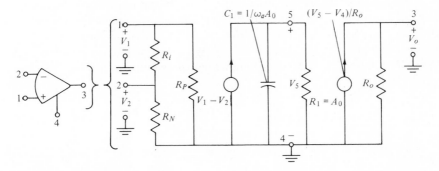

Figure 6.8-2 An operational amplifier model suitable for use in computer-aided circuit-analysis programs.

impedance) is specified as $Z(s)$, then the method requires

$$A(s) = \pm \frac{1}{Z(s)} \tag{4}$$

$A(s)$ is inverting if (4) has a minus sign and noninverting if it has a plus sign. The phase of $Z(j\omega)$ may be expressed as

$$\arg [Z(j\omega)] = \tan^{-1} \frac{|\text{Im } [Z(j\omega)]|}{|\text{Re } [Z(j\omega)]|} = \tan^{-1} Q \tag{5}$$

where Q is the Q factor of the impedance. However, since in practice Q is usually very large, (5) can be written as

$$\arg [Z(j\omega)] \approx \frac{\pi}{2} - \frac{1}{Q} \tag{6}$$

From (4), assuming the plus sign, the phase of $A(j\omega)$ is

$$\arg [A(j\omega)] = -\arg [Z(j\omega)] \approx -\frac{\pi}{2} + \frac{1}{Q} \tag{7}$$

and for the minus sign

$$\arg [A(j\omega)] = \pi - \arg [Z(j\omega)] \approx \frac{\pi}{2} + \frac{1}{Q} \tag{8}$$

Similar relations are obtained if the immittance is an admittance. The choice of the plus or minus sign or whether an impedance or an admittance is used depends on the details of the method (see Sec. 6.3). For example, for low-pass realizations, the internal ladder driving-point immittances are all of the form $1/\Gamma s$, that of an inductor or capacitor. Thus the voltage transfer functions will all be realized by noninverting or inverting integrators.

To illustrate the method consider Fig. 6.3-8a, which is a leapfrog realization of the doubly terminated fifth-order low-pass filter of Fig. 6.3-5a. In the leapfrog circuit it is seen that the inductors are simulated by noninverting integrators consisting of a cascade connection of an inverter and an inverting integrator. Such a cascade is similar to the one shown in Fig. 5.4-9 and analyzed in Sec. 5.4. In like manner the capacitors are simulated by inverting integrators of the type shown in Fig. 5.4-1. The phase shift of the inverting integrator is found from (7) of Sec. 5.4 as approximately

$$\arg [A(j\omega)] \approx \frac{\pi}{2} + \frac{\omega_{RC}}{\omega A_0} - \frac{\omega}{\text{GB}} \tag{9}$$

Combining this with (8) [for the minus sign of (4)] we get

$$Q_C = \frac{1}{\omega_{RC}/\omega A_0 - \omega/\text{GB}} \approx -\frac{\text{GB}}{\omega} \tag{10}$$

where Q_C corresponds to the Q of the simulated capacitor. The negative sign is used to indicate that there is a phase lag from the circuit. This Q is the Q factor of all the simulated capacitors in the realization. The phase shift of the noninverting integrator is given by (24) of Sec. 5.4 as

$$\arg\left[A(j\omega)\right] \approx -\frac{\pi}{2} - \frac{3\omega}{GB} \tag{11}$$

Using (7) and (11) we obtain the Q factor of the simulated inductors as

$$Q_L \approx -\frac{GB}{3\omega} \tag{12}$$

The Q factors obtained in (10) and (12) may be used in (3) to obtain the following result for the leapfrog filter

$$\Delta\alpha \approx 4.34\left(-\frac{4\omega}{GB}\right)\omega\tau(\omega) \tag{13}$$

As an example of the use of this relation, if the configuration of Fig. 6.3-8a is used to realize the filter function shown in Fig. 6.8-1, then the $\Delta\alpha$ at the cutoff frequency is

$$\Delta\alpha(\omega) \approx -180\frac{\omega}{GB} \quad \text{dB} \tag{14}$$

Thus if we assume $\omega/GB = 0.01$, then $\Delta\alpha = -1.8$ dB, which obviously is a rather significant deviation. The performance of a leapfrog realization such as that shown in Fig. 6.3-8a can be improved by careful design of the integrators. For example, if the circuit of Fig. 5.4-8 is used for the inverting integrator, then from (21) of Sec. 5.4 we obtain

$$Q_C \approx \frac{-1}{\omega/GB_1 - \omega/GB_2 + \omega\omega_{RC}/GB_1GB_2 - \omega_{RC}/A_{01}} \approx \frac{-GB_1GB_2}{\omega\omega_{RC}} \tag{15}$$

Similarly if the circuit of Fig. 5.4-10 is used for the noninverting integrator, from (28) of Sec. 5.4 we get

$$Q_L \approx \frac{GB}{\omega} \tag{16}$$

In both the above expressions, we see that the Qs have opposite signs, which will reduce $\Delta\alpha(\omega)$. Assuming that $GB_1 = GB_2 = GB$ and $\omega = \omega_{RC}$, then (15) simplifies to

$$Q_C \approx -\left(\frac{GB}{\omega}\right)^2 \tag{17}$$

Substituting (16) and (17) into (3) we obtain

$$\Delta\alpha(\omega) \approx 4.34\left[\frac{\omega}{GB} - \left(\frac{\omega}{GB}\right)^2\right]\omega\tau(\omega) \tag{18}$$

If ω is at the cutoff frequency, then (18) becomes

$$\Delta\alpha(\omega) \approx 45 \frac{\omega}{\text{GB}}\left(1 - \frac{\omega}{\text{GB}}\right) \tag{19}$$

Comparing this with the previous result by letting $\omega/\text{GB} = 0.01$ we find $\Delta\alpha = 0.446$ dB. This is nearly a factor of 4 decrease in the deviation of the attenuation function, obviously illustrating the direct influence that the choice of integrator circuit has on the characteristics of the resulting realization.

Similar considerations hold for the case when the leapfrog technique is used to realize a bandpass filter. In this case the quantities Q_L and Q_C are the Qs of the second-order blocks. Consequently, the analysis must concentrate on the Q of these networks. The resulting Qs may be substituted into (3) in order to determine the performance of the realization.

6.9 SUMMARY

In this and the preceding chapters we have presented an introduction to the major techniques of active RC filter synthesis. In Chaps. 4 and 5, it will be recalled, we showed how second-order RC-amplifier filter sections could be designed, and how a cascade of these sections could be used to realize high-order filters. In this chapter, however, quite different synthesis methods are used. Here the emphasis is on a direct realization of the entire filter (no matter what its order), rather than the individual realization of the second-order factors of the filter's network function. The methods presented in this chapter are divided into two groups. The first group operates by directly realizing a passive prototype network. In this group, the first two methods, namely the ones using synthetic inductors (Sec. 6.1) and FDNRs (Sec. 6.2), provide a direct simulation of the elements of the prototype network. The synthetic-inductance method is especially well suited to providing realizations of high-pass filters, while the FDNR technique is directly applicable to realizing low-pass and bandpass ones. In Sec. 6.3, another such method, the leapfrog one, is presented. It is quite different in concept. For low-pass realizations it uses integrators as building blocks in an active RC configuration which simulates the voltage and current variables of the prototype network. It may also be used to realize bandpass functions. In this case, the integrators are replaced by second-order RC-amplifier bandpass building blocks of the type discussed in Chaps. 4 and 5. The second group of methods presented in this chapter uses, as a starting point, the actual desired network function, rather than a passive prototype network. The first of these methods is the PRB technique (Sec. 6.4), which, like the leapfrog method, uses second-order stages as building blocks. The second method of this group is the parallel-cascade one (Sec. 6.5). It also operates directly from the network function but it employs GICs and passive RC networks.

Some general comments are appropriate on the two different ways used by the synthesis methods described above for specifying the desired filter characteristics, namely, the use of a prototype passive network and the use of a network function.

In general, more satisfactory results are obtained from the use of a prototype network since, using the appropriate denormalization techniques, the values of the elements of the resulting active realization are obtained without appreciable loss of accuracy. On the other hand, using the network function as a starting point, and using the values of its coefficients as the numerical specifications, is inherently a less accurate procedure. The reason for this, of course, is that each such coefficient represents an interaction of the numerical values of several network elements.

PROBLEMS

6-1 (*Sec. 6.1*) (*a*) Find the transmission parameters for the operational amplifier realization of the current-controlled voltage source shown in Fig. P6-1*a*. Assume that the operational amplifier is ideal, i.e., of infinite gain, zero output impedance, and infinite input impedance.
   (*b*) Repeat for the voltage-controlled current source realization shown in Fig. P6-1*b*.
   (*c*) Repeat for the current-controlled current source realization shown in Fig. P6-1*c*.

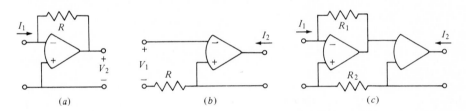

(*a*)   (*b*)   (*c*)

Figure P6-1

6-2 (*Sec. 6.1*) (*a*) Derive the transmission parameters given in (4) of Sec. 6.1.
   (*b*) Repeat for the parameters given in (5) of Sec. 6.1.

6-3 (*Sec. 6.1*) (*a*) Find the transmission parameters for the two-port network shown in Fig. 6.1-2 when the output port is defined at the terminal pair where $Z_1(s)$ is connected. Assume that $Z_1(s)$ is not part of the two-port network.
   (*b*) Repeat for $Z_2(s)$.
   (*c*) Repeat for $Z_3(s)$.

6-4 (*Sec. 6.1*) (*a*) Realize the impedance $Z_{in}(s) = sK$ using the GIN shown in Fig. 6.1-2, and using only single resistors and capacitors for the impedances $Z_1(s)$ through $Z_5(s)$. If the value of the resistors is 1 kΩ and that of the capacitors is 1 μF, find the value of the constant K, and give its units.
   (*b*) Repeat for the impedance $Z_{in}(s) = K/s$.
   (*c*) Repeat for the impedance $Z_{in}(s) = Ks^2$.
   (*d*) Repeat for the impedance $Z_{in}(s) = K/s^2$.
   (*e*) Repeat for the impedance $Z_{in}(s) = K/s^3$.

6-5 (*Sec. 6.1*) Assume that the range of available resistors is from 1 to 100 kΩ, and that of the available capacitors is from 100 pF to 1 μF. Find the range of values of synthetic inductance which can be realized using the GIN shown in Fig. 6.1-2.

6-6 (*Sec. 6.1*) (*a*) Find the network functions $V_{o1}(s)/V_1(s)$ and $V_{o2}(s)/V_1(s)$ for the GIN shown in Fig. 6.1-3 for the case where $Z_2(s) = 1/sC$ and $R_1 = R_3 = R_4 = R_5 = R$. Let $C = 0.1$ μF and $R = 10$ kΩ. Draw the Bode plots of the magnitude of each of the transfer functions.
   (*b*) Repeat for the case where all the impedances are resistors of value R, except for $Z_4(s) = 1/sC$.

6-7 (*Sec. 6.1*)   Use the synthetic-inductance method to design a fourth-order single-terminated high-pass filter with a Butterworth voltage transfer function. The single terminating resistance should be 1000 Ω and the cutoff frequency should be 500 Hz. The prototype filter can be found using the tables in App. A. The GIN used to realize the synthetic inductances should have $Z_2(s) = 1/sC$. The resistors of the synthetic inductors should be equal-valued and all have a value of 10 kΩ.

6-8 (*Sec. 6.2*)   (*a*) For the network shown in Fig. P6-8a, find the open-circuit voltage transfer function $V_2(s)/V_1(s)$ and note the pole and zero locations. Is the transfer function a stable one?
 (*b*) Repeat for the network shown in Fig. P6-8b.

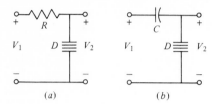

| (*a*) | (*b*) | **Figure P6-8** |

6-9 (*Sec. 6.2*)   If the values of the resistors and capacitors in the circuit shown in Fig. 6.2-2 are constrained to the ranges 1 to 100 kΩ and 100 pF to 1 μF respectively, what range of values for the constant D defined in (1) of Sec. 6.2 can be obtained?

6-10 (*Sec. 6.2*)   Develop the relationships showing how the unnormalized value of the FDNR constant D defined in (1) of Sec. 6.2 is related to a normalized constant D_n by the frequency and impedance denormalization constants Ω_n and z_n defined in Sec. 1.4.

6-11 (*Sec. 6.2*)   Using FDNRs, design a filter which has a fourth-order low-pass Butterworth voltage transfer function. The cutoff frequency should be 5 kHz and the prototype source and load resistors should have a value of 500 Ω. The prototype filter may be found from App. A. All the FDNR resistors should have a value of 7500 Ω. The resulting filter realization should have a dc path to ground from all the operational amplifier input terminals.

6-12 (*Sec. 6.2*)   Using FDNRs, design a realization of the attenuation characteristic shown in Fig. P6-12. A third-order elliptic function should be used to approximate the characteristic. The prototype filter may be designed using App. A. Impedance denormalize the realization by 10^4. All the FDNR capacitors should have a value of 10^{-8} F. The resulting realization should have a dc path to ground from all the operational amplifier input terminals.

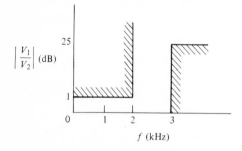

Figure P6-12

6-13 (*Sec. 6.3*)   (*a*) If the first element of a ladder network is a shunt one, and the input excitation is provided by a current source, find a block diagram equivalent to the one shown in Fig. 6.3-3 which simulates the network. Label the voltage and the polarity of each "immittance" block.
 (*b*) Repeat using the diagram shown in Fig. 6.3-4.

6-14 (*Sec. 6.3*)   (*a*) Starting with the circuit shown in Fig. 6.3-5b develop the general filter configuration shown in Fig. 6.3-9a.
 (*b*) Repeat for the configuration shown in Fig. 6.3-9b.

6-15 (*Sec. 6.3*)   Use the leapfrog method to design a filter having a fifth-order low-pass Butterworth voltage transfer function. The cutoff frequency should be 1 krad/s. The filter's terminating resistors should be equal and have a value of 10 kΩ.

6-16 (*Sec. 6.3*)   Derive (10) of Sec. 6.3 starting from the circuit shown in Fig. 6.3-12, and assuming that the operational amplifiers are ideal.

6-17 (*Sec. 6.3*)   Design a sixth-order Chebyshev bandpass filter with a voltage transfer function which has a center frequency of 1000 Hz and an octave bandwidth in which the ripple is 1 dB. The capacitor values should be 10^{-8} F. The filter should have equal-resistance terminations of 1000 Ω. Use the resonator circuit of Fig. 6.3-12 to realize the second-order blocks.

6-18 (*Sec. 6.3*)   (*a*) Develop a table similar to that shown as Table 6.3-1, showing how the universal active filter realization of Fig. 5.3-1 may be used to realize the second-order bandpass blocks required in the leapfrog method. Specifically, the figure should give the values of R_1, R_2, R_3, R_7, and R_8 in terms of the element values shown in Fig. 6.3-11a.

(*b*) Repeat for the element values of Fig. 6.3-11b.

6-19 (*Sec. 6.4*)   Use the PRB technique to design a filter having the specifications given in Prob. 6-17 with $H = 1$. Use Fig. 6.3-12 to realize the second-order blocks.

6-20 (*Sec. 6.4*)   Use the PRB technique to design a tenth-order Butterworth bandpass filter whose transfer function has a center frequency and a bandwidth of 2000π rad/s. The design should have a form similar to that shown in Fig. 6.4-2, that is, it should give the characteristics of the second-order blocks and the values of the feedback resistors. The value of the resistor R_0 should be chosen as 10 kΩ.

6-21 (*Sec. 6.5*)   Use the parallel-cascade method to develop a realization for the following normalized transfer function:

$$\frac{V_2(s)}{V_1(s)} = \frac{s^3 - 2s^2 + 2s - 1}{s^3 + 2s^2 + 2s + 1}$$

Denormalize the realization using the values $\Omega_n = z_n = 10^3$ for the denormalization constants defined in Sec. 1.4. In developing the realization, let $r_{1i} = r_{3i} = 10$ kΩ, and $c_{2i} = c_{4i} = c_i$. Find the value of c_i for the GIC.

6-22 (*Sec. 6.5*)   Use the parallel-cascade method to design a filter having a fourth-order low-pass Butterworth voltage transfer function with a cutoff frequency of 6.28 krad/s. Use an impedance denormalization constant $z_n = 10$ kΩ. Let $r_{1i} = r_{3i} = 10$ kΩ, and $c_{2i} = c_{4i} = c_i$. Give values for c_i for each of the GICs.

6-23 (*Sec. 6.5*)   Use the parallel cascade method to obtain a realization of a voltage transfer function with the poles and zeros shown in Fig. P6-23. Use an impedance denormalization constant $z_n = 10$ kΩ

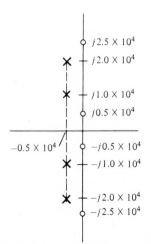

Figure P6-23

for the resulting design. Assume that $r_{1i} = r_{3i} = 1$ kΩ. Let $c_{2i} = c_{4i} = c_i$, and specify the value of c_i for each of the GICs.

6-24 *(Sec. 6.6)* Make plots comparable to the one shown in Fig. 6.6-3 for the elements R_1, R_3, and C_4 of the FDNR realization of Example 6.6-1 (see Fig. 6.6-2b).

6-25 *(Sec. 6.6)* Find $d|N(j1)|/|N(j1)|$ for the FDNR realization of Example 6.6-1 that results from a $+10$ percent change in the values of all the resistors in the circuit.

6-26 *(Sec. 6.6)* Find $d|N(j2)|/|N(j2)|$ for the synthetic-inductor circuit of Example 6.6-2 for a temperature change of 30°C. Assume that all the capacitors have a temperature coefficient of $+250$ ppm/°C and all the resistors have a temperature coefficient of -500 ppm/°C.

6-27 *(Sec. 6.6)* It is desired to compare the sensitivities of passive singly and doubly terminated third-order low-pass Butterworth filters. To do this, make plots of $S_x^{|N(j\omega)|}$ for the elements of the filter shown in Fig. 6.6-6, and compare these with the ones obtained in Example 6.6-1 and Prob. 6-24. Show that the latter plots are the same as the ones for the passive singly terminated filter.

6-28 *(Sec. 6.6)* Find the sensitivities of $|N(j\omega)|$ with respect to R_{11} and C_{11} of a third-order singly terminated low-pass Butterworth filter realized by the leapfrog technique. Compare the results at $\omega = 1$ with those obtained in Example 6.6-3.

6-29 *(Sec. 6.7)* Verify (1) of Sec. 6.7.

6-30 *(Sec. 6.7)* To realize a synthetic inductor let $Y_4(s)$ in Fig. 6.7-1 be a capacitor and all the other elements be resistors. Use the techniques of Sec. 6.7 to show whether $Q(\omega)$ can be independent of GB_1 and GB_2, and if so, what relationships must be satisfied.

6-31 *(Sec. 6.7)* Repeat Example 6.7-1 for the case where $GB_1 = 1.1$ MHz and $GB_2 = 1$ MHz.

6-32 *(Sec. 6.7)* Derive a relation similar to that of (25) of Sec. 6.7 for the case where an FDNR is realized by choosing $Z_1(s) = R_1$, $Z_2(s) = R_2$, $Z_3(s) = 1/sC_3$, $Z_4(s) = R_4$, $Z_5(s) = 1/sC_5$. Determine whether $Q(\omega)$ can be independent of GB_1 and GB_2, and if so, what relationships provide this result.

6-33 *(Sec. 6.7)* If, in an FDNR realization, $G_2 = G_4 = G_5 = G$ and $C_1 = C_3 = C$, plot $Q_{FDNR}(\omega)$ for (a) $GB_1 > GB_2$, (b) $GB_1 = GB_2$, and (c) $GB_1 < GB_2$. Assume $A_{01} = A_{02} = A_0$.

6-34 *(Sec. 6.8)* (a) Find the deviation in attenuation $\Delta\alpha$ at the cutoff frequency of a singly terminated seventh-order Butterworth filter realized by passive elements. Assume that the capacitors are polystyrene and the Qs of the inductors are 100.

(b) Repeat for inductor Qs of 30.

6-35 *(Sec. 6.8)* Find the minimum inductor Q for the filter described in Prob. 6-34 if the maximum value of $\Delta\alpha$ is to be 0.5 dB at the cutoff frequency.

6-36 *(Sec. 6.8)* Plot $\Delta\alpha(\omega)$ over the frequency range of 0 to 1.2 rad/s for a fifth-order low-pass 1-dB-ripple Chebyshev filter realized by passive elements. Use the group delay given in Fig. 6.8-1. Assume that the capacitors are Mylar and that the inductor Qs are 100 and are constant, i.e., that they are not functions of frequency.

6-37 *(Sec. 6.8)* Find the $\Delta\alpha$ at the cutoff frequency of the leapfrog filter of Example 6.3-1 as realized by Fig. 6.3-10b. Assume that the operational amplifiers have a gain bandwidth of 1 MHz.

6-38 *(Sec. 6.8)* Find the deviation in attenuation $\Delta\alpha$ at the cutoff frequency of a fourth-order low-pass Butterworth filter realized by the FDNR method. Assume that the high frequency Q of the FDNR, analyzed by methods similar to those described in Sec. 6.7, is 500 at the cutoff frequency. Since the capacitors are realized by resistors, assume $Q_c = \infty$.

6-39 *(Sec. 6.8)* (a) Use the model for an operational amplifier given in Fig. 6.8-2 and a suitable computer-aided analysis program[1] to obtain the magnitude response for frequencies from 0 to $2f_c$ for the realization of Prob. 6-38. Use operational amplifier parameters $R_i = 10^6$ Ω, $R_N = R_P = \infty$, $C_1 = 0$, $A_0 = 10^5$, and $R_o = 100$ Ω.

(b) Repeat for the case where R_N and R_P are changed to 500 MΩ and ω_a has the value 10 rad/s.

---

[1] For example, see L. W. Nagel and D. O. Pederson, "Simulation Program with Integrated Circuit Emphasis (SPICE)," *Electronics Research Lab. Rep.* ERL-M382, University of California at Berkeley, April 1973.

SEVEN

CONCLUSION

In this chapter we conclude our presentation of active filters by considering two final aspects of the subject. The first of these is an examination of some of the promising directions in which active filter research is now heading. Of the many possible topics that could be treated, we have chosen two which currently appear to have great future potential. The first of these is the development of filters for high-frequency applications. This is the subject of Sec. 7.1. The material includes a treatment of active R filters, i.e., ones which do not use capacitors. The second research topic chosen for discussion is the analog sampled-data filter, also known as the *switched-capacitor filter*. It is a type of filter which is particularly suitable for MOS integration. It is described in Sec. 7.2.

The second important aspect of active filters which is treated in this chapter is the design process. In Sec. 7.3 the design of an active filter to meet a specific industrial application is described in detail. The advantages and disadvantages of some of the different filter types covered in the preceding chapters are brought out in the discussion of the design process.

7.1 HIGH-FREQUENCY FILTERS

In this section we consider the realization of high-frequency filters. The upper limit of the frequency range of active filters is approximately 10 to 20 kHz for most of the techniques that we have studied. The primary cause of this limitation is the finite GB of the operational amplifiers used in the realizations. The effect of the limited GB is to cause the filter poles to shift from their desired locations. In some realizations this shift may eventually result in instability. As the value of Q which is to be realized increases, the frequency limit caused by the GB becomes more severe. Although methods which ideally eliminate the effect of GB on the transfer function were developed (see Sec. 5.7), these methods assume GB matching between amplifiers. In practice, however, the typical GB match among a sample of operational amplifiers such as the 747 might be ±5 percent. Consequently, unless one tunes for a perfect match, the methods of Sec. 5.7 can only be expected to extend the frequency range to approximately 100 kHz. In practice, however, the 100-kHz range is rarely achieved, because of the effects of slew rate. In order to realize such a high-frequency response, the slew rate of the operational amplifiers must be as large as possible and the signal amplitudes must be constrained to low levels.

The above considerations show that a frequency capability of up to 100 kHz for active filters can be achieved when compensation techniques are applied or when operational amplifiers of higher-frequency capabilities are used. In this section we shall investigate a different approach to the realization of high-frequency performance, namely the use of specialized design techniques which can extend the frequency limit. Before pursuing this objective, we should ask if such a result is desirable. Certainly for frequencies above 100 kHz, the inductor is a much more attractive component than it is at lower frequencies. As a matter of fact, for frequencies much over 20 kHz, one must have a good reason for *not* using *RLC* passive filters. The major reason that still favors the active filters approach for these frequencies is that of compatibility with silicon integrated-circuit technology. Until the inductor becomes compatible with this technology, there will still be many situations where active filters capable of 100 kHz or greater frequency performances are useful.

There are two approaches which have shown promise for obtaining active filter response in excess of 100 kHz. The first of these uses the inherent frequency response of the operational amplifiers to achieve frequency discrimination in a circuit consisting only of resistors and operational amplifiers. Such circuits are called *active R filters*. They provide frequency response up to approximately half of the GB of the operational amplifiers.

To see how such filters operate we note that most compensated operational amplifiers have a gain characteristic with a −6 dB/octave slope from the frequency at which the low-frequency pole is located to the frequency corresponding to the gain bandwidth. Recent approaches resulting in high-frequency filters make use of this −6 dB/octave slope in the design of the filter.[1,2,3,4,5]

[1] F. Capparelli and A. Liberatore, "Active Bandpass Networks with Only Resistors as Passive Elements," *Electronic Letters*, vol. 8, no. 2, Jan. 27, 1972, pp. 43–44.

[2] B. Berman and R. Newcomb, "Transistor-Resistor Synthesis of Voltage Transfer Function," *IEEE Trans. Circuit Theory*, vol. CT-20, September 1973, pp. 591–593.

[3] R. Schaumann, "Low Sensitivity High-Frequency Tunable Active Filter without External Capacitors," *IEEE Trans. Circuits and Systems*, vol. CAS-22, January 1975, pp. 39–44.

[4] A. K. Mitra and V. K. Aatre, "Low Sensitivity High Frequency Active R Filters," *IEEE Trans. Circuits and Systems*, vol. CAS-23, no. 11, November 1976, pp. 670–676.

[5] M. Soderstrand, "Active Filters Using Only Resistors and Amplifiers," *Proc. 8th Asilomar Conf. Circuits, Systems, Computers*, December 1974, pp. 675–681.

The method of synthesis is usually straightforward. It simply takes account of the fact that a compensated operational amplifier approximately realizes an integrator. Thus its transfer function may be written as

$$\frac{V_2(s)}{V_1(s)} \approx -\frac{GB}{s} \tag{1}$$

In terms of such an approximation, it is possible to take a filter configuration which uses integrators and replace each integrator with a compensated operational amplifier. The result of doing this to the state-variable filter is shown in Fig. 7.1-1a. Note that the response given in (1) may be obtained either with a minus sign (as shown) or with a plus sign depending on whether the inverting or noninverting terminal of the operational amplifier is used as the input. Using this technique allows us to remove the A_3 amplifier resulting in the circuit shown in Fig. 7.1-1b, where for future convenience the resistors have been renumbered.

For the circuit of Fig. 7.1-1b, if we assume that

$$\frac{V_{\text{BP}}(s)}{V_{\text{HP}}(s)} = \frac{GB_1}{s} \qquad \frac{V_{\text{LP}}(s)}{V_{\text{BP}}(s)} = \frac{GB_2}{s} \tag{2}$$

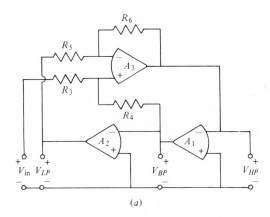

(a)

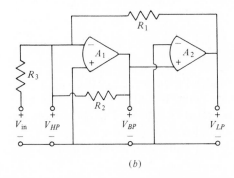

(b)

Figure 7.1-1 A state-variable active R filter.

then, solving for the bandpass transfer function, we obtain

$$\frac{V_{BP}(s)}{V_{in}(s)} = \frac{-s\dfrac{GB_1}{R_3}}{\left(\dfrac{1}{R_1} + \dfrac{1}{R_2} + \dfrac{1}{R_3}\right)s^2 + \dfrac{GB_1}{R_2}s + \dfrac{GB_1 GB_2}{R_1}} \tag{3}$$

Equations (2) and (3) can easily be used to solve for the high-pass and low-pass functions $V_{HP}(s)/V_{in}(s)$ and $V_{LP}(s)/V_{in}(s)$. Design equations can be developed by comparing (3) to the standard second-order bandpass function. Thus we find that

$$\frac{1}{R_1} + \frac{1}{R_2} + \frac{1}{R_3} = 1 \tag{4a}$$

$$\omega_n = \sqrt{\frac{GB_1 GB_2}{R_1}} \tag{4b}$$

$$\frac{\omega_n}{Q} = \frac{GB_1}{R_2} \tag{4c}$$

Solving for R_1, R_2, and R_3 gives

$$R_1 = \frac{GB_1 GB_2}{\omega_n^2} \tag{5a}$$

$$R_2 = \frac{GB_1 Q}{\omega_n} \tag{5b}$$

$$R_3 = \frac{GB_1 GB_2}{GB_1 GB_2 - \omega_n^2 - (\omega_n/Q)GB_2} \tag{5c}$$

For positive values of R_3 we must have

$$\omega_n^2 + \frac{\omega_n}{Q} GB_2 < GB_1 GB_2 \tag{6}$$

For moderate values of Q we have the constraint that $\omega_n < GB_2$. In this case, if ω_n becomes too low, the resistor values become very large, so that ω_n is typically in the range of

$$\frac{GB}{1000} < \omega_n < \frac{GB}{10} \tag{7}$$

From (3) we see that $H_0 = R_2/R_3$ which, under the constraint of (6), is approximately $GB_1 Q/\omega_n$. Thus H_0 is not a free parameter. If control of H_0 is required, another stage to control the passband gain must be added. The active resistor filter approach is also capable of realizing complex zeros or any other form of the second-order transfer function. The techniques are the same as those outlined for the state-variable filter in Sec. 5.2.

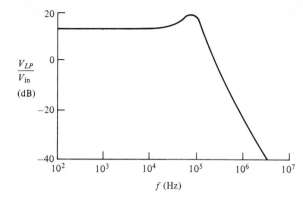

Figure 7.1-2 Response of the filter of Example 7.1-1.

Now let us consider some of the disadvantages of the active R filter. The basic problem is that the GB of the operational amplifiers is an inherent part of the filter response. To see this, note that the Q and ω_n sensitivities to the GB are

$$S_{GB_1}^Q = -\frac{1}{2} \qquad S_{GB_2}^Q = \frac{Q}{2} \qquad S_{GB_1}^{\omega_n} = S_{GB_2}^{\omega_n} = \frac{1}{2} \tag{8}$$

Hence any quantity which causes the GB to change will influence the filter. Thus active R realizations will be particularly sensitive to power supply and temperature variations since both of these quantities affect the GB. In addition, the variation of GB between operational amplifiers will always necessitate individual tuning of these filters. One last problem is that of slew-rate limitations. At high frequencies, even low signal levels may cause slewing problems, and the resultant phase shift may produce instability.

Example 7.1-1 *An Active R Filter* It is desired to use the active R filter method to realize a filter with $f_n = 100$ kHz and $Q = 4/3$. The operational amplifiers to be used have a gain characteristic which is compensated with a -6 dB/octave slope and their gain bandwidth is 1 MHz. From (5) we get $R_1 = 100\ \Omega$, $R_2 = 13.33\ \Omega$, and $R_3 = .0929\ \Omega$. Impedance scaling by 1 kΩ gives $R_1 = 100$ kΩ, $R_2 = 13.33$ kΩ, and $R_3 = .0929$ kΩ. The response of this filter is shown in Fig. 7.1-2. ☐

The second approach to the development of high-frequency active filters is the use of operational amplifiers which are not of the VCVS (*voltage-controlled voltage source*) type to synthesize voltage transfer functions. Typically, operational amplifiers with low input impedance offer improved frequency response. Thus the CCVS (*current-controlled voltage source*) and CCCS (*current-controlled current source*) operational amplifiers may be used. This approach has yielded active filters with a frequency capability of over 500 kHz.

Let us first consider the CCCS operational amplifier. For this we use the symbol shown in Fig. 7.1-3. The amplifier is modeled as shown in Fig. 7.1-4. We

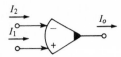

Figure 7.1-3 Symbol for a CCCS operational amplifier.

will assume that the amplifier is constructed so that if R_i is zero, the inputs are at ground potential—in other words, that there is no input common-mode voltage swing. Ideally, of course, R_i is zero and R_o is infinite. To see how the CCCS operational amplifier is used, consider the circuits of Fig. 7.1-5. In Fig. 7.1-5a, a noninverting amplifier is shown, whereas in Fig. 7.1-5b an inverting one is shown. The analysis of these circuits is done by assuming that each amplifier input is a virtual ground. Thus, for Fig. 7.1-5a we have $I_1(s) = I_2(s)$ (analogous to $V_1(s) = V_2(s)$ in the VCVS operational amplifier). This gives

$$\frac{V_o(s)}{V_s(s)} = \frac{Z_2(s)}{Z_1(s)} \tag{9}$$

Likewise, for Fig. 7.1-5b, $I_2(s) = V_1(s)/Z_1(s) + V_o(s)/Z_2(s)$ and $I_1(s) = 0$. Therefore, equating $I_1(s)$ to $I_2(s)$ gives

$$\frac{V_o(s)}{V_s(s)} = -\frac{Z_2(s)}{Z_1(s)} \tag{10}$$

If the gain of the CCCS amplifier is large but not infinite, then we must use a more general relationship between $I_o(s)$ and $I_1(s)$ and $I_2(s)$. This is

$$I_o(s) = A_d(s)[I_1(s) - I_2(s)] \tag{11}$$

where $A_d(s)$ is the differential-mode gain. Using this, for Fig. 7.1-5a, we see that

$$I_o(s) = A_d(s)\left[\frac{V_s(s)}{Z_1(s)} - \frac{V_o(s)}{Z_2(s)}\right] \tag{12}$$

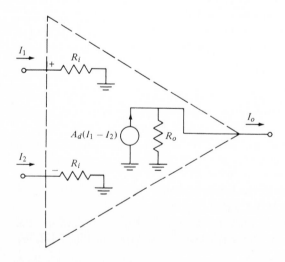

Figure 7.1-4 Model for a CCCS operational amplifier.

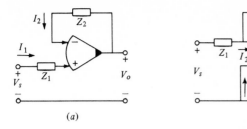

(a)   (b)

Figure 7.1-5 Amplifiers using CCCS operational amplifiers. (a) Noninverting. (b) Inverting.

If we assume that $Z_2(s)$ is the dominant load on the output then

$$I_o(s) = \frac{V_o(s)}{Z_2(s)} \tag{13}$$

Equating (11) and (13) gives

$$\frac{V_o(s)}{V_s(s)} = \frac{Z_2(s)}{Z_1(s)} \left[\frac{A_d(s)}{1 + A_d(s)} \right] \tag{14}$$

It should be noted that the open-loop gain is independent of the feedback network. The reason is that all of $I_o(s)$ flows through $Z_2(s)$ into the input of the amplifier (assuming $R_i = 0$). Thus the feedback is unity and independent of $Z_2(s)$. The advantage of the CCCS is shown by comparing its magnitude response curve with that of a VCVS operational amplifier. The result is shown in Fig. 7.1-6. Here a dominant pole model for $A_d(s)$ has been assumed. The low-frequency gain of the CCCS may be smaller or larger than A_0 depending on the value of $Z_2(0)/Z_1(0)$. Here it has been assumed that this ratio is greater than unity. The plot assumes $Z_1 = R_1$ and $Z_2 = R_2$, and that $(R_2/R_1) > A_0$. Note that the -3-dB bandwidth of the CCCS amplifier occurs at the GB of the VCVS amplifier.

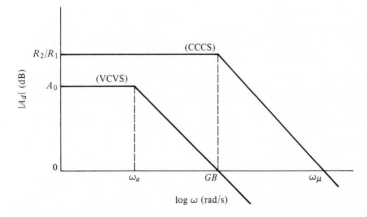

Figure 7.1-6 Magnitude responses for CCCS and VCVS operational amplifiers.

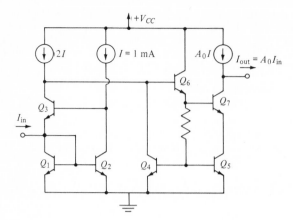

Figure 7.1-7 A CCCS operational amplifier with large GB.

Two problems are encountered in actually applying the technique described above. The first is that a buffer must be used at the output of the CCCS operational amplifier so that all the current $I_o(s)$ flows into the feedback loop. The second is that integrators built using this technique will have a pole at $\omega = $ GB regardless of whether $1/RC$ is above or below GB. Hence, unless GB is increased, no gain in performance over a VCVS operational amplifier is realized. Since A_0 need not be large, it is possible to devise a CCCS amplifier having large GBs. Figure 7.1-7 shows such a realization of a CCCS. The transistors Q_1, Q_2, and Q_3 form a Wilson current source[1] which provides the very low input impedance (less than an ohm) required for the successful application of this technique. The second stage is used to get a current gain of A_0. The current gain is provided by having the base-emitter area of transistor Q_5 be A_0 times that of Q_4. For simplicity differential inputs have not been provided. A CCCS amplifier can be built similar to the one shown in Fig. 7.1-7 using MOS transistors[2]. Figure 7.1-8 shows a resonator configuration realized by the CCCS of Fig. 7.1-7. The buffers, which are emitter followers, are necessary in order to obtain independence between the loop gain and the feedback network. This circuit was used to realize a bandpass second-order filter with design values of $Q_o = 8$ and $f_{no} = 300$ kHz. Analysis of this circuit shows that $\omega_{no} = 1/RC$. Selecting $R = 1$ K gives $C = 530$ pF. The series RC in the feedback path of the damped integrator helps to provide phase lead in the primary feedback loop. This arrangement creates a finite zero on the negative-real axis which would give a high-pass response except for the inherent roll-off of the amplifiers.

It should be noted that the CCCS operational amplifier by itself is not directly responsible for the increased frequency response. The fact that high gains are not

[1] G. R. Wilson, "A Monolithic Junction FET-NPN Operational Amplifier," *Intern. Solid State Circuits Conf., Digest of Technical Papers*, vol. 11, 1968, pp. 20–21.

[2] W. J. Parrish, "An Ion Implanted CMOS Amplifier for High Performance Active Filters," Ph.D. thesis, University of California, Santa Barbara, 1976.

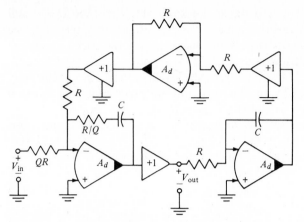

Figure 7.1-8 A resonator filter using the operational amplifier of Fig. 7.1-7.

required is also a factor, since this allows the amplifier gain bandwidths to be much larger than would be the case for the normal operational amplifier. In addition, the low input impedance eliminates the effect of capacitive parasitics. A further extension of this technique is shown in Fig. 7.1-9 where the gain of each CCCS is -1. Thus we see that (14) can be written as

$$\frac{V_o(s)}{V_s(s)} = -\frac{1}{2}\frac{Z_2(s)}{Z_1(s)} \tag{15}$$

The simplicity of this circuit is self-evident. It was used to realize a second-order bandpass filter with $Q_o = 8.5$ and $f_{no} = 650$ kHz. Values of $R = 1$ kΩ, $C = 245$ pF,

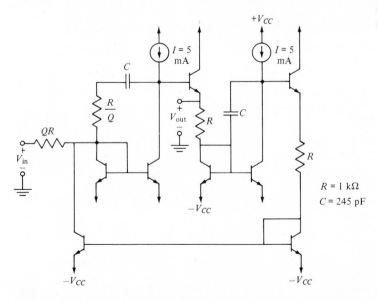

Figure 7.1-9 A second-order filter using CCCS operational amplifiers with gains of -1.

and $I = 5$ mA were used. The experimental results obtained were $Q = 7$ and $f_n = 634$ kHz, and a center frequency gain of about -2 dB. Much of the difference between the specifications and the performance is due to the current gain not being exactly -1 because of transistor matching. This points out that the design is strongly dependent on matching of the transistors, resulting in a large sensitivity to the current gain of the CCCSs. In order to avoid this problem it is desirable to have the gain of the amplifier sufficiently large so that $A_d/(1 + A_d) \approx 1$.

Another version of the approach to high-frequency active filters described above is to use the CCVS operational amplifier instead of the CCCS one. Using such an amplifier the voltage transfer function for an integrator is

$$\frac{V_o(s)}{V_s(s)} = \pm \frac{\omega_{RC}}{s} \frac{1}{1 + \dfrac{1}{sCR_d(s)}} \tag{16a}$$

where

$$R_d(s) = R_0 \frac{\omega_a}{s + \omega_a} \tag{16b}$$

At high frequencies, however, $R_d(s) \approx \omega_a R_0/s$. Thus the frequency dependence of $R_d(s)$ and the integrating capacitor cancel to give

$$\frac{V_o(s)}{V_s(s)} = \pm \frac{\omega_{RC}}{s} \frac{1}{1 + 1/\omega_a R_0 C} \approx \pm \frac{\omega_{RC}}{s} \tag{17}$$

A realization of the resonator using a noninverting CCVS integrator cascaded with an inverting CCVS integrator is shown in Fig. 7.1-10. The transfer function is

$$\frac{V_{out}(s)}{V_{in}(s)} = \frac{-(R_1/R_4)\omega_1 s}{s^2 + \omega_1 s + 1/R_2 R_3 C_1 C_2} \tag{18}$$

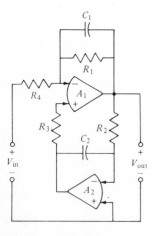

Figure 7.1-10 A resonator filter using CCVS operational amplifiers.

where $\omega_1 = 1/R_1 C_1$. The design equations can be found as

$$\omega_n = \sqrt{\frac{1}{R_2 R_3 C_1 C_2}} \tag{19a}$$

$$Q = R_1 \sqrt{\frac{C_1}{R_2 R_3 C_2}} \tag{19b}$$

and

$$H_0 = \frac{R_1}{R_4}$$

Ideally, the input impedances of the CCVS operational amplifiers are zero. If this is not true, then a frequency dependence of the amplifier upon the realization is observed. In this case, in Fig. 7.1-10 the damped integrator A_1 requires a new analysis for phase shift because of the input resistance. The transfer function from the output of A_2, V_{o2}, to the output of A_1, V_o can be found as

$$\frac{V_o}{V_{o2}} = \frac{GB_1(s + k\omega_1)}{s^2 + [k\omega_1 + (R_3 GB_1/R_i)]s + [\omega_a \omega_1 k + (R_3/R_i)GB_1 \omega_1]} \tag{20}$$

where $k = 1 + R_1/R_i$, $\omega_1 = 1/R_1 C_1$, $GB_1 = \omega_a R_0/R_3$, and R_i is the input resistance of the CCVS operational amplifier and R_0 is the low-frequency gain of $R_d(s)$. If $\omega = \omega_{no}$ and $\omega_1 = \omega_{no}/Q$, then the phase shift of (20) can be approximated as

$$\theta_1(j\omega) \approx -\frac{\pi}{2} + \frac{Q_o}{k} - \frac{R_i}{R_3}\frac{\omega_{no}}{GB_1} + \frac{kR_i}{R_3 Q_o}\frac{\omega_a}{GB_1} + \frac{1}{Q_o} \tag{21}$$

The total loop phase shift is

$$\theta_T(j\omega) = \frac{R_i}{R_2}\frac{\omega_{no}GB_2}{\omega_2(\omega_2 + GB_2)} + \frac{Q_o R_i}{R_1 + R_i} + \frac{R_i + R_1}{R_3 Q_o}\frac{\omega_a}{GB_1} + \frac{1}{Q_o} - \frac{R_i}{R_3}\frac{\omega_{no}}{GB_1} \tag{22}$$

Thus it may be shown that (see Sec. 5.7)

$$Q \approx \frac{1}{\dfrac{1}{Q_o} + Q_o\dfrac{R_i}{R_i + R_1} + \dfrac{R_i}{R_2}\dfrac{\omega_{no}GB_2}{\omega_2(\omega_2 + GB_2)} + \dfrac{\omega_a}{Q_o GB_1}\dfrac{R_1 + R_i}{R_3} - \dfrac{R_i}{R_3}\dfrac{\omega_{no}}{GB_1}} \tag{23}$$

As R_i becomes small, (23) simplifies to

$$Q \approx \frac{1}{\dfrac{1}{Q_o} + Q_o\dfrac{R_i}{R_1} + \dfrac{R_i}{R_2}\dfrac{1}{1 + \omega_{no}/GB_2} + \dfrac{\omega_a}{GB_1} - \dfrac{R_i}{R_3}\dfrac{\omega_{no}}{GB_1}} \tag{24}$$

where $\omega_2 = \omega_{no}$. Since $GB_1 = \omega_a R_0/R_3$ and $R_1/R_2 = Q_o$, (24) reduces to

$$Q \approx \frac{1}{1/Q_o + 2R_i/R_2 + R_3/R_0 - (R_i/R_3)(\omega_{no}/GB_1)} \tag{25}$$

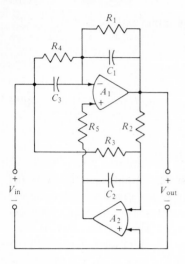

Figure 7.1-11 A biquadratic filter using CCVS operational amplifiers.

If we assume that $R_2 = R_3 = R$ and that $R_3 \ll R_0$, then the design frequency for this realization must satisfy the following relationship (when Q_o is large):

$$\omega_{no} < 2\text{GB}_1 \tag{26}$$

Thus, the circuit shown in Fig. 7.1-10 has a potential for high-frequency operation. As is always true, this potential can only be realized if slew-rate limitations can be avoided.

A general filter configuration based on Fig. 7.1-10 is shown in Fig. 7.1-11. Its transfer function is

$$\frac{V_{out}(s)}{V_{in}(s)} = \frac{-C_3\left(s^2 + \dfrac{s}{R_4 C_3} + \dfrac{1}{R_3 R_5 C_2 C_3}\right)}{C_1\left(s^2 + \dfrac{s}{R_1 C_1} + \dfrac{1}{R_2 R_5 C_1 C_2}\right)} \tag{27}$$

The design equations follow the form of those given for the circuit of Fig. 5.2-6. Although it cannot be used to generate right-half plane zeros, this is still a very useful circuit.

7.2 ANALOG SAMPLED-DATA FILTERS

One of the advantages of active RC filters, as opposed to passive RLC ones, is that they have the potential for being constructed using integrated circuit (IC) technology. Unfortunately, this potential has not been realized on a widespread commercial basis. The primary reason is that in active filters, the RC products must be accurately defined. This implies that the absolute values of the resistors and capacitors be closely controlled, a condition not found in current IC technology.

In addition, standard IC resistors and capacitors have poor linearity and temperature characteristics and require a large area compared to active devices such as an operational amplifier. Most active RC filters which have been completely integrated require careful trimming techniques[1] and as a result are too costly for commercial production.

One approach which makes active filters more amenable to IC technology is the active R methods discussed in the preceding section. Although these methods avoid the need for capacitors, their dependence on resistor tolerances and operational amplifier gain bandwidths is still a serious problem. Another disadvantage of this approach is that resistors typically take more chip area than capacitors; thus this method is not very suitable for IC technology, although some interesting results have been reported.[2]

A filter which has good potential for being compatible with both the requirements of active filters and with IC technology is called the *analog sampled-data filter*.[3,4] The basic theory underlying the operation of such a filter may be understood by considering the switched capacitor shown in Fig. 7.2-1. This circuit performs the function of a resistor.[5] To see this, assume that the switch is initially at position 1 so that capacitor C is charged to V_1 volts. Now let the switch be thrown to position 2 so that the capacitor C is discharged (or charged) to V_2. The amount of charge that flows into (or from) the voltage source V_2 is $C(V_1 - V_2)$. If the switch is thrown back and forth every T_c seconds, then the current flow into the voltage source V_2 is

$$i = \frac{C\,\Delta V}{\Delta T} = \frac{C(V_1 - V_2)}{T_c - 0} = \frac{C(V_1 - V_2)}{T_c} \tag{1}$$

[1] J. Friend et al., "Star: An Active Biquadratic Filter Section," *IEEE Trans. Circuits and Systems*, vol. CAS-22, no. 2, February 1975, pp. 115–121.

[2] K. Tan and P. Gray, "Higher-Order Monolithic Analog Filters Using Bipolar/JFET Technology," *Proc. IEEE Intern. Solid-State Circuits Conf.*, February 1978, pp. 80–81.

[3] D. Fried, "Analog Sampled Data Filters," *IEEE J. Solid-State Circuits*, vol. SC-7, August 1972, pp. 302–303.

[4] W. Kuntz, "A New Sample-and-Hold Device and Its Application to the Realization of Digital Filters" (letter), *Proc. IEEE*, vol. 56, November 1968, pp. 2092–2093.

[5] B. Hosticka, R. Brodersen, and P. Gray, "MOS Samples Data Recursive Filters Using Switched Capacitor Integrators," *IEEE J. Solid-State Circuits*, vol. SC-12, no. 6, December 1977, pp. 600–608.

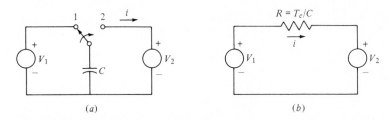

(a) (b)

Figure 7.2-1 A switched-capacitor equivalent of a resistor.

Thus, this circuit performs the same function as a resistor of value T_c/C ohms as shown. If the switching rate $f_c = 1/T_c$ is much larger than the signal frequencies of interest, the sampling time of the signal which occurs in this circuit can be ignored and the switched capacitor can then be considered as a direct replacement for a conventional resistor. On the other hand, if the switch rate and signal frequencies are of the same order, then sampled-data techniques are required for analysis.[1] In such a case, as in any sampled-data system, the frequency content of the input signal should be band-limited below $f_c/2$.

The reason the switched capacitor approach is useful can be seen by examining the RC product of a resistor designated R_1 and a capacitor designated C_2. If the resistor is replaced by a switched capacitor, then $R_1 = T_c/C_1$. Consequently, the RC product or time constant τ_{RC} is

$$\tau_{RC} = R_1 C_2 = T_c \frac{C_2}{C_1} \tag{2}$$

In MOS technology, the switched capacitor of Fig. 7.2-1a can be implemented as shown in Fig. 7.2-2a. The two MOSFETs are operated as switches from a two-phase nonoverlapping clock (ϕ_1 and ϕ_2) at a frequency f_c as shown in Fig. 7.2-2b.

[1] A. Oppenheim and R. Schafer, *Digital Signal Processing*, Prentice-Hall, Inc., Englewood Cliffs, N.J., 1975.

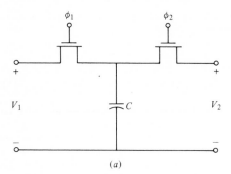

(a)

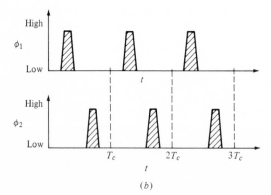

(b)

Figure 7.2-2 An MOS implementation of a switched capacitor.

When ϕ_1 is high, C is connected to V_1. When ϕ_2 is high, C is connected to V_2. Otherwise C is connected to neither V_1 nor V_2. For a given clock rate, the value of τ_{RC} of (2) is determined by the ratio of capacitors. Since the *relative* values of capacitors C_1 and C_2 are determined by photolithographic methods, it is possible to achieve a high precision in the capacitance ratio. It has been shown that the error in such ratios can be less than 0.1 percent using standard MOS processing techniques.[1] Since the MOS capacitor very closely approaches ideal capacitor characteristics, much better stability and linearity are obtained than is possible with diffused resistors. In addition the ratio of two MOS capacitors has very little temperature dependence. Therefore we conclude that the switched-capacitor resistor of Fig. 7.2-1a makes possible the design of precise, stable active RC filters which can be fully integrated using MOS technology.

A basic circuit used in many active RC filter configurations is the inverting integrator shown in Fig. 7.2-3a. If the resistor R_1 in this circuit is replaced by the switched capacitor of Fig. 7.2-1a, then the sampled-data integrator of Fig. 7.2-3b results. When the switch is in position 1, C_1 is charged to the value of $V_{\rm in}$. If the switch is now moved to position 2, then the charge of C_1 is all transferred to C_2. All the charge is transferred because the operational amplifier forces the voltage across C_1 to be zero. The net charge on C_2 is equal to the previous charge on C_2 minus the charge transferred to C_2 by C_1. The charge transferred to C_2 by C_1 is subtracted because of the inverting configuration. At time $(n-1)T_c$, the input charge on C_1 is

$$q_1[(n-1)T_c] = C_1\, v_{\rm in}[(n-1)T_c] \qquad (3)$$

and the charge on C_2 is

$$q_2[(n-1)T_c] = C_2\, v_{\rm out}[(n-1)T_c] \qquad (4)$$

At the time nT_c, the charge $q_1[(n-1)T_c]$ has been transferred to C_2. Therefore, the charge on C_2 at $t = nT_c$ is

$$q_2(nT_c) = q_2[(n-1)T_c] - q_1[(n-1)T_c] \qquad (5)$$

[1] J. L. McCreary and P. R. Gray, "All-MOS Charge Redistribution Analog-to-Digital Conversion Techniques—Part I," *IEEE J. Solid-State Circuits*, vol. SC-10, December 1975, pp. 371–379.

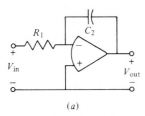

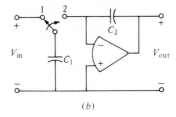

(a) (b)

Figure 7.2-3 A sampled-data inverting integrator.

Substituting (3) and (4) into (5) results in

$$v_{out}(nT_c) = v_{out}[(n-1)T_c] - \frac{C_1}{C_2} v_{in}[(n-1)T_c] \qquad (6)$$

Taking the Laplace transform of both sides of (6) we obtain

$$V_{out}(s)e^{s(nT_c)} = V_{out}(s)e^{s(n-1)T_c} - \frac{C_1}{C_2} V_{in}(s)e^{s(n-1)T_c} \qquad (7)$$

Multiplying both sides of the above by e^{-snT_c} gives

$$V_{out}(s) = V_{out}(s)e^{-sT_c} - \frac{C_1}{C_2} V_{in}(s)e^{-sT_c} \qquad (8)$$

The factor e^{-sT_c} is recognized as a time delay of T_c seconds. Solving for $V_{out}(s)/V_{in}(s)$ gives

$$\frac{V_{out}(s)}{V_{in}(s)} = \frac{-(C_1/C_2)e^{-sT_c}}{1 - e^{-sT_c}} = \frac{-C_1/C_2}{e^{sT_c} - 1} \qquad (9)$$

If $sT_c \ll 1$, then $e^{sT_c} \approx 1 + sT_c$ and (9) becomes

$$N_1(s) = \frac{V_{out}(s)}{V_{in}(s)} \approx \frac{-C_1/C_2}{sT_c} = \frac{-1}{s\dfrac{T_c}{C_1}C_2} \qquad (10)$$

The transfer function of the conventional integrator of Fig. 7.2-3a is given as

$$N_2(s) = \frac{V_{out}(s)}{V_{in}(s)} = \frac{-1}{sR_1C_2} \qquad (11)$$

If $s = j\omega$, the frequency response of (10) and (11) may be expressed as

$$N_1(j\omega) \approx \frac{-1}{j\omega(T_c/C_1)C_2} = -\frac{\omega_o}{j\omega} \qquad (12)$$

where ω_o is the unity-gain frequency of the integrator, and

$$N_2(j\omega) = \frac{-1}{j\omega R_1 C_2} = -\frac{\omega_o}{j\omega} \qquad (13)$$

Thus we see that if $\omega \ll 1/T_c$ or $f \ll f_c/2\pi$, R_1 in Fig. 7.2-3a has been approximated by T_c/C_1 in Fig. 7.2-3b, and this figure is a direct replacement for Fig. 7.2-3a. We note with interest that the integrator time constant can be easily varied by varying the clock frequency.

The operation of the sampled-data integrator can be further illustrated through the use of Fig. 7.2-4. The input is assumed to be a sinusoid of frequency f_o which is 10 times smaller than the clock frequency and which has a peak amplitude of 1 V. V_{in} is sampled in the middle of every clock cycle, and this value of v_{in} is held on C_1 for a half clock period until it is transferred to C_2 according to (5).

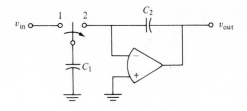

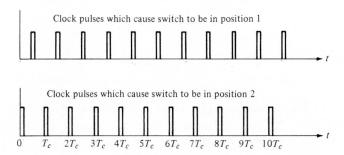

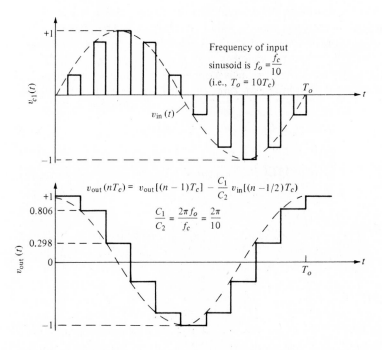

Figure 7.2-4 The operation of the sampled-data integrator.

The output is held for a clock period until a new sample is transferred to the capacitor C_2. In this example, the sinusoidal frequency of f_o implies that the gain of the integrator is unity from (12). Therefore, we see that $\omega(T_c/C_1)C_2$ of (12) is equal to unity when $\omega = \omega_o$. Thus

$$\frac{C_1}{C_2} = \omega_o T_c = \frac{2\pi f_o}{f_c} = \frac{2\pi}{10} = \frac{\pi}{5} \tag{14}$$

In (6) we see that only $\pi/5$ of the input voltage is subtracted from the output. If we assume that $v_{out}(0) = 1$ V (an arbitrary choice), then we may calculate the value of v_{out} at T_c as follows. It is seen that $v_{in}(T_c/2) = \sin(18°) = 0.309$. Therefore

$$v_{out}(T_c) = v_{out}(0) - \frac{\pi}{5}v_{in}(T_c/2) = 1 - 0.194 = 0.806$$

This operation can be repeated to obtain the waveforms of Fig. 7.2-4. Normally, the input frequency would be much less than f_c in order to reduce the output distortion.

Many of the inverting applications that we have examined such as the resonator circuit of Fig. 5.2-4 require the integrator to be damped as shown in Fig. 7.2-5a. The transfer function of this circuit for $N_1(s) = V_{out}(s)/V_{in}(s)$ is

$$N_1(j\omega) = -\frac{R_3/R_1}{1 + j\omega R_3 C_2} \tag{15}$$

Figure 7.2-5b shows the sampled-data equivalent circuit where the resistors R_1 and R_3 of Fig. 7.2-5a have been replaced with switched capacitors. The phasing of the switches is important and is indicated in this circuit by the arrows. The interpretation of these arrows is that the switches are simultaneously connected to the input and output and then simultaneously connected together to the inverting input of the operational amplifier. At time $(n - 1)T_c$ the various charges on the capacitors are

$$q_1[(n - 1)T_c] = C_1 v_{in}[(n - 1)T_c] \tag{16}$$

$$q_2[(n - 1)T_c] = C_2 v_{out}[(n - 1)T_c] \tag{17}$$

and

$$q_3[(n - 1)T_c] = C_3 v_{out}[(n - 1)T_c] \tag{18}$$

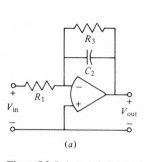 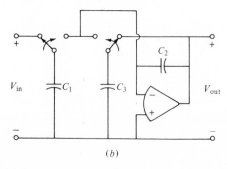

(a)   (b)

Figure 7.2-5 A sampled-data damped integrator.

At time nT_c the switches have changed position to give

$$q_2(nT_c) = q_2[(n-1)T_c] - q_3[(n-1)T_c] - q_1[(n-1)T_c] \tag{19}$$

Substituting (16) through (18) into (19) results in

$$v_{\text{out}}(nT_c) = v_{\text{out}}[(n-1)T_c] - \frac{C_3}{C_2}v_{\text{out}}[(n-1)T_c] - \frac{C_1}{C_2}v_{\text{in}}[(n-1)T_c] \tag{20}$$

Converting from the time domain to the s domain and solving for $N_2(s) = V_{\text{out}}(s)/V_{\text{in}}(s)$ give the following transfer function:

$$N_2(s) = \frac{-C_1/C_2}{e^{sT_c} - 1 + C_3/C_2} \tag{21}$$

Replacing s by $j\omega$ and assuming that $e^{j\omega T_c} \approx 1 + j\omega T_c$ give the desired result of

$$N_2(j\omega) \cong \frac{-(C_1/T_c)(T_c/C_3)}{1 + j\omega(T_c/C_3)C_2} \tag{22}$$

where $\omega T_c \ll 1$. From (15) and (22) it is clear that $R_1 = T_c/C_1$ and $R_3 = T_c/C_3$.

Another basic component of active RC filters is the differential integrator shown in Fig. 7.2-6a. The differential integrator is capable of integrating the difference between two signals. A continuous realization of a differential integrator is given in Fig. 7.2-6b. Analysis of this circuit yields

$$V_{\text{out}}(s) = -\frac{\omega_o}{s}[V_1(s) - V_2(s)] \tag{23}$$

where $\omega_o = 1/R_1C_2$. Although every resistor of Fig. 7.2-6b could be replaced by a switched capacitor, there is a more efficient sampled-data realization which is

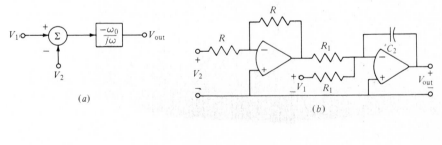

(a)

(b)

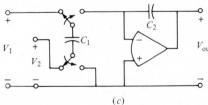

(c)

Figure 7.2-6 A sampled-data differential integrator.

given in Fig. 7.2-6c. Following the same procedure as before we may express $V_{out}(s)$ of Fig. 7.2-6c as

$$V_{out}(s) \approx \frac{-1}{s(T_c/C_1)C_2}[V_1(s) - V_2(s)] \tag{24}$$

where, for $s = j\omega$, $\omega T_c \ll 1$.

The use of switched capacitors to replace resistors can be easily applied to active RC networks. Although each resistor can be replaced directly with a switched capacitor, it is more efficient to replace functional blocks such as that of Fig. 7.2-6b by the circuit of Fig. 7.2-6c. As an example, consider the resonator filter shown in Fig. 7.2-7a. This is exactly the same as the circuit of Fig. 5.2-4 except that $R_3 = R_4$. The low-pass and bandpass transfer function can be found

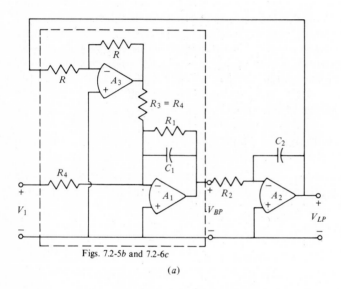

Figs. 7.2-5b and 7.2-6c

(a)

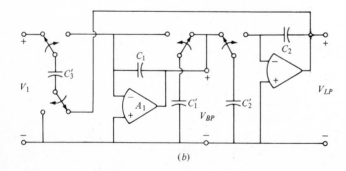

(b)

Figure 7.2-7 A sampled-data resonator filter.

from (19) and (20) of Sec. 5.2 as

$$\frac{V_{BP}(s)}{V_1(s)} = \frac{-\dfrac{s}{R_3 C_1}}{s^2 + \dfrac{s}{R_1 C_1} + \dfrac{1}{R_2 R_3 C_1 C_2}} \tag{25}$$

and

$$\frac{V_{LP}(s)}{V_1(s)} = \frac{\dfrac{1}{R_2 R_3 C_1 C_2}}{s^2 + \dfrac{s}{R_1 C_1} + \dfrac{1}{R_2 R_3 C_1 C_2}} \tag{26}$$

The only constraint introduced by requiring $R_3 = R_4$ is that the gain at ω_n of the low-pass realization will be 1. It can be shown that the combination of Fig. 7.2-5b and Fig. 7.2-6c will realize the dashed portion of Fig. 7.2-7a. If Fig. 7.2-3b is used for the remaining inverting integrator, then the analog sampled-data filter of Fig. 7.2-7b results. The capacitors which represent resistors are primed and have the subscript corresponding to the resistor they replace. The transfer function of Fig. 7.2-7b may be found by applying the previous techniques to the components of this figure. For example, the transfer function of the circuit within the dashed box can be found as

$$V_{BP}(s) \approx \frac{C_3'/C_1'}{1 + sT_c(C_1/C_1')}[V_{LP}(s) - V_1(s)] \tag{27}$$

if for $s = j\omega$, $\omega T_c \ll 1$. Combining this result with that for the inverting integrator gives

$$\frac{V_{BP}(s)}{V_1(s)} \approx \frac{-s\dfrac{C_3'}{T_c}\dfrac{1}{C_1}}{s^2 + s\dfrac{C_1'}{T_c}\dfrac{1}{C_1} + \dfrac{C_2' C_3'}{T_c T_c}\dfrac{1}{C_1 C_2}} \tag{28}$$

The low-pass transfer function is

$$\frac{V_{LP}(s)}{V_1(s)} \approx \frac{(C_2'/T_c)(C_3'/T_c)(1/C_1 C_2)}{s^2 + s\dfrac{C_1'}{T_c}\dfrac{1}{C_1} + \dfrac{C_2' C_3'}{T_c T_c}\dfrac{1}{C_1 C_2}} \tag{29}$$

Example 7.2-1 *An Analog Sampled-Data Resonator Filter* It is desired to design a low-pass and bandpass second-order resonator filter having $Q = 20$ and $f_n = 1$ kHz using sampled-data techniques. The clock frequency is to be 100 kHz. Let the values of the capacitor ratios be $\alpha_1 = C_1'/C_1$ and $\alpha_2 = C_2'/C_2$. Choosing $C_1' = C_2' = C_3'$ in (29) gives

$$\alpha_1 \alpha_2 = \omega_n^2 T_c^2 \tag{30}$$

and

$$\alpha_1 = \frac{\omega_n T_c}{Q} \tag{31}$$

From the specifications $\alpha_1 = 0.00314$ and $\alpha_2 = 1.256$. ☐

It is clear from (30) and the above example that the product $\alpha_1 \alpha_2$ must be equal to $4\pi^2$ times the ratio $f_n : f_c$ and that having values of α close to unity would imply that the clock frequency is approximately 40 times the pole frequency of the filter. As the filter frequency approaches the clock frequency it becomes necessary to reexamine the filter with the use of z transforms.[1] The distinction between an analog filter and a sampled-data analog filter is illustrated in Fig. 7.2-8 for the previous example. As the frequency approaches f_c, the frequency response goes back up rather than going to zero as $\omega \to \infty$, as in analog filters. This effect is called *aliasing*. It may be necessary to use an additional analog filter in cascade with the output of a sampled-data analog filter to correct for it. If $f_n \ll f_c$, the additional filter is generally unnecessary.

Another important consideration is to avoid half-cycle delays. For example, in Fig. 7.2-7*b* when V_{BP} changes to a new value at some value of time, it will maintain that value for a full clock period. The C_2' capacitor switch can be phased to either sample the new V_{BP} value as soon as it appears or to wait for half a clock cycle. It can be shown[2] that if one chooses to sample at $\frac{1}{2}$ cycle later, an excess phase lag occurs when the condition $f_n \ll f_c$ is not satisfied. Therefore it is good practice to phase the switches so that when the output voltage of an integrator changes, this value is transferred on to the next circuit as soon as possible.

The sensitivities of sampled-data analog filters should be identical with those of their analog filter counterparts until f_n begins to approach f_c. In order to investigate the sensitivities of the various α's with respect to f_n/f_c it is necessary to avoid using the assumption that $e^{sT_c} \approx 1 + sT_c$. In order to analyze this problem a transformation is made between the s domain and the z domain. The transformation is

$$z = e^{sT_c} \tag{32}$$

[1] Oppenheim and Schafer, op. cit.

[2] G. Jacobs, D. Allstot, R. Brodersen, and P. Gray, "Design Considerations for MOS Switched Capacitor Ladder Filters," *Proc. IEEE Intern. Conf. Circuits and Systems*, May 1978, pp. 324–329.

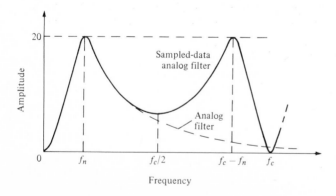

Figure 7.2-8 The effect of aliasing in sampled-data filters.

Thus z^{-1}, which equals e^{-sT_c}, is seen to correspond with a delay of T_c seconds. Returning to (9) we see that an inverting integrator as used in Fig. 7.2-7a can be expressed as

$$\frac{V_{LP}(z)}{V_{BP}(z)} = \frac{-C'_2/C_2}{z - 1} = \left(-\frac{C'_2}{C_2}\right)\frac{z^{-1}}{1 - z^{-1}} \tag{33}$$

The dashed portion of Fig. 7.2-7 can be expressed as

$$V_{BP}(s) = \frac{-C'_3/C_1}{e^{sT_c} - 1 + C'_1/C_1}[V_1(s) - V_{LP}(s)] \tag{34}$$

Transforming from the s domain to the z domain gives

$$V_{BP}(z) = \frac{-C'_3/C_1}{z - 1 + C'_1/C_1}[V_1(z) - V_{LP}(z)] \tag{35}$$

Combining (33) and (35) results in

$$\frac{V_{BP}(z)}{V_1(z)} = \frac{-(C'_3/C_1)(z - 1)}{z^2 - (2 - C'_1/C_1)z + [1 + (C'_3/C_1)(C'_2/C_2) - C'_1/C_1]} \tag{36}$$

Equation (36) can be simplified by letting $C'_3 = C'_1$ and using $\alpha_1 = C'_1/C_1$ and $\alpha_2 = C'_2/C_2$, resulting in

$$\frac{V_{BP}(z)}{V_1(z)} = \frac{-\alpha_1(z - 1)}{z^2 - (2 - \alpha_1)z + (1 + \alpha_1\alpha_2 - \alpha_1)} \tag{37}$$

The low-pass transfer function can be obtained by multiplying (37) by (33) to get

$$\frac{V_{LP}(z)}{V_1(z)} = \frac{\alpha_1\alpha_2}{z^2 - (2 - \alpha_1)z + (1 + \alpha_1\alpha_2 - \alpha_1)} \tag{38}$$

The location of a pair of complex poles in the s plane may be written as

$$s_1, s_2 = -\frac{\omega_n}{2Q} \pm j\frac{\omega_n}{2Q}\sqrt{4Q^2 - 1} \tag{39}$$

In the z plane these poles are written as

$$z_1, z_2 = Re^{\pm j\theta} \tag{40}$$

Substituting (39) into (32) and comparing with (40) yields the following formulas:

$$f_n = \frac{f_c}{2\pi}\sqrt{\theta^2 + \ln^2(R)} \tag{41}$$

and

$$Q = \frac{-\pi f_n}{f_c \ln(R)} \tag{42}$$

The standard second-order denominator of a complex pole pair in the z domain may be expressed as

$$D(z) = (z - z_1)(z - z_2) = (z - R \cos \theta - jR \sin \theta)(z - R \cos \theta + jR \sin \theta)$$

$$= z^2 - 2R \cos \theta z + R^2 \tag{43}$$

The relationships of (41) through (43) will permit a more exact design formula to be developed for α_1 and α_2 compared to the formulas of (30) and (31) when $e^{sT_c} \approx 1 + sT_c$. Comparing the denominator of (37) or (38) to (43) yields

$$\alpha_1 = 2 \left[1 - e^{-\pi f_n / f_c Q} \cos \left(\frac{\pi f_n}{f_c} \sqrt{4 - \frac{1}{Q^2}} \right) \right] \tag{44}$$

and

$$\alpha_2 = 1 + \frac{1}{\alpha_1} \left(e^{-2\pi f_n / Q f_c} - 1 \right) \tag{45}$$

Substituting the values of Example 7.2-1 into these formulas yields $\alpha_1 = 0.00708$ and $\alpha_2 = 0.5567$. These values differ from the ones found in the example by a factor of approximately 2. If the clock frequency is increased to 1 MHz (or f_n changed to 100 Hz) the values of α_1 and α_2 from (30) and (31) become 0.00031416 and 0.1256 respectively, whereas (44) and (45) give α_1 and α_2 as 0.0035356 and 0.11607 respectively. Consequently, even though the filter frequency is much less than the clock frequency, certain realizations, such as the one of Fig. 7.2-7b, obviously become very sensitive to the ratio of f_n / f_c.

The sensitivities of f_n to α_1 and α_2 can be determined from (44) and (45), assuming that $Q \gg 1$ and $f_n \ll f_c$. Thus we find

$$S_{\alpha_1}^{f_n} = \frac{\partial f_n}{\partial \alpha_1} \frac{\alpha_1}{f_n} \approx \frac{f_c}{2\pi f_n} \sin \left(\frac{\pi f_n}{f_c} \right) \tag{46}$$

$$S_{\alpha_2}^{f_n} = \frac{f_c}{2\pi f_n Q} \frac{S_{\alpha_1}^{f_n}}{1 - S_{\alpha_1}^{f_n}} \approx 0 \tag{47}$$

$$S_{\alpha_1}^{Q} \approx 0 \tag{48}$$

$$S_{\alpha_2}^{Q} \approx 2\pi \frac{f_n Q}{f_c} \tag{49}$$

As f_c / f_n becomes much less than 1, the sensitivity of f_n with respect to α_1 approaches $\frac{1}{2}$. On the other hand, the sensitivity of Q with respect to α_2 can become large (if Q is large) as f_c / f_n approaches 1. This means that f_n / f_c must be very small to maintain low sensitivities with large Qs.

Sampled-data circuits can also be used to realize the state-variable filter. One way of accomplishing this is shown in Fig. 7.2-9. Note that the replacement of resistors R_3, R_4, R_5, and R_6 with capacitors does not change the ideal transfer function of the summing differential amplifier of Fig. 5.2-2. Replacement of R_1 and R_2 by switched capacitors yields the result of Fig. 7.2-9. The design equations of Sec. 5.2 may be used for this circuit if R_i is replaced by $1/C_i$, where $i = 3, 4, 5, 6$.

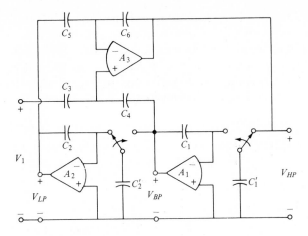

Figure 7.2-9 A sampled-data state-variable filter.

Unfortunately, the practical aspects of operational amplifiers make this realization impractical. One reason is that the lack of dc feedback causes the dc offsets at the amplifier outputs to be excessive. In general, because of the number of resistors required, the state-variable circuit is less suitable to realization as a sampled-data filter than the resonator. A simplified realization of the state-variable filter by sampled-data techniques has been given in the literature.[1]

The sampled-data realization method is also suitable for other second-order filters such as the infinite-gain ones of Sec. 5.1. An example of the realization of a low-pass infinite-gain filter and its fabrication using MOS technology can be found in the literature.[1]

The passive network simulation methods of Chap. 6 can also be realized by sampled-data techniques. For example, the leapfrog method of Sec. 6.3 is very compatible with the sampled-data realization method and offers excellent sensitivity characteristics. To see this, note that the basic element of a leapfrog filter is the integrator. Since Fig. 7.2-6c realizes a difference integrator, the leapfrog circuit of Fig. 6.3-2 may be used as a prototype. The transfer function of an intermediate stage of the leapfrog structure is given by (1) of Sec. 6.3 and is expressed in general as

$$V_j(s) = T'_j(s)[V_i(s) - V_k(s)] \tag{50}$$

where the voltages V_i, V_j, and V_k progress consecutively down the ladder from left to right and may be either current analogs or actual voltages. If in Fig. 7.2-6c we replace V_2 by V_i, V_1 by V_j, and V_{out} by V_k, then from (23) we can write that

$$V_j(s) = \frac{1}{s\left(\dfrac{T_c}{C'_j}\right)C_j}[V_i(s) - V_k(s)] \tag{51}$$

[1] Hosticka et al., op. cit.

where for $s = j\omega$, $\omega T_c \ll 1$, and C_1 and C_2 of Fig. 7.2-6c have been replaced by C_j' and C_j. Since $T_j'(s)$ of (50) is typically normalized to an impedance level of 1 Ω and a cutoff frequency of 1 rad/s, we must either denormalize (50) or normalize (51) before we can compare the two expressions. Assuming that the cutoff frequency is specified, then the simplest approach is to make a normalization of (51), which amounts to normalizing the clock frequency to a value T_c' given as

$$T_c' = T_c \Omega_n \tag{52}$$

Here it has been assumed that (50) is normalized to 1 rad/s. Therefore α_j, the ratio of C_j' to C_j, can be designed from the following expression

$$\alpha_j = s[T_c \Omega_n T_j'(s)] = \frac{C_j'}{C_j} \tag{53}$$

Since the ladder structure is always terminated in a load resistor, the last integrator must be damped. Also, note that the last integrator does not need to be differential since there is no feedback path. Figure 7.2-6c can still be used for these conditions by connecting V_1 to V_{out}. Using the notation of (51), V_j is equal to V_k. Thus (51) can be written as

$$\frac{V_j(s)}{V_i(s)} = \frac{V_k(s)}{V_i(s)} = \frac{1}{s\dfrac{T_c}{\alpha_j} + 1} \tag{54}$$

If the ladder filter is of mth order, then the transfer function of the last stage can be written in general as

$$\frac{V_m(s)}{V_{m-1}(s)} = T_m(s) = \frac{1}{R_{Ln} + sL_{mn}} = \frac{1}{G_{Ln} + sC_{mn}} \tag{55}$$

where the subscripts n refer to the normalized values of a component. If we choose values of R_{Ln} as 1 Ω or G_{Ln} as 1 Ω then we may use (54) and (55) to solve for α_m of the last stage as

$$\alpha_m = \frac{T_c \Omega_n}{L_{mn}} = \frac{T_c \Omega_n}{C_{mn}} \tag{56}$$

where the last (right-hand) elements of the ladder filter are a 1-Ω resistor in series with a normalized inductance of L_{mn} or a 1-Ω resistor in parallel with a normalized capacitance of C_{mn}.

If the ladder is doubly terminated, then it will also have a source resistance R_S. While the source resistance (and the load resistance) could be external to the realization let us consider what modifications need to be made to the first (left-most) stage in order to include the source resistance R_S. The two possible source configurations are of a Thevenin or Norton type. The transfer function of the first stage is

$$V_2(s) = T_1'(s)[V_1(s) - V_3(s)] = \frac{V_1(s) - V_3(s)}{R_{Sn} + sL_{1n}} = \frac{V_1(s) - V_3(s)}{G_{Sn} + sC_{1n}} \tag{57}$$

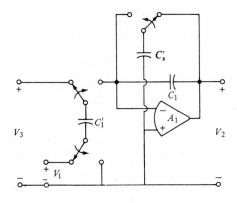

Figure 7.2-10 A sampled-data differencing damped integrator.

Unfortunately, it is necessary to difference the input with the output of stage 2, so that Fig. 7.2-6c is not directly applicable. However, a differencing damped integrator may be developed by simulating a damping resistor in Fig. 7.2-6c with a switched capacitor. The result is illustrated in Fig. 7.2-10. Since this circuit is identical to the circuit contained in the dashed box of Fig. 7.2-7a, we may use the results of (27) to write that

$$V_2(s) = \frac{1}{sT_c(C_1/C_1') + C_s'/C_1'} [V_1(s) - V_3(s)] \tag{58}$$

where $\alpha_1 = C_1'/C_1$ and $\alpha_S = C_S'/C_1'$. Normalizing T_c and comparing (57) with (58) result in

$$\alpha_1 = \frac{C_1}{C_1'} = \frac{C_{1n}}{T_c\Omega_n} = \frac{L_{1n}}{T_c\Omega_n} \tag{59}$$

and

$$\alpha_S = \alpha_1 R_{Sn} = \alpha_1 G_{Sn} \tag{60}$$

Example 7.2-2 *Sampled-Data Realization of Example 6.3-1* It is desired to develop a sampled-data realization of the low-pass third-order filter designed in Example 6.3-1. We assume that the clock frequency is 100 kHz and the cutoff frequency of the filter is 1 kHz ($\Omega_n = 2000\pi$). Using a right-to-left element numbering scheme (similar to that shown in Fig. 6.3-10), (53) gives

$$\alpha_3 = \frac{4000\pi}{3} \times 10^{-5} = 0.0419$$

and

$$\alpha_2 = \frac{6000\pi}{4} \times 10^{-5} = 0.0471$$

If $R_{Ln} = R_L = 1\ \Omega$, then (55) gives

$$\alpha_1 = 4000\pi \times 10^{-5} = 0.1257$$

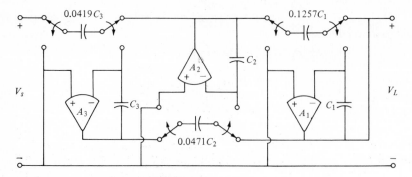

Figure 7.2-11 A sampled-data leapfrog filter.

Figure 7.2-11 shows a realization of this filter using sampled-data techniques. It is interesting to note that the values of C_1, C_2, and C_3 are arbitrary and may be selected based on other considerations having to do with the **MOS** technology or other factors. ☐

Several other examples of using sampled-data techniques to realize a leapfrog structure can be found in the literature. Among these examples are a fourth-order doubly terminated Butterworth low-pass filter,[1] a fifth-order doubly terminated Chebyshev low-pass filter,[2] and a third-order doubly terminated elliptic low-pass filter.[3] A technique for realizing zeros through a switched-capacitor integrator/summer has also been given.[4]

Since the sampled-data integrator is a basic element in the realization techniques of this section, let us examine its performance in more detail. Equation (9) represents the transfer function in the s domain of a realization of a sampled-data integrator. If $V_{out}(s)/V_{in}(s) = H(s)$, and s is replaced by $j\omega$, then (9) becomes

$$H(j\omega) = \frac{-C_1/C_2}{e^{j\omega T_c} - 1} = \left(-\frac{C_1}{C_2}\right)\frac{e^{-j\omega T_c/2}}{e^{j\omega T_c/2} - e^{-j\omega T_c/2}} \tag{61}$$

Using the definition of (12) for ω_o and using exponential relations for trigonometric functions result in

$$H(j\omega) = -(\omega_o T_c)\frac{e^{-j\omega T_c/2}}{2j \sin(\omega T_c/2)} = -\frac{\omega_o}{j\omega}\left[\frac{\omega T_c e^{-j\omega T_c/2}}{2\sin(\omega T_c/2)}\right]$$

$$= -\frac{\omega_o}{j\omega}\left[\frac{\pi f}{f_c}\frac{e^{-j\pi f/f_c}}{\sin(\pi f/f_c)}\right] \tag{62}$$

[1] Jacobs et al., op. cit.

[2] D. Allstot, R. Brodersen, and P. Gray, "Fully-Integrated High-Order NMOS Sampled-Data Ladder Filters," *Proc. Intern. Solid-State Circuits Conf.*, February 1978, pp. 82–83.

[3] Ibid.

[4] Ibid.

Equation (62) shows that the actual integrator function consists of the product of the ideal function and a nonideal factor. In the ladder structure it was found that $\omega_o = C_j'/T_c C_j = \alpha_j/T_c$. Therefore $\omega_o/\omega = \alpha_j f_c(2\pi f)$. In Example 7.2-2, if $f = f_{\text{cutoff}}$, which was 100 times less than f_c, then $\omega_o/\omega = 50\alpha_j/\pi$, which is the ideal gain of the integrator at f_{cutoff}. However, the term in the brackets of (62) is equal to $1.00016/-1.8°$. The 1.8° of phase lag can have a serious effect in high-Q realizations. If the clock frequency is increased to 1 MHz, then the bracket term of (61) becomes $1.0000016/0.18°$. The delay term of $e^{-j\omega f/f_c}$ can be removed if the output of the integrator is sampled by the next stage as soon as it changes to a new value. This point was previously discussed, and the realization of Fig. 7.2-11 shows the proper switch phasing to remove the delay term in the parentheses of (62).

A sampled-data integrator was recently proposed[1,2] which doubles the sample rate of the integrator and achieves a zero at half the sample frequency. This reduces aliasing distortion. Figure 7.2-12 shows the configuration. At time $(n-1)T_c$ we assume the switches are in the position shown. Thus C_1 has a voltage of $v_1[(n-1)T_c]$ and C_2 has a voltage of $v_2[(n-1)T_c]$. During the next cycle the switches change position and the voltage on C_2 is

$$v_2(nT_c) = v_2[(n-1)T_c] - \frac{C_1}{C_2}v_1(nT_c) - \frac{C_1}{C_2}v_1[(n-1)T_c] \tag{63}$$

The transfer function of this integrator is given as

$$N(s) = \frac{V_2(s)}{V_1(s)} = -\frac{C_1}{C_2}\left(\frac{1 + e^{-sT_c}}{1 - e^{-sT_c}}\right) \tag{64}$$

Replacing s by $j\omega$ and C_1/C_2 by $\omega_o T_c$, (64) becomes

$$N(j\omega) = -\frac{\omega_o}{j\omega}\left[\frac{\omega T_c}{\tan(\omega T_c/2)}\right] = -\frac{2\omega_o}{j\omega}\left[\frac{\omega T_c/2}{\tan(\omega T_c/2)}\right] \tag{65}$$

The ideal transfer function is seen to become $-2\omega_o/j\omega$ because of the fact that the frequency sampling rate has been effectively doubled. It is seen that there is no delay term for this integrator.

[1] G. C. Temes and I. A. Young, "An Improved Switched-Capacitor Integrator." *Electronics Letters*, vol. 14, no. 9, Apr. 27, 1978, pp. 287–288.

[2] C. K. Sutton, W. K. Jenkins, and T. N. Trick, "New Structures for Switched Capacitor Sampled Data Filters," *Proc. 1978 Midwest Symp. Circuits and Systems*, August 1978, pp. 169–173.

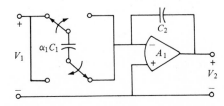

Figure 7.2-12 A sampled-data integrator with reduced aliasing.

There are other limitations for the sampled-data realization of active filters that will be briefly mentioned. The operational amplifier is one of the primary sources of performance limitation. The slew rate of the amplifier must be sufficiently high to allow the output to change rapidly from one sample value to the next. A reasonable transition time from one level to another for a 100 kHz clock is approximately 1 μs. Therefore an amplifier with a slew rate of 1 V/μs could have output voltage changes of less than 1 V without experiencing slew-rate limitations. Since the amplifier is switching at the clock rate, it is important to have a good transient response with minimum settling time. If the Q of the filter is large, the effect of finite gain of the operational amplifier can become a serious limiting factor.[1] Finally the thermal noise contributions of the amplifier and the switches must be considered. In addition to the fact that the amplifier should have low noise characteristics even in continuous active RC filters,[2] the switched capacitors which represent resistors generate a thermal rms noise given as

$$V_C(\text{rms}) = \sqrt{\frac{kT}{C}} \tag{66}$$

where kT is Boltzmann's constant times the temperature. It is seen that the switching noise can be minimized by making C large.

The use of sampled-data techniques for implementing active RC filters is a recent development and should result in an expanded use of active filters commercially. In addition the active filter realized by sampled-data techniques has several interesting properties, among which is the ability to reprogram the filter using the clock frequencies. Another is that since the switched capacitor is passive in nature, it does not create any stability problems apart from those which are inherent in active RC filters.

7.3 EXAMPLES OF ACTIVE FILTER APPLICATIONS

In this section we present two examples of active filters which were designed to meet specific industrial requirements. The objective of these examples is to show how to bring together the information from the previous chapters in such a way as to produce a successful practical realization. The examples stress different aspects of filter design and should serve as guidelines for approaching an actual filter design problem.

The first example involves a low-pass filter with the attenuation specification given in Fig. 7.3-1. This filter had been previously realized by the passive network shown in Fig. 7.3-2 and successfully used as a component of a large system. A decision to hybridize the system required the simultaneous design of an active filter to replace the passive one.

---

[1] Hosticka et al., op. cit.

[2] F. Trofimenkoff, D. Treleaven, and L. Bruton, "Noise Performance of RC-Active Quadratic Filter Sections," *IEEE Trans. Circuit Theory*, vol. CT-20, no. 5, September 1973, pp. 524–532.

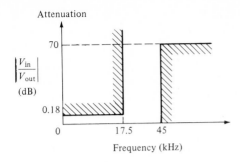

Attenuation

$\left|\dfrac{V_{in}}{V_{out}}\right|$ (dB)

70

0.18

0   17.5   45

Frequency (kHz)

Figure 7.3-1 A low-pass filter attenuation specification.

Three prototype designs were built and evaluated. They were:

Design 1: A fifth-order elliptic filter using the FDNR technique.

Design 2: A seventh-order Chebyshev filter using the leapfrog technique.

Design 3: A fifth-order elliptic filter using a cascade of two second-order blocks and a first-order one.

Application of the nomographs in Figs. 2.2-5 and 2.3-5 shows that either a fifth-order elliptic function with a passband ripple of 0.18 dB or a seventh-order Chebyshev one with 0.1-dB ripple will meet the specifications shown in Fig. 7.3-1. For design 1, using a standard table of elliptic filter approximations,[1] we obtain the passive network of Fig. 7.3-3. The $RLC:CRD$ transformation of Sec. 6.2 produces the circuit shown in Fig. 7.3-4. For this, at zero frequency we find

$$\frac{V_{out}}{V_{in}} = \frac{R_B}{R_A + R_B + 4.48062} \tag{1}$$

If we assume $R_A = 100\ \Omega$ and equate (1) to 0.5, we get $R_B = 104.4062\ \Omega$. The FDNRs are designed by letting the impedances of Fig. 6.1-2a be $Z_1(s) = 1/sC$, $Z_2(s) = R$, $Z_3(s) = R$, $Z_4(s) = R_4$, $Z_5(s) = 1/sC$. Thus $D = C^2 R_4$. Selecting $C = 1.0\ F$ and $R = 1.0\ \Omega$, the value of R_4 required to realize D_2 is $1.30215\ \Omega$. Similarly the value of R_4 required to realize D_4 is $1.21921\ \Omega$. The normalized design is shown in Fig. 7.3-5. The final realization is obtained by frequency denormalizing for a cutoff frequency of 17.5 kHz, and impedance denormalizing by 1000. Thus the final element values are found by multiplying all the resistor values

---

[1] A. I. Zverev, *Handbook of Filter Synthesis*, John Wiley & Sons, Inc., New York, 1967.

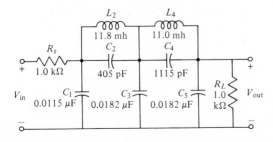

Figure 7.3-2 A passive filter realization for the specifications of Fig. 7.3-1.

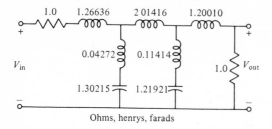

Ohms, henrys, farads

Figure 7.3-3 A passive elliptic filter realization.

by 10^3 and the capacitor values by 9.0946×10^{-9}. The actual measured frequency response of the resulting circuit is shown in Fig. 7.3-6. Note that the stopband characteristics do not meet the specifications. Subsequent experimental measurements showed that the level of attenuation obtained was actually dependent upon the input signal level, suggesting a nonlinear effect such as that of slew rate.

The FDNRs were tested in a manner similar to that described in Sec. 6.2. Using (11) of that section we find

$$\frac{V_{o2}(s)}{V_1(s)} = 1 - sR_4C \tag{2}$$

The chosen values for R_4 and C yield break frequencies of 13.439 kHz for D_2 and 14.353 kHz for D_4. In attempting to measure these curves it was necessary to first reduce the breakpoint for reasons that will shortly become apparent. To do this, values of $R_4 = 107$ kΩ and $C = 10^{-9}$ F were used to give a breakpoint at 1487 Hz. Figure 7.3-7 gives the result of the measurement. In this figure, one of the serious problems associated with FDNR realizations is apparent, namely, the output voltage of the operational amplifiers increases with frequency, causing a "jump" in operating conditions which is caused by slew-rate limitations. The two curves are for two different values of the compensation capacitor C_p (LM101A operational amplifiers were used). Since the "jump" effect is sensitive to the amplitude of the input signal level, this accounts for the sensitivity of the filter response to the signal level. Also, in Fig. 7.3-6, note that neither of the elliptic

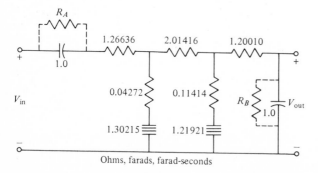

Ohms, farads, farad-seconds

Figure 7.3-4 The $RLC:CRD$ transformation of the filter of Fig. 7.3-3 using FDNRs.

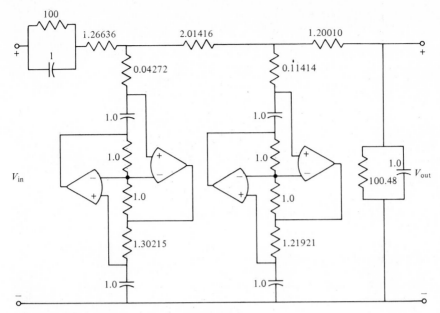

Figure 7.3-5 A realization of the filter of Fig. 7.3-4.

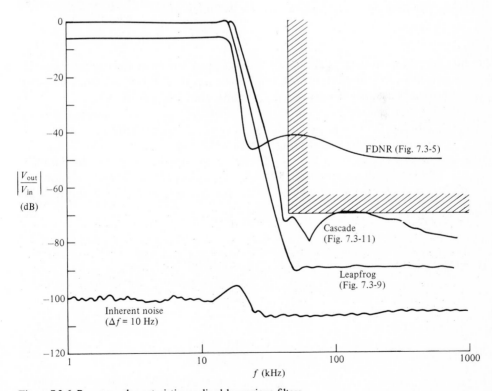

Figure 7.3-6 Response characteristics realized by various filters.

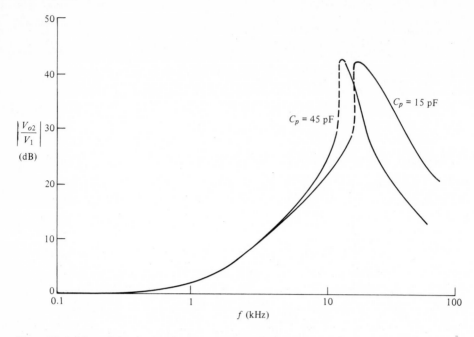

Figure 7.3-7 Effect of slew-rate limitations on FDNR characteristics.

function's stopband zeros are apparent. This gives further evidence that the series connected FDNR/R elements are not "resonating." The use of higher-slew-rate operational amplifiers improved the performance, but not enough to meet the specifications.

For design 2, the leapfrog circuit was chosen and the Chebyshev approximation was used. The low-pass prototype network was taken from standard tables and is shown in Fig. 7.3-8.[1] A single-terminated structure was used to eliminate the need for a source resistor. A leapfrog realization of this network is shown in Fig. 7.3-9. The element values for an impedance level of 23.72 kΩ and a cutoff

---

[1] This realization was obtained from table 13-2 of L. Weinberg, *Network Analysis and Synthesis*, McGraw-Hill Book Company, New York, 1962; reprinted by R. E. Krieger Publishing Co., Huntington, N.Y., 1975.

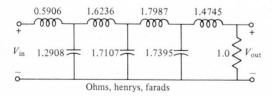

Ohms, henrys, farads

Figure 7.3-8 A seventh-order Chebyshev filter.

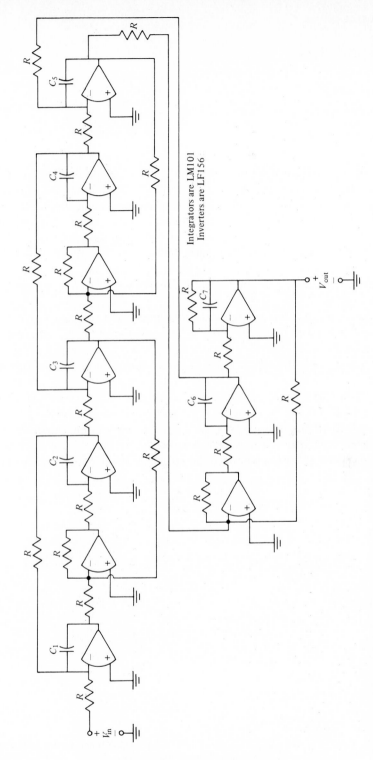

Integrators are LM101
Inverters are LF156

Figure 7.3-9 A leapfrog realization for the filter of Fig. 7.3-8.

383

frequency of 17.5 kHz are

$$R = 23.72 \text{ k}\Omega \qquad C_4 = 656 \text{ pF}$$
$$C_1 = 226 \text{ pF} \qquad C_5 = 690 \text{ pF}$$
$$C_2 = 464 \text{ pF} \qquad C_6 = 667 \text{ pF}$$
$$C_3 = 623 \text{ pF} \qquad C_7 = 565 \text{ pF}$$

The LF156 operational amplifier was used for the inverters and the LM101 for the integrators. The frequency characteristics of the resulting filter are shown in Fig. 7.3-6. Obviously the leapfrog filter easily satisfies the specifications. As shown in the figure, the stopband attenuation is limited at about -90 dB. This limitation is due to the noise level of the filter, which is significant because of the large number of active devices. A disadvantage of the leapfrog realization is that it requires 10 operational amplifiers, 7 precision capacitors, and 20 equal-valued resistors. The large number of components makes this approach undesirable for hybridization.

For design 3, a cascade of two second-order blocks and one first-order block was used to realize the elliptic filter function. The sensitivities of this approach are not as good as those of the other two designs; however, since the number of cascaded stages is small, the results are acceptable. The normalized pole locations for the filter are shown in Fig. 7.3-10. The circuit selected for the second-order block is the resonator of Fig. 7.1-11. This circuit uses a minimum of operational amplifiers and yet has the good frequency response which is necessary for the specified stopband attenuation. The first-order stage is chosen as a simple RC network. The poles and zeros are grouped as indicated in Fig. 7.3-10. The design equations are generated for the second-order stages by letting R_4 of Fig. 7.1-11 be

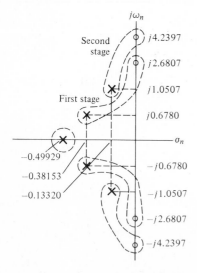

Figure 7.3-10 Normalized pole and zero locations for a fifth-order elliptic filter.

infinite. Thus (27) of Sec. 7.1 becomes

$$\frac{V_{out}(s)}{V_{in}(s)} = -\frac{C_3}{C_1} \frac{s^2 + \dfrac{C_1 R_2}{C_3 R_3} \dfrac{1}{R_2 R_5 C_1 C_2}}{s^2 + \dfrac{s}{R_1 C_1} + \dfrac{1}{R_2 R_5 C_1 C_2}} \tag{3}$$

The general second-order elliptic transfer function can be written as

$$\frac{V_o(s)}{V_s(s)} = -H_0 \frac{s^2 + \omega_z^2}{s^2 + (\omega_p/Q_p)s + \omega_p^2} \tag{4}$$

Therefore, equating (3) to (4) gives

$$\omega_p = \sqrt{\frac{1}{R_2 R_5 C_1 C_2}} \tag{5}$$

$$Q_p = R_1 \sqrt{\frac{C_1}{R_2 R_5 C_2}} \tag{6}$$

$$H_0 = \frac{R_2}{R_3} \tag{7}$$

and

$$\omega_z = \omega_p \sqrt{H_0 \frac{C_1}{C_3}} \tag{8}$$

The design of the filter begins by frequency denormalizing the individual stage transfer functions. Consequently, we get

Stage 1

$$H_1(s) = \frac{s^2 + 8.68825 \times 10^{10}}{s^2 + 83903s + 0.73176 \times 10^{10}} \tag{9a}$$

$$f_{p1} = 13.614 \text{ kHz} \qquad Q_{p1} = 1.0195 \qquad f_{z1} = 46.912 \text{ kHz} \tag{9b}$$

Stage 2

$$H_2(s) = \frac{s^2 + 2.17323 \times 10^{11}}{s^2 + 29292s + 1.35618 \times 10^{10}} \tag{10a}$$

$$f_{p2} = 18.534 \text{ kHz} \qquad Q_{p2} = 3.97566 \qquad f_{z2} = 74.194 \text{ kHz} \tag{10b}$$

Stage 3

$$H_3(s) = \frac{54900}{s + 54900} \tag{11a}$$

$$f_{p3} = 8738 \text{ Hz} \tag{11b}$$

The design of each of the second-order stages is developed from (5) through (8). The operational amplifiers selected are the LM 3900 which is a quad CCVS operational amplifier on a single monolithic chip. In order to have dc coupling, the LM3900 operational amplifiers are powered from ± 15 V (rather than $+ 30$ V). Consequently, it is necessary to add a resistance from the plus terminal of amplifier A_2 in Fig. 7.1-11 to ground. The dc equation for amplifier A_1 with R_4 infinite is

$$\frac{V_{out}(0) + 15}{R_1} = \frac{V_{o2}(0) + 15}{R_5} \tag{12}$$

Similarly for the A_2 amplifier

$$\frac{V_{in}(0) + 15}{R_3} + \frac{V_{out}(0) + 15}{R_2} = \frac{15}{R} \tag{13}$$

where R is the value of the resistance added from the plus terminal of amplifier A_2 to ground. The fact that the input terminals of the LM3900s are at -15 V potential is incorporated into (12) and (13). If $V_{in}(0) = 0$, $V_{o2}(0) = 0$, and $V_{out}(0) = 0$, then

$$R_1 = R_5 \qquad \text{and} \qquad \frac{1}{R} = \frac{1}{R_2} + \frac{1}{R_3} \tag{14}$$

In order to keep the bias current flowing into the LM3900s less than 500 μA, R_1, R_2, R_5, and R should be greater than 30 kΩ. The normal approach of letting $C_1 = C_2$ results in values of R which are just under 30 kΩ. Consequently, the following design procedure is developed to give larger values of R:

1. Select C_2 as a convenient value.
2. Calculate $C_1 = 2Q_p^2 C_2$.
3. Set $R_1 = R_5 = Q_p/\omega_p C_1$.
4. Set $R_2 = R_3 = 2R_1$.
5. Choose $R = R_2/2 = R_1$.
6. Find $C_3 = H_0 C_1 (\omega_p/\omega_z)^2$.

The first stage design gives $C_2 = 40$ pF, $C_1 = 83$ pF, $R_1 = R_5 = 143$ kΩ, $R_2 = R_3 = 286$ kΩ, $R = 143$ kΩ, and $C_3 = 7$ pF. The second stage design results in $C_2 = 20$ pF, $C_1 = 632$ pF, $R_1 = R_5 = 54$ kΩ, $R_2 = R_3 = 108$ kΩ, $R = 54$ kΩ, and $C_3 = 39$ pF. The last stage is a simple buffered RC network. The resulting filter is shown in Fig. 7.3-11. Its frequency response is shown in Fig. 7.3-6. The tuning of the second-order stages is accomplished as follows:

1. Check the outputs of each amplifier to make sure the dc levels are achieved.
2. With C_3 removed, use C_1 to tune f_p.
3. Use R_1 to tune Q_p.
4. Reinsert C_3 and use C_3 to tune f_z.

The measured inherent noise of the filter is shown in Fig. 7.3-6. It is considerably lower than that of the leapfrog realization of design 2. The components required are 11 resistors, 7 capacitors, and 5 operational amplifiers. However, from a hybridization viewpoint there are only two operational amplifier chips. The values of components are compatible with hybridization techniques. The effects of slew rate were observed when input levels higher than 0.5 V rms were used. In the specific filter application, however, the preceding stage was a limiter set for 0.3 V rms. Thus, no problem with slew rate was encountered.

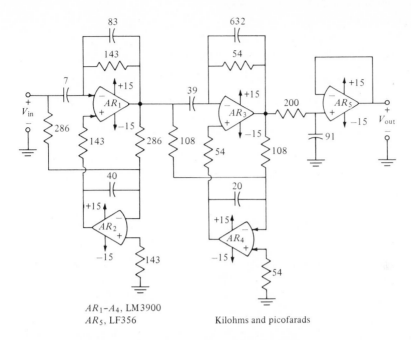

AR_1-A_4, LM3900
AR_5, LF356

Kilohms and picofarads

Figure 7.3-11 A cascade realization of a fifth-order elliptic filter.

The second example involves the use of the leapfrog technique to design an eight-pole Butterworth bandpass filter. The center frequency is 3000 Hz and the bandwidth is 600 Hz. The second-order blocks are realized by the UAF described in Sec. 5.3. The sensitivity performance of the realization is examined experimentally by varying individual stage f_ps and Q_ps by ± 10 percent.

A low-pass fourth-order Butterworth prototype filter is found from App. A and is shown in Fig. 7.3-12a. Transforming this to a bandpass prototype with a center frequency of 1 rad/s and a bandwidth of 0.2 rad/s yields the circuit of Fig. 7.3-12b. The leapfrog configuration of Fig. 6.3-4 and a modification of the UAF shown in Fig. 5.3-2 to obtain a resonator are used to synthesize the filter. The modified circuit is shown in Fig. 7.3-13. The various normalized transfer functions that must be realized by the UAFs are summarized below:

$$H_1(s) = \frac{-1}{Y_1(s)} = \frac{-7.5185s}{s^2 + 1} \tag{15a}$$

$$H_2(s) = \frac{1}{Z_2(s)} = \frac{7.8860s}{s^2 + 1} \tag{15b}$$

$$H_3(s) = \frac{-1}{Y_3(s)} = \frac{-5.4120s}{s^2 + 1} \tag{15c}$$

$$H_4(s) = \frac{1}{Z_4(s)} = \frac{1.9135s}{s^2 + 1.9135s + 1} \tag{15d}$$

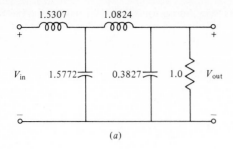

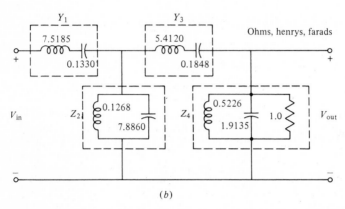

Figure 7.3-12 (*a*) A fourth-order low-pass prototype filter. (*b*) An eighth-order bandpass filter.

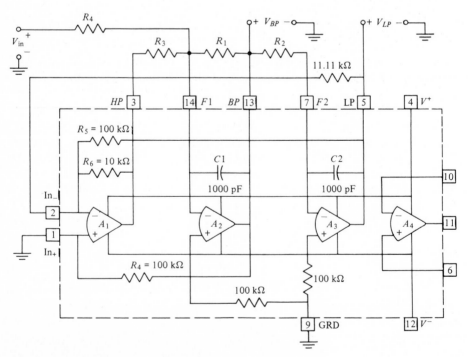

Figure 7.3-13 A modification of the UAF to produce a resonator filter.

The element values in Fig. 7.3-13 are:

$H_1(s)$:
$$R_1 = \infty \qquad R_4 = 7.5185\ \Omega \qquad C = 1\ \text{F} \qquad R = 1\ \Omega \qquad (16a)$$

$H_2(s)$:
$$R_1 = \infty \qquad R_4 = 7.8860\ \Omega \qquad C = 1\ \text{F} \qquad R = 1\ \Omega \qquad (16b)$$

$H_3(s)$:
$$R_1 = \infty \qquad R_4 = 5.4120\ \Omega \qquad C = 1\ \text{F} \qquad R = 1\ \Omega \qquad (16c)$$

$H_4(s)$:
$$R_1 = 1.9135\ \Omega \qquad R_4 = 1.9135\ \Omega \qquad C = 1\ \text{F} \qquad R = 1\ \Omega \qquad (16d)$$

To obtain the required unnormalized value of $C = 1000$ pF provided in the UAF, the capacitors must be denormalized by a value of 10^9. Since the required frequency denormalization is 6000π, the impedance denormalization must be 53,052. Thus all resistor values are multiplied by 53,052 and all capacitor values are divided by 10^9. The resulting realization is shown in Fig. 7.3-14. The UAFs and the pin numbers correspond to those shown in Fig. 7.3-13.

The sensitivity of the realization was examined with respect to changes in f_{pi} and Q_{pi}, where i corresponds to the ith UAF stage. The results are tabulated in Tables 7.3-1 and 7.3-2. Table 7.3-1 gives the center frequency f_o of the realization for a given change in f_{pi} or Q_{pi}. Table 7.3-2 gives the experimentally determined sensitivities of f_o with respect to the various f_{pi} and Q_{pi}. The fact that the transfer function is more dependent upon $S_{x_i}^{\omega_{pi}}$ than $S_{x_i}^{Q_i}$ is borne out in these results.[1]

---

[1] L. T. Bruton, "High-Frequency Limitations of GIC-Derived Ladder Structures and Two-Integrator Loop (Coupled Biquad) Ladder Structures," *Proc. 8th Asilomar Conf. Circuits, Systems, Computers*, December 1974.

Table 7.3-1 Detuning procedure for Fig. 7.3-14

Stage no.	Parameter	% change	Resultant overall f_o
1	f_{p1}	$+10\%$	3.146 kHz
1	f_{p1}	-10%	2.877 kHz
2	f_{p2}	$+10\%$	2.987 kHz
2	f_{p2}	-10%	3.001 kHz
3	f_{p3}	$+10\%$	3.088 kHz
3	f_{p3}	-10%	2.892 kHz
4	f_{p4}	$+10\%$	2.893 kHz
4	f_{p4}	-10%	3.008 kHz
All	f_{pi}	$+10\%$	3.343 kHz
All	f_{pi}	-10%	2.743 kHz
4	Q_{p4}	$+10\%$	2.998 kHz
4	Q_{p4}	-10%	2.998 kHz

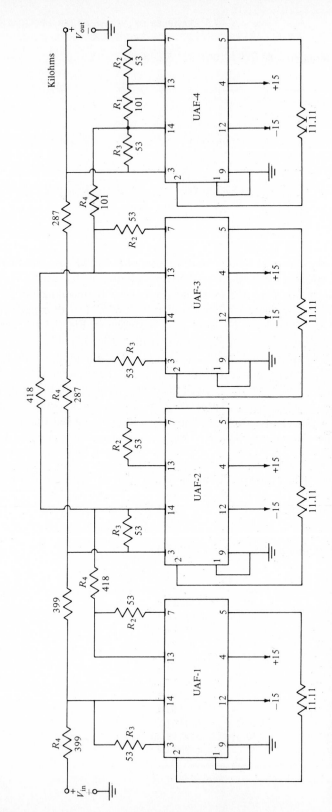

Figure 7.3-14 A leapfrog realization of the eighth-order bandpass filter of Fig. 7.3-12b.

Table 7.3-2 Results of sensitivity analysis on Fig. 7.3-14

Parameter	With respect to	Change	Sensitivity
Overall f_o	Δf_{p1}	$\pm 10\%$	$+0.428$
Overall f_o	Δf_{p2}	$\pm 10\%$	-0.023
Overall f_o	Δf_{p3}	$\pm 10\%$	$+0.317$
Overall f_o	Δf_{p4}	$\pm 10\%$	-0.042
Overall f_o	All Δf_p	$\pm 10\%$	$+0.897$
Overall f_o	ΔQ_{p4}	$\pm 10\%$	$\simeq 0$

PROBLEMS

7-1 (*Sec. 7.1*)   Use the active R circuit shown in Fig. 7.1-1b to design an inverting bandpass filter with $Q = 5$ and $f_n = 100$ kHz. What value of H is realized?

7-2 (*Sec. 7.1*)   Show that if Z_1 and Z_2 of Fig. 7.1-5a become R and $1/sC$ respectively, and that if $A_d(s) = A_0 \omega_a/(s + \omega_a)$, the closed-loop magnitude response has a pole at $-GB$.

7-3 (*Sec. 7.1*)   Use the circuit of Fig. 7.1-10 to obtain a second-order low-pass realization having $Q = 5$ and $f_n = 20$ kHz.

7-4 (*Sec. 7.1*)   Repeat Prob. 7-3 for a second-order bandpass realization.

7-5 (*Sec. 7.1*)   Use the circuit of Fig. 7.1-11 to obtain a realization similar to that of Prob. 7-3, but having complex zeros at $\pm j100\pi$ krad/s.

7-6 (*Sec. 7.1*)   The application of the operational amplifier's open-loop transfer function to create a synthetic inductance is shown in Fig. P7-6a. Find $|Z_{in}(j\omega)|$ and arg $[Z_{in}(j\omega)]$ of this circuit if the open-loop transfer function of the operational amplifier is given in Fig. P7-6b.

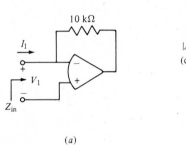
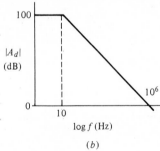

(a)   (b)

Figure P7-6

7-7 (*Sec. 7.1*)   The frequency response of Fig. P7-7 was observed to resonate at 12.7 kHz with a Q of 158. Assuming that the operational amplifier has the open-loop transfer function given in Fig. P7-6b, find V_{out}/V_{in}, find the value of Q and ω_n, and compare with the observed values.

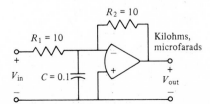

Figure P7-7

7-8 (*Sec. 7.2*)   Show that $v_{\text{out}}(2T_c)$ of Fig. 7.2-4 is equal to 0.298 V.

7-9 (*Sec. 7.2*)   Find the transfer function of (21) of Sec. 7.2 if the switches of Fig. 7.2-5b are oppositely phased, i.e., if C_3 is connected to the inverting input of the operational amplifier when C_1 is connected to V_{in}.

7-10 (*Sec. 7.2*)   Develop (24) of Sec. 7.2.

7-11 (*Sec. 7.2*)   Replace the positive-gain low-pass realization of Fig. 4.2-4 with a switched-capacitor equivalent circuit. Assume that $K = 1$ so that R_A and R_B of Fig. 4.2-4 are infinite and zero respectively. Express α_1 znd α_2 in terms of the filter specifications T_c, ω_n, and Q, using $\alpha_1 = C_1/C_2$ and $\alpha_2 = C_3/C_4$, and letting C_1 and C_3 be the switched capacitor replacements for R_1 and R_3 respectively.

7-12 (*Sec. 7.2*)   (*a*) Use the resonator realization of Fig. 7.2-7b to realize a bandpass second-order filter with a Q of 100 and a resonant frequency f_n of 100 Kz. Assume the clock frequency is 100 kHz and give the values of α_1 and α_2 as defined in (36) of Sec. 7.2.

(*b*) Repeat this design if $f_n = 5$ kHz and the clock frequency is still 100 kHz.

7-13 (*Sec. 7.2*)   Use the state-variable circuit of Fig. 7.2-9 to realize a second-order bandpass filter with a Q of 10 and $f_n = 1$ kHz. Give the values of α_1, α_2, α_3, and α_4 where $\alpha_1 = C'_1/C_1$, $\alpha_2 = C'_2/C_2$, $\alpha_3 = C'_4/C'_3$, and $\alpha_4 = C'_5/C'_6$. Assume that $\alpha_1 = \alpha_2$, $\alpha_4 = 1$, and $T_c = 10^{-5}$ s.

7-14 (*Sec. 7.2*)   Use the sampled-data method to realize a fourth-order low-pass Butterworth voltage transfer function. The cutoff frequency should be 5 kHz. The source and load resistors should both have a value of 500 Ω. The prototype filter may be found in App. A. Give a schematic of your realization showing the values of all capacitors. Use a clock frequency of 100 kHz.

7-15 (*Sec. 7.2*)   Use the sampled-data method to design a realization of the attenuation characteristic shown in Fig. P6-12. A third-order elliptic function should be used to approximate the characteristic. The prototype filter may be designed using App. A. The source and load resistors should have a value of 1000 Ω. The clock frequency is 100 kHz. Give a schematic of your realization and show the values of all capacitors.

7-16 (*Sec. 7.2*)   Show that the sampled-data circuit of Fig. P7-16 realizes an inverting circuit which integrates V_2 and sums it with V_1 by finding $V_3(s)$ as a function of $V_1(s)$ and $V_2(s)$. Assume that $sT_c \ll 1$.

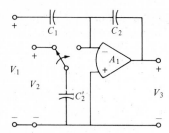

Figure P7-16

BIBLIOGRAPHY

Additional information on many of the topics treated in this book can be obtained from the volumes listed below. Considering first the material on approximation contained in Chap. 2 (and also the subject of passive network synthesis), the classic text, by Guillemin, has a precision of expression not often found. In addition it gives the reader an outstanding insight into the mathematics of circuit synthesis techniques. Other volumes of note in this area are the one by Van Valkenburg, which is especially recommended for its clear treatment of introductory concepts, and the one by Balabanian, which contains an especially detailed treatment of RC networks. Tabulations of various network approximations and filters which realize them may be found in Christian and Eisenmann, Craig, Saal, and Zverev. Considering next the material on sensitivity contained in Chap. 3, the volume by Geher provides an outstanding reference. Finally, for the material on active filters contained in Chaps. 4 through 6, Budak, Daryanani, Lindquist, Sedra and Brackett, and Temes and Mitra provide modern treatments of many of the topics in this general subject area. In addition, Huelsman (1976) has compiled an edited collection of the most significant original research papers which represent the benchmarks in the evolution of the active filter field.

Balabanian, N.: *Network Synthesis*, Prentice-Hall, Inc., Englewood Cliffs, N.J., 1958.

Blinchikoff, Herman J., and Anatol I. Zverev: *Filtering in the Time and Frequency Domains*, John Wiley & Sons, Inc., New York, 1976.

Budak, Aram: *Passive and Active Network Analysis and Synthesis*, Houghton Mifflin Company, Boston, 1974.

Calahan, Donald A.: *Modern Network Synthesis*, vols. 1 and 2, Hayden Book Company, New York, 1964.

Cauer, Wilhelm: *Synthesis of Linear Communication Networks*, vols. I and II, McGraw-Hill Book Company, New York, 1958.

Christian, Erich, and Egon Eisenmann: *Filter Design Tables and Graphs*, Transmission Networks International, Inc., Knightdale, N.C., 1977.

Craig, J. W.: *Design of Lossy Filters*, The MIT Press, Cambridge, Mass., 1970.

Daniels, Richard W.: *Approximation Methods for Electronic Filter Design with Applications to Passive, Active, and Digital Networks*, McGraw-Hill Book Company, New York, 1974.

Daryanani, Gobind: *Principles of Active Network Synthesis and Design*, John Wiley & Sons, Inc., New York, 1976.

Geffe, Philip R.: *Simplified Modern Filter Design*, John F. Rider, Publisher, Inc., New York, 1963.

Geher, K.: *Theory of Network Tolerances*, Akademiai Kiado, Budapest, 1971.

Guillemin, Ernst A.: *Synthesis of Passive Networks*, John Wiley & Sons, Inc., New York, 1957.

Haykin, S. S.: *Synthesis of RC Active Filter Networks*, McGraw-Hill Publishing Company, Ltd., London, 1969.

Hazony, Dov: *Elements of Network Synthesis*, Reinhold Publishing Corporation, New York, 1963.

Hilburn, John L., and David E. Johnson: *Manual of Active Filter Design*, McGraw-Hill Book Company, New York, 1973.

Huelsman, Lawrence P.: *Active Filters: Lumped, Distributed, Integrated, Digital, and Parametric*, McGraw-Hill Book Company, New York, 1970.

———: *Active RC Filters: Theory and Application*, Dowden, Hutchinson & Ross, Stroudsburg, Pa., 1976.

———: *Theory and Design of Active RC Circuits*, McGraw-Hill Book Company, New York, 1968.

Humpherys, DeVerl S.: *The Analysis, Design, and Synthesis of Electrical Filters*, Prentice-Hall, Inc., Englewood Cliffs, N.J., 1970.

Johnson, David E.: *Introduction to Filter Theory*, Prentice-Hall, Inc., Englewood Cliffs, N.J., 1976.

——— and John L. Hilburn: *Rapid Practical Designs of Active Filters*, John Wiley & Sons, Inc., New York, 1975.

Karni, Shlomo: *Network Theory: Analysis and Synthesis*, Allyn and Bacon, Inc., New York, 1966.

Lindquist, Claude S.: *Active Network Design*, Steward & Sons, Long Beach, Calif., 1977.

Lubkin, Yale J.: *Filter Systems and Design: Electrical, Microwave, and Digital*, Addison-Wesley Publishing Company, Inc., Reading, Mass., 1970.

Mitra, Sanjit K.: *Analysis and Synthesis of Linear Active Networks*, John Wiley & Sons, Inc., New York, 1969.

Moschytz, George S.: *Linear Integrated Networks: Design*, Van Nostrand Reinhold Co., New York, 1975.

———: *Linear Integrated Networks: Fundamentals*, Van Nostrand Reinhold Co., New York, 1976.

Newcomb, Robert W.: *Active Integrated Circuit Synthesis*, Prentice-Hall, Inc., Englewood Cliffs, N.J., 1968.

Saal, R.: *The Design of Filters Using the Catalogue of Normalized Low-Pass Filters*, Telefunken G.M.B.H., Backnang/Wurtt., West Germany, 1963.

Sedra, Adel S., and Peter O. Brackett: *Filter Theory and Design: Active and Passive*, Matrix Publishers, Inc., Champaign, Ill., 1977.

Spence, Robert: *Linear Active Networks*, John Wiley & Sons, Inc., New York, 1970.

Su, Kendall L.: *Active Network Synthesis*, McGraw-Hill Book Company, New York, 1965.

———: *Time Domain Synthesis of Linear Networks*, Prentice-Hall, Inc., Englewood Cliffs, N.J., 1971.

Temes, Gabor C., and Sanjit K. Mitra: *Modern Filter Theory and Design*, John Wiley & Sons, Inc., New York, 1973.

Tuttle, David F., Jr.: *Network Synthesis*, vol. I, John Wiley & Sons, Inc., New York, 1958.

Van Valkenburg, M. E.: *Introduction to Modern Network Synthesis*, John Wiley & Sons, Inc., New York, 1960.

Vlach, Jiri: *Computerized Approximation and Synthesis of Linear Networks*, John Wiley & Sons, Inc., New York, 1969.

Wait, John V., Lawrence P. Huelsman, and Granino A. Korn: *Introduction to Operational Amplifier Theory and Applications*, McGraw-Hill Book Company, New York, 1975.

Weinberg, Louis: *Network Analysis and Synthesis*, McGraw-Hill Book Company, New York, 1962; R. E. Krieger Publishing Co., Huntington, N.Y., 1975.

Williams, Arthur B.: *Active Filter Design*, Artech House, Inc., Dedham, Mass., 1975.

Zverev, Anatol I.: *Handbook of Filter Synthesis*, John Wiley & Sons, Inc., New York, 1967.

PASSIVE LOW-PASS FILTER REALIZATIONS

In this appendix we present tables for passive filter realizations of various types of low-pass network functions. The first type of filter is the single-resistance-terminated lossless-ladder network. The form of the realization for a voltage-source excitation is shown in Figs. A-1a (for even-order functions) and A-1b (for odd-order functions). For a current-source excitation the realizations have the form shown in Figs. A-2a (for even-order functions) and A-2b (for odd-order functions). The element values for (normalized) Butterworth, Chebyshev ($\frac{1}{2}$- and 1-dB ripple), and Thomson filters are given in Table A-1.

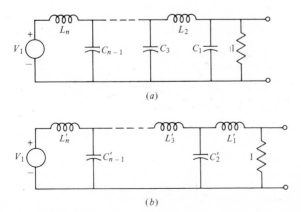

(a)

(b)

Figure A-1 Network configuration for Table A-1 (voltage-source excitation). (a) Even. (b) Odd.

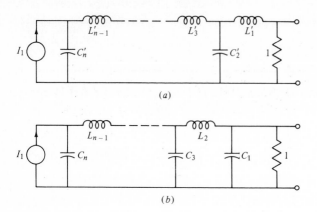

Figure A-2 Network configuration for Table A-1 (current-source excitation). (*a*) Even. (*b*) Odd.

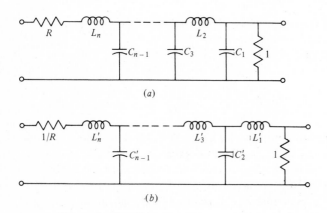

Figure A-3 Network configuration for Tables A-2 and A-3. (*a*) Even. (*b*) Odd.

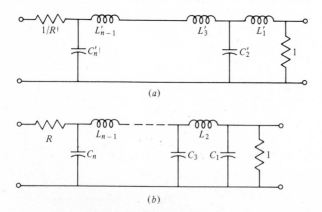

Figure A-4 Network configuration for Tables A-2 and A-3. (*a*) Even. (*b*) Odd.

Table A-1 Element values for low-pass single-resistance-terminated lossless-ladder realizations*

Elements in Figs. A-1a (even) and A-2b (odd)

n	C_1	L_2	C_3	L_4	C_5	L_6	C_7	L_8	C_9	L_{10}
2	0.7071	1.4142								
3	0.5000	1.3333	1.5000		Butterworth					
4	0.3827	1.0824	1.5772	1.5307	(1 rad/s bandwidth)					
5	0.3090	0.8944	1.3820	1.6944	1.5451					
6	0.2588	0.7579	1.2016	1.5529	1.7593	1.5529				
7	0.2225	0.6560	1.0550	1.3972	1.6588	1.7988	1.5576			
8	0.1951	0.5776	0.9370	1.2588	1.5283	1.7287	1.8246	1.5607		
9	0.1736	0.5155	0.8414	1.1408	1.4037	1.6202	1.7772	1.8424	1.5628	
10	0.1564	0.4654	0.7626	1.0406	1.2921	1.5100	1.6869	1.8121	1.8552	1.5643
2	0.7014	0.9403			$\frac{1}{2}$-dB ripple Chebyshev					
3	0.7981	1.3001	1.3465		(1 rad/s bandwidth)					
4	0.8352	1.3916	1.7279	1.3138						
5	0.8529	1.4291	1.8142	1.6426	1.5388					
6	0.8627	1.4483	1.8494	1.7101	1.9018	1.4042				
7	0.8686	1.4596	1.8675	1.7371	1.9712	1.7254	1.5982			
8	0.8725	1.4666	1.8750	1.7508	1.9980	1.7838	1.9571	1.4379		
9	0.8752	1.4714	1.8856	1.7591	2.0116	1.8055	2.0203	1.7571	1.6238	
10	0.8771	1.4748	1.8905	1.7645	2.0197	1.8165	2.0432	1.8119	1.9816	1.4539
2	0.9110	0.9957			1-dB ripple Chebyshev					
3	1.0118	1.3332	1.5088		(1 rad/s bandwidth)					
4	1.0495	1.4126	1.9093	1.2817						
5	1.0674	1.4441	1.9938	1.5908	1.6652					
6	1.0773	1.4601	2.0270	1.6507	2.0491	1.3457				
7	1.0832	1.4694	2.0437	1.6736	2.1192	1.6489	1.7118			
8	1.0872	1.4751	2.0537	1.6850	2.1453	1.7021	2.0922	1.3691		
9	1.0899	1.4790	2.0601	1.6918	2.1583	1.7213	2.1574	1.6707	1.7317	
10	1.0918	1.4817	2.0645	1.6961	2.1658	1.7306	2.1803	1.7215	2.1111	1.3801
2	0.3333	1.0000			Thomson (1-s delay at dc)					
3	0.1667	0.4800	0.8333							
4	0.1000	0.2899	0.4627	0.7101						
5	0.0667	0.1948	0.3103	0.4215	0.6231					
6	0.0476	0.1400	0.2246	0.3005	0.3821	0.5595				
7	0.0357	0.1055	0.1704	0.2288	0.2827	0.3487	0.5111			
8	0.0278	0.0823	0.1338	0.1806	0.2227	0.2639	0.3212	0.4732		
9	0.0222	0.0660	0.1077	0.1463	0.1811	0.2129	0.2465	0.2986	0.4424	
10	0.0182	0.0541	0.0886	0.1209	0.1549	0.1880	0.2057	0.2209	0.2712	0.4161
n	L_1'	C_2'	L_3'	C_4'	L_5'	C_6'	L_7'	C_8'	L_9'	C_{10}'

Elements in Figs. A-1b (odd) and A-2a (even)

* Reprinted by permission from L. Weinberg, *Network Analysis and Synthesis*, McGraw-Hill Book Company, New York, 1962; R. E. Krieger Publishing Co., Huntington, N.Y., 1975.

A second type of filter realization is the double-resistance-terminated lossless-ladder one. The form of the realization is shown in Figs. A-3a (for even-order functions) and A-3b (for odd-order functions). An alternate realization form is shown in Figs. A-4a (for even-order functions) and A-4b (for odd-order functions). The element values for (normalized) Butterworth, Chebyshev ($\frac{1}{2}$- and 1-dB ripple), and Thomson network functions are given in Table A-2 for the equal-resistance case. In this case, no solution exists for even-order Chebyshev functions. Table A-3 gives the element values for the case where the load and source resistances are related by a factor of two. In this case no solution exists for even-order 1-dB-ripple Chebyshev functions.

Table A-2 Element values for low-pass double-resistance-terminated lossless-ladder realizations with $R = 1\ \Omega^*$

Elements in Figs. A-3a (even) and A-4b (odd)

n	C_1	L_2	C_3	L_4	C_5	L_6	C_7	L_8	C_9	L_{10}
2	1.4142	1.4142								
3	1.0000	2.0000	1.0000			Butterworth				
4	0.7654	1.8478	1.8478	0.7654		(1 rad/s bandwidth)				
5	0.6180	1.6180	2.0000	1.6180	0.6180					
6	0.5176	1.4142	1.9319	1.9319	1.4142	0.5176				
7	0.4450	1.2470	1.8019	2.0000	1.8019	1.2470	0.4450			
8	0.3902	1.1111	1.6629	1.9616	1.9616	1.6629	1.1111	0.3902		
9	0.3473	1.0000	1.5321	1.8794	2.0000	1.8794	1.5321	1.0000	0.3473	
0	0.3129	0.9080	1.4142	1.7820	1.9754	1.9754	1.7820	1.4142	0.9090	0.3129
3	1.5963	1.0967	1.5963			$\frac{1}{2}$-dB ripple Chebyshev				
5	1.7058	1.2296	2.5408	1.2296	1.7058	(1 rad/s bandwidth)				
7	1.7373	1.2582	2.6383	1.3443	2.6383	1.2582	1.7373			
9	1.7504	1.2690	2.6678	1.3673	2.7239	1.3673	2.6678	1.2690	1.7504	
3	2.0236	0.9941	2.0236			1-dB ripple Chebyshev				
5	2.1349	1.0911	3.0009	1.0911	2.1349	(1 rad/s bandwidth)				
7	2.1666	1.1115	3.0936	1.1735	3.0936	1.1115	2.1666			
9	2.1797	1.1192	3.1214	1.1897	3.1746	1.1897	3.1214	1.1192	2.1797	
2	1.5774	0.4226				Thomson (1-s delay at dc)				
3	1.2550	0.5528	0.1922							
4	1.0598	0.5116	0.3181	0.1104						
5	0.9303	0.4577	0.3312	0.2090	0.0718					
6	0.8377	0.4116	0.3158	0.2364	0.1480	0.0505				
7	0.7677	0.3744	0.2944	0.2378	0.1778	0.1104	0.0375			
8	0.7125	0.3446	0.2735	0.2297	0.1867	0.1387	0.0855	0.0289		
9	0.6678	0.3203	0.2547	0.2184	0.1859	0.1506	0.1111	0.0682	0.0230	
0	0.6305	0.3002	0.2384	0.2066	0.1808	0.1539	0.1240	0.0911	0.0557	0.0187
n	L_1'	C_2'	L_3'	C_4'	L_5'	C_6'	L_7'	C_8'	L_9'	C_{10}'

Elements in Figs. A-3b (odd) and A-4a (even)

* Reprinted by permission from L. Weinberg, *Network Analysis and Synthesis*, McGraw-Hill Book Company, New York, 1962; R. E. Krieger Publishing Co., Huntington, N.Y., 1975.

The form of the filter for double-resistance-terminated lossless-ladder realizations of elliptic low-pass functions is given in Fig. A-5. An alternate realization form is shown in Fig. A-6. The element values for various values of passband ripple and for various orders are given in Tables A-4 and A-5. Table A-4 is for equal-resistance-terminated realizations for even-order (case c) and odd-order filters (0.1- and 1.0-dB passband ripple). Table A-5 is for case b (unequal-resistance terminations) for even-order filters (0.1- and 1.0-dB passband ripple).

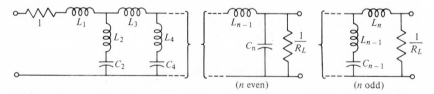

Figure A-5 Network configuration for Tables A-4 and A-5.

Table A-3 Element values for low-pass double-resistance-terminated lossless-ladder realizations with $R = \frac{1}{2}\ \Omega^*$

Elements in Figs. A-3a (even) and A-4b (odd)

n	C_1	L_2	C_3	L_4	C_5	L_6	C_7	L_8	C_9	L_{10}
					Butterworth					
2	3.3461	0.4483			(1 rad/s bandwidth)					
3	3.2612	0.7789	1.1811							
4	3.1868	0.8826	2.4524	0.2175						
5	3.1331	0.9237	3.0510	0.4955	0.6857					
6	3.0938	0.9423	3.3687	0.6542	1.6531	0.1412				
7	3.0640	0.9513	3.5532	0.7512	2.2726	0.3536	0.4799			
8	3.0408	0.9558	3.6678	0.8139	2.6863	0.5003	1.2341	0.1042		
9	3.0223	0.9579	3.7426	0.8565	2.9734	0.6046	1.7846	0.2735	0.3685	
10	3.0072	0.9588	3.7934	0.8864	3.1795	0.6808	2.1943	0.4021	0.9818	0.0825
					$\frac{1}{2}$-dB ripple Chebyshev					
2	1.5132	0.6538			(1 rad/s bandwidth)					
3	2.9431	0.6503	2.1903							
4	1.8158	1.1328	2.4881	0.7732						
5	3.2228	0.7645	4.1228	0.7116	2.3197					
6	1.8786	1.1884	2.7589	1.2403	2.5976	0.7976				
7	3.3055	0.7899	4.3575	0.8132	4.2419	0.7252	2.3566			
8	1.9012	1.2053	2.8152	1.2864	2.8479	1.2628	2.6310	0.8063		
9	3.3403	0.7995	4.4283	0.8341	4.4546	0.8235	4.2795	0.7304	2.3719	
10	1.9117	1.2127	2.8366	1.2999	2.8964	1.3054	2.8744	1.2714	2.6456	0.8104
					1-dB ripple Chebyshev					
3	3.4774	0.6153	2.8540		(1 rad/s bandwidth)					
5	3.7211	0.6949	4.7448	0.6650	2.9936					
7	3.7916	0.7118	4.9425	0.7348	4.8636	0.6757	3.0331			
9	3.8210	0.7182	5.0013	0.7485	5.0412	0.7429	4.9004	0.6797	3.0495	
					Thomson					
2	2.6180	0.1910			(1 s delay at dc)					
3	2.1156	0.2613	0.3618							
4	1.7893	0.2461	0.6127	0.0530						
5	1.5686	0.2217	0.6456	0.1015	0.1393					
6	1.4102	0.1999	0.6196	0.1158	0.2894	0.0246				
7	1.2904	0.1821	0.5797	0.1171	0.3497	0.0542	0.0735			
8	1.1964	0.1676	0.5395	0.1135	0.3685	0.0683	0.1684	0.0142		
9	1.1202	0.1558	0.5030	0.1081	0.3580	0.0744	0.2195	0.0336	0.0453	
10	1.0569	0.1460	0.4710	0.1024	0.3586	0.0763	0.2456	0.0450	0.1100	0.0925
n	L'_1	C'_2	L'_3	C'_4	L'_5	C'_6	L'_7	C'_8	L'_9	C'_{10}

Elements in Figs. A-3b (odd) and A-4a (even)

* Reprinted by permission from L. Weinberg, *Network Analysis and Synthesis*, McGraw-Hill Book Company, New York, 1962; R. E. Krieger Publishing Co., Huntington, N.Y., 1975.

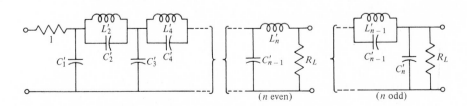

Figure A-6 Network configuration for Tables A-4 and A-5.

Table A-4 Element values for elliptic low-pass double-resistance-terminated lossless-ladder realizations with $R_L = 1\ \Omega$ (case C for even cases)*

Elements in Fig. A-5

0.1-dB passband ripple

n	ω_s	K_P	L_1 (C_1')	C_2 (L_2')	L_2 (C_2')	L_3 (L_3')	C_4 (C_3')	L_4 (L_4')	L_5 (C_5')	C_6 (L_6')	L_6 (C_6')	L_7 (L_7')	C_8 (C_7')	L_8 (L_8')	L_9 (C_9')	C_{10} (L_{10}')
3	1.05	1.748	.35550	.15374	5.39596	.35550										
	1.10	3.374	.44626	.26993	2.70353	.44626										
	1.20	6.691	.57336	.44980	1.30805	.57336										
	1.50	14.848	.77031	.74561	.47797	.77031										
	2.00	24.010	.89544	.93759	.20697	.89544										
4	1.05	3.284	.00442	.17221	4.93764	1.01224	.84445									
	1.10	6.478	.17279	.32758	2.30386	1.04694	.89415									
	1.20	12.085	.37139	.56638	1.02294	1.11938	.92440									
	1.50	23.736	.62815	.94009	.40730	1.24711	.93518									
	2.00	36.023	.77554	1.17646	.17957	1.33373	.93382									
5	1.05	13.841	.70813	.76630	.73572	1.12761	.20138	4.38116	.04985							
	1.10	20.050	.81296	.92418	.49338	1.22645	.37193	2.13500	.29125							
	1.20	28.303	.91441	1.06516	.31628	1.38201	.60131	1.09329	.52974							
	1.50	43.415	1.02789	1.21517	.15134	1.63179	.93525	.44083	.81549							
	2.00	58.901	1.08758	1.29322	.07317	1.79387	1.14330	.20038	.97720							
6	1.05	18.727	.44177	.71651	.90905	.83142	.36274	2.44680	.80463	.99857						
	1.10	26.230	.57630	.88798	.61282	.97304	.59060	1.35666	.94305	1.01181						
	1.20	36.113	.70984	1.06266	.39136	1.15974	.87407	.76185	1.09176	1.02462						
	1.50	54.202	.86595	1.27403	.18554	1.43306	1.27235	.33007	1.28253	1.03317						
	2.00	72.761	.95131	1.39297	.08926	1.60132	1.51866	.15421	1.39521	1.03621						
7	1.05	30.470	.91937	1.07659	.34220	1.09623	.40518	2.20850	.84335	.50342	1.51827	.41098				
	1.10	39.357	.98821	1.16726	.24374	1.27743	.59720	1.35681	1.04029	.67801	.96669	.58282				
	1.20	50.963	1.05029	1.24672	.16124	1.48377	.82869	.81542	1.28723	.87428	.58918	.75395				
	1.50	72.129	1.11593	1.33554	.07857	1.75687	1.15174	.37160	1.63827	1.12502	.26822	.95588				
	2.00	93.809	1.14910	1.37979	.03822	1.92026	1.35221	.17692	1.85664	1.27023	.12694	1.06720				
8	1.05	36.268	.68105	1.01040	.47466	.91103	.48838	1.83667	.67613	.70502	1.12616	.98462	1.04901			
	1.10	46.939	.77921	1.13273	.33839	1.08668	.70667	1.15081	.86562	.92698	.74468	1.09627	1.06140			
	1.20	59.639	.82950	1.25079	.22404	1.28997	.97154	.69915	1.09601	1.17713	.46698	1.21321	1.07163			
	1.50	83.807	.87790	1.38443	.10930	1.56051	1.34331	.32099	1.41679	1.50420	.21807	1.35777	1.08093			
	2.00	108.575	.91037	1.45551	.05321	1.72464	1.57463	.15324	1.61419	1.69718	.10439	1.44027	1.08484			
9	1.05	47.276	1.02597	1.21654	.20583	1.29803	.60675	1.36728	.76114	.44746	2.01083	.94136	.74312	.84407	.63916	
	1.10	58.707	1.07226	1.27741	.14773	1.46403	.79046	.92320	1.00154	.63575	1.28473	1.14956	.89544	.57633	.77015	
	1.20	73.629	1.11295	1.33139	.09815	1.64257	.99964	.58858	1.29050	.86538	.78945	1.39229	1.05143	.36877	.89697	
	1.50	100.842	1.15593	1.38761	.04800	1.86765	1.27611	.27927	1.69055	1.18960	.36523	1.72037	1.23694	.17437	1.04130	
	2.00	128.717	1.17976	1.41168	.02339	1.99976	1.44055	.13971	1.93638	1.30225	.17487	1.91900	1.33887	.08371	1.11842	
10	1.05	53.576	.82096	1.17319	.30389	1.09174	.66563	1.26609	.76760	.54466	1.65398	.76756	.97842	.69517	1.11421	1.08154
	1.10	66.262	.89544	1.26213	.21933	1.25556	.87195	.85540	.86007	.76133	1.06638	.95500	1.18228	.48206	1.20554	1.09096
	1.20	82.830	.96461	1.34441	.14648	1.44189	1.10918	.54607	1.11622	1.03051	.65893	1.17011	1.39629	.31255	1.29689	1.10032
	1.50	113.056	1.03982	1.43372	.07190	1.67758	1.42540	.25986	1.47272	1.42505	.30603	1.45794	1.65959	.14962	1.40489	1.10919
	2.00	144.023	1.07857	1.47977	.03516	1.81552	1.61465	.12563	1.69286	1.66641	.14673	1.63110	1.80819	.07224	1.46428	1.11311

Elements in Fig. A-6

Elements in Fig. A-5

1.0-dB passband ripple

n	ω_s	K_P	L_1	C_2	L_2	L_3	C_4	L_4	L_5	C_6	L_6	L_7	C_8	L_8	L_9	C_{10}
3	1.05	8.134	1.05507	.25223	3.28904	1.05507										
	1.10	11.480	1.22525	.37471	1.94752	1.22525										
	1.20	16.209	1.42450	.52544	1.11977	1.42450										
	1.50	25.176	1.69200	.73340	.48592	1.69200										
	2.00	34.454	1.85199	.85903	.22590	1.85199										
4	1.05	11.322	.63708	.35277	2.41039	1.11522	1.39953									
	1.10	15.942	.80935	.54042	1.40015	1.18107	1.45001									
	1.20	22.293	1.00329	.77733	.79634	1.26621	1.49217									
	1.50	34.179	1.25675	1.11431	.34362	1.38981	1.53225									
	2.00	46.481	1.40677	1.32367	.15960	1.46762	1.55071									
5	1.05	24.135	1.56191	.67560	.81300	1.55460	.26584	3.31881	.88528							
	1.10	30.471	1.69691	.77511	.58827	1.79892	.39922	1.98907	1.12109							
	1.20	38.757	1.82812	.87005	.38720	2.09095	.56347	1.16672	1.38094							
	1.50	53.875	1.77687	.97694	.18824	2.49161	.79362	.51950	1.71889							
	2.00	69.360	2.05594	1.03392	.09152	2.73567	.93561	.24486	1.91939							
6	1.05	29.113	1.07458	.80116	.57746	.92735	.51753	1.71498	.92186	1.60511						
	1.10	36.680	1.22059	.94235	.41284	1.10900	.75718	1.05819	1.01676	1.64682						
	1.20	46.571	1.37146	1.08633	.26628	1.32610	1.05110	.63354	1.12484	1.68498						
	1.50	64.661	1.55425	1.25876	.18779	1.62529	1.46557	.28655	1.26961	1.72482						
	2.00	83.221	1.65661	1.35450	.09179	1.80860	1.72376	.13586	1.35729	1.74424						
7	1.05	40.026	1.82156	.86343	.42668	1.67632	.34381	2.60271	1.23696	.46779	1.63392	1.22362				
	1.10	49.816	1.91040	.92662	.30705	1.93579	.42016	1.68753	1.55276	.59277	1.10699	1.41994				
	1.20	61.422	1.99168	.98474	.20446	2.22804	.64444	1.04856	1.92724	.73012	.70551	1.62539				
	1.50	82.588	2.07882	1.04761	.10016	2.61372	.87393	.48973	2.44021	.90483	.33349	1.77717				
	2.00	104.268	2.12329	1.07993	.04884	2.84446	1.01638	.23538	2.75306	1.00567	.16034	2.01924				
8	1.05	46.727	1.34673	1.00922	.47521	1.08540	.54692	1.64007	.70773	.86533	.91752	1.00315	1.71154			
	1.10	56.958	1.46597	1.10092	.34817	1.28842	.75883	1.07170	.90030	1.11345	.61997	1.07811	1.75961			
	1.20	70.098	1.68346	1.18748	.23559	1.52224	.96907	.66997	1.12753	1.39664	.39359	1.15904	1.79429			
	1.50	94.266	1.71869	1.28337	.11791	1.83710	1.36932	.31689	1.43764	1.77189	.18513	1.26147	1.83033			
	2.00	119.034	1.79131	1.33358	.05807	2.02800	1.58946	.15185	1.62645	1.99561	.08878	1.32075	1.84787			
9	1.05	57.736	1.95471	.95672	.26172	1.94887	.46951	1.76694	1.12605	.35392	2.54224	1.40079	.63099	.94407	1.47897	
	1.10	69.167	2.01503	.99876	.18875	2.18068	.60062	1.21501	1.47339	.48929	1.66927	1.71691	.73764	.69962	1.63779	
	1.20	84.089	2.06867	1.03833	.12585	2.43022	.74985	.78464	1.88531	.65376	1.04499	2.06810	.84652	.45803	1.79580	
	1.50	111.302	2.12470	1.07895	.06173	2.74001	.94696	.37634	2.45079	.88531	.49076	2.53625	.97584	.22103	1.97957	
	2.00	139.176	2.15275	1.09942	.03012	2.92871	1.06415	.18236	2.79680	1.02979	.23642	2.81740	1.04688	.10706	2.07916	
10	1.05	64.036	1.52461	1.11270	.32041	1.31260	.68239	1.23499	.69671	.59634	1.51065	.77127	1.18327	.57482	1.06945	1.79994
	1.10	76.722	1.62236	1.17101	.23621	1.51411	.87139	.85595	.92029	.82679	.98969	.94303	1.42284	.40056	1.12735	1.83423
	1.20	93.289	1.71564	1.22500	.15651	1.77787	1.08551	.55798	1.18808	1.10906	.61761	1.13757	1.68092	.25963	1.18655	1.86522
	1.50	123.515	1.81756	1.28114	.08056	2.02843	1.36690	.27098	1.55972	1.50076	.28886	1.39429	2.00434	.12388	1.25749	1.87727
	2.00	154.482	1.87378	1.30964	.03972	2.19972	1.53360	.13227	1.78908	1.76138	.13884	1.54779	2.18935	.05966	1.29688	1.91286
			C'_1	L'_2	C'_2	C'_3	L'_4	C'_4	C'_5	L'_6	C'_6	C'_7	L'_8	C'_8	C'_9	L'_{10}

Elements in Fig. A-6

* Computed using a program written by David Jose Baezlopez, "Sensitivity and Synthesis of Elliptic Functions," Ph.D. dissertation, University of Arizona, Tucson, 1978.

Table A-5 Element values for elliptic low-pass double-resistance-terminated lossless-ladder realizations for case B*

Elements in Fig. A-5

0.1-dB passband ripple
$R_L = 0.73781\ \Omega$

n	ω_s	K_P	L_1	C_2	L_2	L_3	C_4	L_4	L_5	C_6	$L_6{'}$	L_7	C_8	L_8	L_9	C_{10}
4	1.05	4.465	.15780	.18091	4.73822	1.20743	.82637									
	1.10	8.308	.33411	.33438	2.28333	1.26881	.84827									
	1.20	14.387	.53773	.55478	1.12558	1.36980	.85261									
	1.50	26.320	.79962	.88310	.43628	1.53677	.84068									
	2.00	38.697	.95051	1.08631	.19517	1.64684	.83004									
6	1.05	20.307	.57153	.65752	1.01346	.92972	.32584	2.72744	1.03524	.88809						
	1.10	27.889	.70783	.81703	.57992	1.10484	.51890	1.54640	1.19779	.88523						
	1.20	37.827	.84264	.98082	.43111	1.32791	.75659	.88144	1.37708	.87992						
	1.50	55.966	.99836	1.17887	.20248	1.64500	1.08849	.38623	1.61158	.87198						
	2.00	74.548	1.08280	1.28970	.09690	1.84134	1.29301	.18123	1.75160	.86710						
8	1.05	37.529	.79899	.92906	.52853	1.02243	.42917	2.09084	.78906	.59418	1.34317	1.24821	.89367			
	1.10	47.686	.89538	1.04944	.37264	1.22051	.61878	1.31484	1.01703	.77697	.89307	1.38901	.89115			
	1.20	60.949	.98606	1.16509	.24409	1.44606	.84957	.79988	1.29323	.98124	.56277	1.53864	.88795			
	1.50	85.138	1.08612	1.29596	.11768	1.74572	1.17448	.36725	1.67684	1.24564	.26411	1.72539	.88347			
	2.00	109.915	1.13830	1.36543	.05694	1.92553	1.37705	.17531	1.91249	1.40030	.12672	1.83237	.88075			
10	1.05	54.608	.92833	1.09111	.33387	1.21271	.58502	1.44313	.74613	.46770	1.92642	.91319	.80993	.84618	1.41635	.89491
	1.10	67.307	.99839	1.18097	.23833	1.39449	.76903	.97094	.90183	.65631	1.24697	1.14180	.97201	.59039	1.53570	.89383
	1.20	83.887	1.06237	1.26448	.15755	1.59212	.98322	.61709	1.28687	.88676	.77257	1.40608	1.14191	.38429	1.65727	.89178
	1.50	114.123	1.13092	1.35538	.07560	1.84367	1.26943	.29213	1.69639	1.21247	.35973	1.76038	1.34883	.18475	1.80184	.88893
	2.00	145.095	1.16590	1.40227	.03721	1.98983	1.44131	.14083	1.94870	1.41617	.17269	1.97377	1.46362	.08940	1.88151	.88719
			$C_1{'}$	$L_2{'}$	$C_2{'}$	$C_3{'}$	$L_4{'}$	$C_4{'}$	$C_5{'}$	$L_6{'}$	$C_6{'}$	$C_7{'}$	$L_8{'}$	$C_8{'}$	$C_9{'}$	$L_{10}{'}$

Elements in Fig. A-6

Elements in Fig. A-5

n	ω_s	K_P	L_1	C_2	L_2	L_3	C_4	L_4	L_5	C_6	L_6	L_7	C_8	L_8	L_9	C_{10}
4	1.05	13.243	.95111	.26779	3.20104	1.90749	.80699									
	1.10	18.140	1.16239	.39958	1.91077	2.05228	.80907									
	1.20	24.700	1.40135	.56068	1.11374	2.23453	.80633									
	1.50	36.771	1.71483	.78307	.49201	2.49368	.79924									
	2.00	49.156	1.90048	.91820	.23091	2.65459	.79441									
6	1.05	30.730	1.40432	.58067	1.14761	1.37588	.31837	2.79144	1.79883							
	1.10	38.342	1.55906	.69149	.80335	1.66832	.45609	1.75937	1.99786							
	1.20	48.285	1.73631	.80659	.52424	2.01190	.62218	1.07185	2.22816							
	1.50	66.425	1.93461	.94611	.25229	2.47740	.85305	.49283	2.53990							
	2.00	85.008	2.04359	1.02402	.12205	2.75884	.99547	.23540	2.72966							
8	1.05	47.987	1.67197	.76069	.64615	1.55530	.34069	2.56605	1.17579	.48610	1.64181	2.03669	.82473			
	1.10	58.146	1.79301	.84446	.46307	1.83957	.48544	1.67598	1.51273	.61510	1.12807	2.21830	.82355			
	1.20	71.408	1.90884	.92585	.30716	2.16030	.65000	1.04547	1.91110	.75902	.72753	2.41639	.82209			
	1.50	95.597	2.03847	1.01816	.14980	2.58395	.88106	.48955	2.45568	.94522	.34805	2.66846	.82006			
	2.00	120.374	2.10671	1.06721	.07286	2.83733	1.02492	.23554	2.78753	1.05416	.16833	2.81453	.81883			
10	1.05	65.067	1.83096	.86998	.41873	1.81829	.45061	1.87358	1.10022	.35991	2.50341	1.35196	.63152	1.08524	2.24376	.82561
	1.10	77.767	1.92046	.93288	.30171	2.07454	.58273	1.28236	1.45156	.49556	1.65147	1.67847	.74584	.76942	2.40166	.82484
	1.20	94.346	2.00313	.99155	.20092	2.35262	.73518	.82529	1.86693	.66055	1.03714	2.05233	.86549	.50703	2.56480	.82390
	1.50	124.583	2.09263	1.05564	.09835	2.70617	.93910	.39489	2.44690	.89307	.48639	2.50089	1.01109	.24638	2.76130	.82261
	2.00	155.554	2.13863	1.08880	.04792	2.91150	1.06145	.19123	2.80119	1.03824	.23556	2.85037	1.09249	.11976	2.87052	.82181
			C'_1	L'_2	C'_2	C'_3	L'_4	C'_4	C'_5	L'_6	C'_6	C'_7	L'_8	C'_8	C'_9	L'_{10}

1.0-dB passband ripple
$R_L = 0.37598\ \Omega$

Elements in Fig. A-6

* Computed using a program written by David Jose Baezlopez, "Sensitivity and Synthesis of Elliptic Functions," Ph.D. dissertation, University of Arizona, Tucson, 1978.

403

PROPERTIES OF OPERATIONAL AMPLIFIERS

In this appendix we present a brief introduction to the properties of operational amplifiers. An operational amplifier is essentially a controlled source in which the forward gain parameter has become very large. In normal use, feedback is applied around it to make the transfer function independent of the forward gain parameter. There are two types of operational amplifiers which are presently available for use. They are the VCVS (*voltage-controlled voltage source*) and CCVS (*current-controlled voltage source*) operational amplifiers. These operational amplifiers are represented by the symbols shown in Fig. B-1*a* (for the VCVS type) and B-1*b* (for the CCVS type). The output voltages are

$$V_o(s) = A_d[V_1(s) - V_2(s)] + A_c \frac{V_1(s) + V_2(s)}{2} \tag{1}$$

for the VCVS operational amplifier, and

$$V_o(s) = R_d[I_1(s) - I_2(s)] + R_c \frac{I_1(s) + I_2(s)}{2} \tag{2}$$

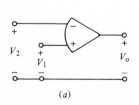

(a)

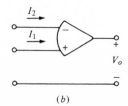

(b)

Figure B-1 Symbols for operational amplifiers. (*a*) VCVS. (*b*) CCVS.

for the CCVS operational amplifier. The parameters A_d and R_d are called the *differential-mode gains*, while A_c and R_c are called the *common-mode gains*. The *common-mode rejection ratio* (CMRR) is defined as

$$\text{CMRR} = \frac{A_d}{A_c} \tag{3}$$

for the VCVS operational amplifier, and

$$\text{CMRR} = \frac{R_d}{R_c} \tag{4}$$

for the CCVS operational amplifier. Ideally, the output resistance for both amplifiers should be zero. The input resistance of the VCVS operational amplifier should be infinite while the input resistance of the CCVS one should be zero. If the forward gain of an operational amplifier is sufficiently large so that it may be considered infinite, then the concept of a null port may be used to analyze a circuit containing it. A *null port* is simply a port, i.e., a pair of terminals, at which both the voltage and current are simultaneously zero. It can be shown that the input terminal pair of either the VCVS or the CCVS operational amplifier realizes a null port as the differential-mode gain approaches infinity. As an example, in Fig. B-2a, a noninverting finite-gain voltage amplifier circuit is shown. The variables $V_i(s)$ and $I_i(s)$ identify the null port. Assuming ideal characteristics, we find that

$$\frac{V_o(s)}{V_s(s)} = 1 + \frac{R_2}{R_1} \tag{5}$$

Similarly, in Fig. B-2b an inverting finite-gain voltage amplifier circuit is shown. Again $V_i(s)$ and $I_i(s)$ define a null port. For this circuit we obtain

$$\frac{V_o(s)}{V_s(s)} = -\frac{R_2}{R_1} \tag{6}$$

A higher level of complexity for the modeling of operational amplifiers occurs when the forward or differential-mode gain is not sufficiently large to use the null

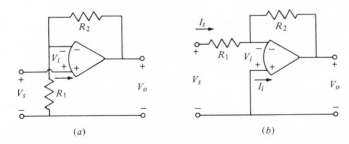

(a)   (b)

Figure B-2 Finite-gain voltage amplifiers. (a) Noninverting. (b) Inverting.

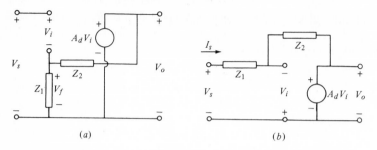

(a) (b)

Figure B-3 Models for finite-gain amplifiers. (a) Noninverting. (b) Inverting.

port concept. In this case the general model of the noninverting configuration is shown in Fig. B-3a, where the forward operational amplifier gain is given as

$$V_o(s) = A_d V_i(s) \tag{7}$$

The overall voltage gain $A(s)$ is

$$A(s) = \frac{V_o(s)}{V_s(s)} = \frac{A_d}{1 + A_d Z_1/(Z_1 + Z_2)} \tag{8}$$

The input and output impedances of the circuit retain their infinite and zero values respectively. Similarly the general model for the inverting configuration is shown in Fig. B-3b. The voltage gain $A(s)$ is

$$A(s) = \frac{V_o(s)}{V_s(s)} = \frac{-A_d Z_2/(Z_1 + Z_2)}{1 + A_d Z_1/(Z_1 + Z_2)} \tag{9}$$

For this circuit the output impedance is zero, but the input impedance is not infinite. It is found to be

$$Z_{in}(s) = \frac{V_s(s)}{I_s(s)} = \frac{Z_1}{1 + Z_1/(Z_1 + Z_2 + A_d Z_2)} \tag{10}$$

This finite input impedance of the inverting configuration causes difficulties in inverting finite-gain RC amplifier filters and is one of the reasons such filters are not widely used.

A still higher level of complexity for the modeling of operational amplifiers occurs when their input and output resistances are considered. For the noninverting finite-gain case, a circuit is shown in Fig. B-4a. For this the input and output impedances are

$$Z_{in}(s) = \frac{V_s(s)}{I_s(s)} = \frac{R_i(Z_1 + Z_2 + R_o) + Z_1(R_o + Z_2) + A_d R_i Z_1}{Z_1 + Z_2 + R_o} \tag{11a}$$

$$Z_{out}(s) = \frac{V_o(s)}{I_o(s)} = \frac{R_o[R_i(Z_1 + Z_2) + Z_1 Z_2]}{R_i(Z_1 + Z_2 + R_o) + Z_1(R_o + Z_2) + A_d R_i Z_1} \tag{11b}$$

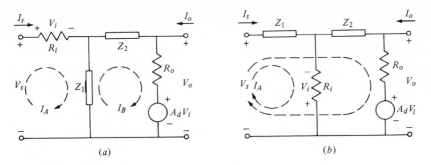

Figure B-4 Models for finite-gain amplifiers. (*a*) Noninverting. (*b*) Inverting.

A model for the inverting configuration is shown in Fig. B-4*b*. For this,

$$Z_{in}(s) = \frac{V_s(s)}{I_s(s)} = Z_1 + \frac{R_o R_i + Z_2 R_i}{Z_2 + R_o + R_i + A_d R_i} \tag{12a}$$

$$Z_{out}(s) = \frac{V_o(s)}{I_o(s)} = R_o \frac{Z_1 Z_2 + R_i R_1 + R_i Z_2}{R_i(Z_1 + Z_2 + R_o) + Z_1(R_o + Z_2) + A_d R_i Z_1} \tag{12b}$$

The voltage gain for the noninverting configuration is

$$\frac{V_o(s)}{V_s(s)}\bigg|_{I_o(s)=0} = A(s) = \frac{R_o Z_1 + A_d R_i(Z_1 + Z_2)}{R_i(Z_1 + Z_2 + R_o) + Z_1(R_o + Z_2) + A_d Z_1 R_i} \tag{13}$$

For the inverting configuration we find

$$\frac{V_o(s)}{V_s(s)}\bigg|_{I_o(s)=0} = A(s) = \frac{R_o R_i - A_d R_i Z_2}{R_i(Z_1 + R_o + Z_2) + Z_1(Z_2 + R_o) + A_d R_i Z_1} \tag{14}$$

As an example of the above relations, let the parameters of an operational amplifier be $R_i = 10^6 \ \Omega$, $R_o = 10^2 \ \Omega$, and $A_d = 10^5$. If the feedback impedances in Fig. B-4 have the values $Z_1 = 1 \ k\Omega$, and $Z_2 = 10 \ k\Omega$, then for the noninverting configuration from (11) we find $Z_{in} = 9.010 \times 10^9 \ \Omega$ and $Z_{out} = 0.011 \ \Omega$. Similarly, for the inverting configuration from (12) we obtain $Z_{in} = 1000.1 \ \Omega$ and $Z_{out} = 0.011 \ \Omega$. It should be noted that values of the order of those calculated in the preceding example are not obtainable on a practical basis (except for Z_{in} of the inverting realization). One reason is that the maximum input impedance of an operational amplifier used in a single input connection cannot exceed the common-mode impedance R_{icm} to ground. This impedance is modeled as shown in Fig. B-5. R_{icm} is typically caused by the resistance from the base to the collector of a transistor operating in the normal region. Thus, typical values are in the 10 to 100 MΩ range depending upon the degree of reverse bias and the value of the collector current. The maximum value of Z_{in} thus is 100 MΩ or less. Similarly the output impedance will be limited to about 1 Ω due to contact resistance and the finite resistance of the conductors used in the operational amplifier.

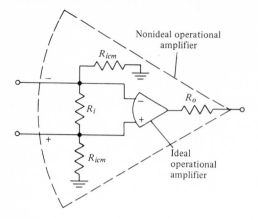

Figure B-5 Model for common-mode impedance.

Circuits which use the CCVS operational amplifier may be analyzed in exactly the same manner as illustrated in this section for the VCVS operational amplifier. A model of the CCVS operational amplifier which accounts for the nonideal impedance levels is shown in Fig. B-6. Ideally R_i and R_o are zero. It is seen that common-mode input impedances are meaningless for the CCVS operational amplifier.

Most monolithic operational amplifiers use direct (dc) coupling between their various internal stages. Such coupling introduces the possibility of dc errors in the overall amplifier characteristics. There are two sources of such error, namely the bias currents and the offset voltages. The *bias currents* may be modeled by two current sources I_{B1} and I_{B2} as shown in Fig. B-7. In this figure, the solid-line symbol represents an operational amplifier with no dc errors while the dashed-line symbol includes the error sources. The values of these sources represent the bias currents required for proper operation of the operational amplifier. These currents are small, typically in the 100-nA range. For operational amplifiers which are designed using FETs (*field effect transistors*) in the input, the bias currents are

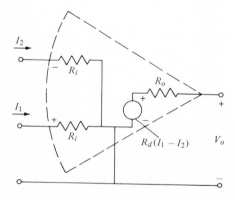

Figure B-6 Model of the CCVS operational amplifier.

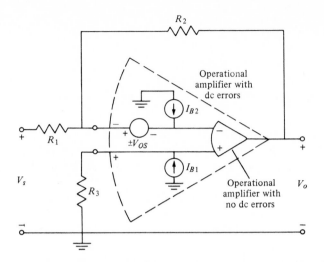

Figure B-7 Model for bias currents and offset voltage.

even smaller, typically having values of about 10 pA. The difference between I_{B1} and I_{B2} is called the *offset current*. It is defined as

$$I_{os} = |I_{B1} - I_{B2}| \tag{15}$$

The offset current is typically 5 to 10 percent of the bias current. It is the result of inequalities in the current gain of the two sides of the input differential stage of the operational amplifier.

The second source of dc error in an operational amplifier is the *offset voltage*. When an operational amplifier is operated with no differential input, a voltage will appear at the output even though the input voltage is zero. This effect is modeled by the offset voltage source V_{os} in Fig. B-7. The offset voltage accounts for any transconductance variation of the operational amplifier input stages. Typically, values of offset voltage are 1 to 5 mV. The combined effects of the two types of dc errors give

$$V_o(s) = -\frac{R_2}{R_1} V_s(s) \pm \left(1 + \frac{R_2}{R_1}\right) V_{os} + \left(1 + \frac{R_2}{R_1}\right) R_3 I_{B1} - R_2 I_{B2} \tag{16}$$

Since, in general, the polarities to be used with V_{os} and I_{os} are not predictable, (16) must be treated from a worst-case viewpoint. The dc errors will be minimized if the resistances to ground from the inverting and noninverting inputs of the operational amplifier are equal. Since in active filter applications the operational amplifier is always used with feedback, the dc errors do not usually cause serious problems.

Conventional direct-coupled operational amplifiers require the use of a split power supply providing both a positive and a negative polarity output. Recently, however, single-supply operational amplifiers have been introduced. These are available in both VCVS and CCVS types. A circuit symbol for the single-supply VCVS is shown in Fig. B-8. The voltages at the inverting and noninverting inputs are shown as V^- and V^+ respectively, while the supply voltage is labeled as V_{cc}.

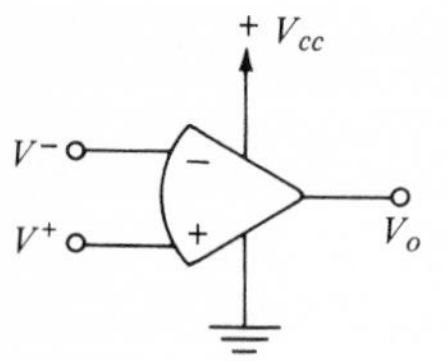

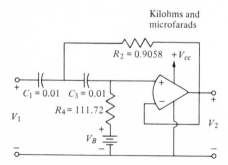

Figure B-8 Symbol for single-supply VCVS operational amplifier.

Unlike split-supply operational amplifiers for which dc coupling is readily used, single-supply operational amplifiers must have a biased input. In practice the bias is selected so that under zero signal conditions the output voltage will be midway between the power supply voltage V_{cc} and ground. As an example, consider the use of a single-supply VCVS operational amplifier to realize a unity-gain high-pass realization (see Example 4.3-3). Figure B-9 shows the realization with an additional source V_B being used to provide the bias capability. If V_B is set to $V_{cc}/2$, then the dc value of V_2 will also be $V_{cc}/2$. V_B can be obtained directly from the power supply V_{cc} by using a voltage divider of the form shown in Fig. B-10. In using such a network one must make certain that it does not affect the dc characteristics of the circuit. In this case, since $\omega_n \approx 10^4$ rad/s, we calculate from Fig. B-10 that the ac source impedance of V_B is approximately 100 Ω, which is insignificant when placed in series with R_4, which is 111 kΩ.

Now let us consider the ac properties of operational amplifiers. One of the most important characteristics of an operational amplifier, especially with regard to its application in active filters, is its frequency response. As frequency increases, the gain of an operational amplifier decreases due to the bandwidth limitations of its solid-state devices. The more the gain decreases, the more the overall filter realization becomes dependent upon the open-loop gain of the amplifier. In addition, if excessive phase lag is introduced into the open-loop transfer function, a shift in the desired pole locations is produced which can lead to instability.

The actual frequency response of an operational amplifier is a complex expression. A simplified approximation that works well consists of three negative-real poles.[1] Such a model predicts that when feedback is applied around it, instabil-

---

[1] G. E. Tobey, J. D. Graeme, and L. P. Huelsman, *Operational Amplifiers—Design and Applications,* McGraw-Hill Book Company, New York, 1971, chap. 5.

Figure B-9 Realization for a unity-gain high-pass filter.

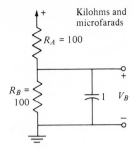

Figure B-10 Obtaining V_B from a voltage divider.

ity can result. Thus, frequency compensation must be applied to most operational amplifiers before they are usable in active filters or in almost any other application. The objective of most compensation schemes is to achieve an operational amplifier forward gain function having the form

$$A_d(s) = \frac{A_0 \omega_a}{s + \omega_a} = \frac{GB}{s + \omega_a} \tag{17}$$

where A_0 is the dc gain, ω_a is the bandwidth, and GB is the gain-bandwidth product. A magnitude plot of (17) is given in Fig. B-11 where typical values are $A_0 = 10^5$ and $\omega_a = 10$ rad/s. If we let $s = jGB$ in (17), we see that GB is the radian frequency at which the magnitude of $A_d(j\omega)$ is unity. Discussions of how frequency compensation is accomplished may be found in the literature.[1] The actual frequency response of most compensated operational amplifiers has a second pole at or above $\omega = GB$, as shown by the dashed -12 dB/octave line in Fig. B-11. Usually, however, this pole may be ignored.

Another characteristic of an operational amplifier on which active filter performance is dependent is the *slew rate*, i.e., the ability to follow a rapidly changing input signal. This effect is difficult to model since it is a nonlinear one. In active

[1] James Roberge, *Operational Amplifiers—Theory and Practice*, John Wiley & Sons, Inc., New York, 1975.

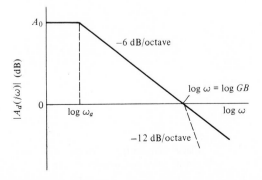

Figure B-11 Magnitude of the forward gain of an operational amplifier.

filter applications, it can cause an amplitude-dependent phase lag in the feedback loop. For example, in some filters, as the input signal frequency is swept from low to high values, it is possible for the circuit to break into instability as the frequency approaches the resonant frequency. If, however, the amplitude of the input signal is reduced, a stable sweep can be performed. Such an instability is caused by the fact that while the signal frequency is being increased, the resonance in the filter characteristic also increases the signal amplitude. The operational amplifier's slew-rate limitation thus creates enough phase lag to produce oscillation. To see why the slope of the output voltage of an operational amplifier must be limited to a maximum value (the slew rate), consider the frequency-compensated operational amplifier model shown in Fig. B-12.[1] In this model the source labeled $f(V_i)$ is a VCCS (*voltage-controlled current source*) whose dependence on an input voltage is given by Fig. B-13. Here SR is the slew rate. The output voltage for Fig. B-12 can be written as

$$V_o(s) = \frac{R\omega_a}{s + \omega_a} f(V_i) \tag{18}$$

where

$$\omega_a = \frac{1}{RC} \tag{19}$$

For values of V_i less than δ, (18) simplifies to

$$\frac{V_o(s)}{V_i(s)} = g_m R \frac{\omega_a}{s + \omega_a} = \frac{A_0 \omega_a}{s + \omega_a} \tag{20}$$

which is equivalent to (17). Thus $g_m R = A_0$, the small-signal low-frequency (or dc) gain. For frequencies above ω_a, (20) reduces to

$$\frac{V_o(s)}{V_i(s)} \simeq \frac{A_0 \omega_a}{s} = \frac{GB}{s} \tag{21}$$

[1] J. Solomon, "The Monolithic Op Amp: A Tutorial Study," *IEEE J. Solid-State Circuits*, vol. SC-9, no. 6, December 1974, pp. 314–332.

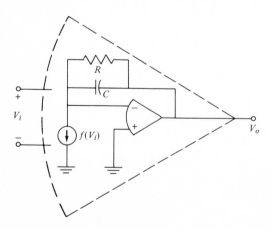

Figure B-12 Model of a frequency-compensated operational amplifier.

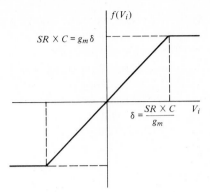

Figure B-13 Characteristic for source $f(V_i)$ of Fig. B-12.

If $V_i(s)$ is a step of V volts, then the output time-domain waveform is expressed as

$$v_o(t) = \text{GB} \times V \times t \qquad (22)$$

The slope of the output is seen to be proportional, among other things, to the amplitude of V_i. However, if the magnitude of V_i is greater than δ, then the maximum value of $f(V_i)$ is given as

$$f(V_i)_{\text{max}} = C \left. \frac{dv_o}{dt} \right|_{\text{max}} = \text{SR} \times C \qquad (23)$$

It is apparent from (23) that the maximum slope of the output voltage is limited to SR, the slew rate of the operational amplifier.

Another characteristic of operational amplifiers which influences active filter performance is noise. This effectively limits the minimum signal level. To obtain large values of signal attenuation in the stopband of a filter characteristic, low noise levels are necessary. The noise characteristics of an operational amplifier can be modeled as shown in Fig. B-14. In this figure the quantities $e_n^2(\omega)$ and $i_n^2(\omega)$ are called *voltage and current noise spectral densities*. Their units are mean square volts and mean square amperes respectively. Their values are also frequently given in rms (root mean square) units. Note that, as indicated in the figure, noise sources have no reference polarity. Typical noise spectral densities for an operational amplifier are shown in Fig. B-15. The increase of noise spectral density at decreasing frequencies is usually referred to as a *1/f effect*.[1] Another source of

---

[1] C. D. Motchenbacher and F. C. Fitchen, *Low-Noise Electronic Design*, John Wiley & Sons, Inc., New York, 1973.

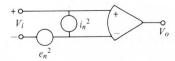

Figure B-14 Model for noise characteristics of an operational amplifier.

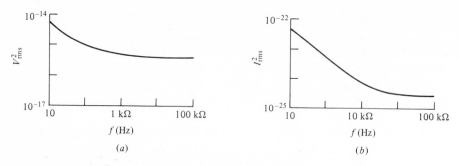

Figure B-15 Noise spectral densities for an operational amplifier.

noise in active filters is the resistive components. Two equivalent noise models for a resistor of value R are shown in Fig. B-16. For these

$$e_R^2(\omega) = 4kTR \text{ volts}^2/\text{Hz} \qquad i_R^2(\omega) = \frac{4kT}{R} \text{ amps}^2/\text{Hz} \qquad (24)$$

where k is the Boltzmann constant of 1.38×10^{-23} Ws/°K. Examples of noise calculations may be found in the literature.[1,2]

Effective design techniques can considerably minimize the effects of noise in active filters. For example, the signal level of the filter should be chosen as large as possible to give the maximum dynamic range. The inherent noise voltage is typically of the order of -100 dBV (100 dB below 1 V) for a 10-Hz bandwidth. For each decade increase in frequency, a 10-dB increase in noise results. For example, if a system has a noise floor of -100 dBV in a 10-Hz bandwidth, then the inherent noise voltage would be -70 dBV over a 10-kHz bandwidth. If we compare the relative contributions of resistors and amplifiers to overall noise, the amplifier is the primary culprit. Thus, the most effective way to decrease output noise levels is to select a low noise operational amplifier. In most applications $e_n^2(\omega)$ is more important than $i_n^2(\omega)$, so that in practice it is important to select an amplifier with a low value of $e_n^2(\omega)$. If it is important to minimize $1/f$ noise, then a

[1] F. N. Trofimenkoff, D. H. Treleaven, and L. T. Bruton, "Noise Performance of RC-Active Quadratic Filter Sections," *IEEE Trans. Circuit Theory*, vol. CT-20, no. 5, September 1973, pp. 524–532.

[2] L. T. Bruton and D. H. Treleaven, "Electrical Noise in Low-Pass FDNR Filters," *IEEE Trans. Circuit Theory*, vol. CT-20, no. 2, March 1973, pp. 154–158.

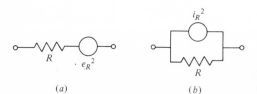

Figure B-16 Noise models for a resistor of value R.

JFET-input operational amplifier should be selected, since the breakpoint in the $e_n^2(\omega)$ curve is less for JFETs than for bipolar junction transistors.

There are some additional miscellaneous parameters of operational amplifiers which deserve mention, since they are important in active filter applications. The first of these concerns the dc error previously characterized by V_{os} and I_{os}. In practice these parameters are dependent upon temperature. Typical values are $dV_{os}/dT \approx 6$ mV/°C and $dI_{os}/dT \approx 2$ pA/°C. Fortunately, the feedback used in active filters will minimize these effects in most realizations. Another temperature-dependent parameter is GB. A typical temperature coefficient of GB is approximately -2000 ppm/°C. A final parameter is the power supply rejection ratio (PSRR). The PSRR is a measure of the ratio of ripple at the output of the amplifier to ripple at the power supply output. In general the PSRR should be greater than A_0 to prevent instability problems.

Index

Index

Page numbers in italic indicate tables.

THE
UNIVERSITY SHOP
VILLANOVA,PA.

05/28/86

MERCH.   4   1.95
BOOKS   45.95
SUBTTL   47.90
TAX 1   0.12
TOTAL   48.02
CASH   50.00
CHANGE   1.98

       2   #ITEMS

0171A 1   14:19
KEEP FOR RETURNS